Materials in sports equipment

Materials in
sports equipment

Edited by

Mike Jenkins

CRC Press
Boca Raton Boston New York Washington, DC

WOODHEAD PUBLISHING LIMITED
Cambridge England

Published by Woodhead Publishing Limited, Abington Hall, Abington
Cambridge CB1 6AH, England
www.woodhead-publishing.com

Published in North America by CRC Press LLC, 2000 Corporate Blvd, NW
Boca Raton FL 33431, USA

First published 2003, Woodhead Publishing Ltd and CRC Press LLC
© 2003, Woodhead Publishing Limited
The authors have asserted their moral rights.

British Library Cataloguing in Publication Data
A catalogue record for this book is available from the British Library.

Library of Congress Cataloging in Publication Data
A catalog record for this book is available from the Library of Congress.

Woodhead Publishing ISBN 1 85573 599 7
CRC Press ISBN 0-8493-1766-5
CRC Press order number: WP1766

Typeset by SNP Best-set Typesetter Ltd., Hong Kong
Printed by TJ International, Cornwall, England

Contents

Preface

Examination of the sporting goods sector in the UK reveals a market of significant size and growth potential. In 1998, the Sports Industries Federation evaluated consumer expenditure in the sports market sector. It was estimated at over £12b, with sports equipment contributing in the region of 28% to this market. A number of growth areas were also highlighted, including golf, team sports and fitness equipment. The growth of golfing activity has been most significant, increasing from £115m in 1995 to £180m in 1998. The growth of the sports sector as a whole is in accordance with increased participation in sporting activity and this has significant financial implications for society.

In terms of the financial costs of healthcare for the population, participation in sport and exercise can reduce the costs to the state in later life. In 1985, Australian research predicted savings of A$400m in the year 2000 if 50% of the population participated in some sporting activity. In addition, there are many opportunities for business diversification as new sports proliferate. There is also a significant financial incentive for entering the market if the following example is considered. In 1971, a 'professional' tennis racket cost £11. In 1998, based on inflation, the racket should have cost £48; it actually cost £230.

The driving force for a consumer to pay £230 for a tennis racket is the benefit that advanced materials bring to the product, most often in terms of increased stiffness and reduced weight. Therefore, in recent years the use of advanced materials in sport has increased and there has also been a corresponding increase in athletic performance. The purpose of this book is to detail the use of materials in sport and to discuss the relationship between materials selection, materials processing and design in a sporting context. To achieve this, a wide range of sports and equipment is considered including golf, tennis, body protection and mountaineering. The content of each chapter is aimed primarily at academics and manufacturers. However, it is also likely to be of interest to postgraduate students working in the area of sports engineering. To ensure this, each chapter is supported by numerous literature and internet-based resources.

Finally, I would like to extend my thanks to each author for their valuable contributions to this work.

Mike Jenkins

Contributors

*indicates main point of contact

Chapter 1
Dr M. Jenkins
Metallurgy and Materials
School of Engineering
University of Birmingham
Edgbaston
Birmingham
B15 2TT
UK
Tel: +44 (0) 121 414 2841
Fax: +44 (0) 121 414 5232
Email: m.j.jenkins@bham.ac.uk

Chapters 2 & 4
Dr N. J. Mills
Metallurgy and Materials
School of Engineering
University of Birmingham
Edgbaston
Birmingham
B15 2TT
UK
Tel: +44 (0) 121 414 5185
Fax: +44 (0) 121 414 5232
Email: n.j.mills@bham.ac.uk

Chapter 3
Dr C. Walker
University of Strathclyde

Mechanical Engineering
James Weir Building
75 Montrose Street
Glasgow
G1 1XJ
UK
Tel: +44 (0) 141 548 2657
Fax: +44 (0) 141 552 5105
Email:
cwalker@mecheng.strath.ac.uk

Chapter 5
Dr J. Macari Pallis
Cislunar Aerospace, Inc.
PO Box 320768
San Francisco
California 94132
USA
Tel: +1 415 681 9619
Fax: +1 415 681 9163
Email: deke@cislunar.com

Dr R. D. Mehta
Sports Aerodynamics Consultant
209 Orchard Glen Court
Mountain View
California 94043
USA
Tel: +1 650 960 0587
Fax: +1 650 903 0746
Email: rabi44@aol.com

Chapter 6
Dr M. Strangwood
Metallurgy and Materials
School of Engineering
University of Birmingham
Edgbaston
Birmingham
B15 2TT
UK
Tel: +44 (0) 121 414 5169
Fax: +44 (0) 121 414 5232
Email: m.strangwood@bham.ac.uk

Chapter 7
Dr H. Dong
Surface Engineering Group
Metallurgy and Materials
School of Engineering
University of Birmingham
Edgbaston
Birmingham
B15 2TT
UK
Tel: +44 (0) 121 414 5197
Fax: +44 (0) 121 414 7373
Email: h.dong.20@bham.ac.uk

Chapter 8
Professor R. Cross
School of Physics A28
University of Sydney
New South Wales 2006
Australia
Tel: +61 (0) 2 9351 2545
Fax: +61 (0) 2 9351 7727
Email: cross@physics.usyd.edu.au

Chapter 9
Mr H. Lammer* & Mr J. Kotze
R&D Racquetsport
HEAD SPORT AG
Wuhrkopfweg 1
A-6921

Kennelbach
Austria
Tel: +43 (0) 5574 608 780
Fax: +43 (0) 5574 608 795
Email: h.lammer@head.com
                J.kotze@head.com

Chapter 10
Dr J. M. Morgan
Department of Mechanical
   Engineering
University of Bristol
Queen's Building
University Walk
Bristol
BS8 1TR
UK
Tel.: +44 (0) 117 928 9900
Fax: +44 (0) 117 929 4423
Email: john.morgan@bristol.ac.uk

Chapter 11
Dr J. R. Blackford
School of Mechanical Engineering
University of Edinburgh
Sanderson Building
The King's Buildings
Edinburgh
EH9 3JL
Scotland
UK
Tel: +44 (0) 131 650 5677
Fax: +44 (0) 131 667 3677
Email: jane.blackford@ed.ac.uk

Chapter 12
Dr H. Casey
1322 Cibola Circle
Santa Fe
New Mexico
87501
USA
Email: hughcasey@msn.com

Chapter 13
Associate Professor A. J. Subic
Department of Mechanical and
Manufacturing Engineering
RMIT University
PO Box 71
Bundoora
VIC 3083
Australia
Tel: +61 3 9925 6080
Fax: +61 3 9925 6108
E-mail:
aleksandar.subic@rmit.edu.au

Dr A. J. Cooke*
Cooke Associates
Tudor Cottage
Stonebridge Lane
Fulbourn

Cambridge
CB1 5BW
UK
Tel: +44 (0) 1223 882072
Fax: +44 (0) 1223 881442
Email:
alison@cookeassociates.com

Chapter 14
Dr J. Macari Pallis
Cislunar Aerospace, Inc.
PO Box 320768
San Francisco
California 94132
USA
Tel: +1 415 681 9619
Fax: +1 415 681 9163
Email: deke@cislunar.com

1
Introduction

M. JENKINS

University of Birmingham, UK

1.1   Factors determining sports performance

Sports performance is determined by a number of factors. Some originate from the human element of the sport, such as the physiological and psychological state of the competitor, while others originate from the equipment used by the athlete, which includes the design and materials used in the production of the item. Two sports that clearly illustrate these ideas are the triple jump and the pole vault.

If the world records for these sports are examined over the last 100 years, it is clear that the jump lengths and heights have increased, as shown in Figs 1.1 and 1.2. The factors that affect performance in the triple jump are dominated by the human elements, with increases in performance deriving from improved fitness and psychological condition in the athlete. In addition, advances in high speed video and motion analysis equipment have resulted in a deeper understanding of the biomechanics of the event, which in turn has developed the technique of the athlete.

Evidence of the human factors can be found in the event of the pole vault; however, these factors are accompanied by additional elements, such as the design and materials used in the construction of the vaulting pole. Closer inspection of the trend shown in Fig. 1.2 reveals what is almost a step change in the world record in the post-war years. The explanation for this occurrence requires an investigation of the evolution of the construction of the vaulting poles over the last 100 years.

1.2   Materials, processing and design in the pole vault

The regulations that define the construction of vaulting poles are very liberal.[1] As a result, the design has evolved significantly. Poles were originally made from solid sections of wood, but rapidly developed through the use of long lengths of bamboo. The incorporation of this natural material

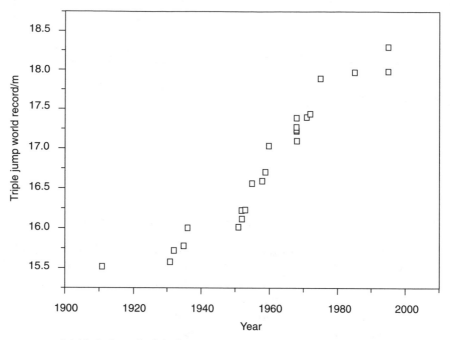

1.1 Variation of triple jump world record with time.

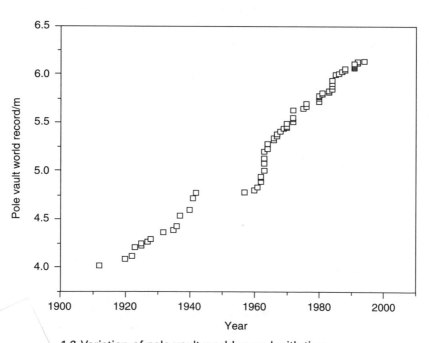

1.2 Variation of pole vault world record with time.

dramatically reduced the weight of the poles, but also limited the length of poles that could be made. In the post-war years, two innovations of materials technology occurred. In the 1950s, aluminium poles were introduced, but they were followed rapidly in the 1960s by glass fibre composites. The introduction of glass fibre poles corresponds to the step change in world record identified above, but the explanation of why this occurred requires analysis of the biomechanics of the event.[2]

In essence, the athlete aims to generate the maximum possible kinetic energy before the launch. When the athlete leaves the ground, kinetic energy is transformed into potential energy, which in part determines the height of the jump. Therefore, the faster the athlete runs in the sprint stage of the vault, the higher the jump height. However, there is another dimension to the event, one that derives from the materials used in the production of the pole.

If the event is examined using high speed video equipment, it is clear that a high degree of bending in the pole develops following the launch. This is made possible due to the selection of glass fibre composite as a construction material. The properties of the material include relatively high strength, intermediate stiffness and low density. Intermediate stiffness enables the pole to bend, but relatively high strength ensures that the pole will not break. If a vault with a bamboo or aluminium pole is examined photographically, there is little evidence of significant bending, due to increased bending stiffnesses in these poles. The consequence of the glass fibre composite material selection is that the stored elastic strain energy is increased because of the decreased radii of curvatures made possible as a result of high strength and low bending stiffness. Since glass fibre composites generally exhibit relatively low densities, the speed on the athlete prior to launch can be correspondingly increased, resulting in increased kinetic energy. The release of the stored strain energy, coupled with the transformation of kinetic to potential energy, helps to propel the athletes to ever-increasing heights, as shown by the step change in Fig. 1.2.

There is also a materials processing aspect to the performance of the vaulting pole. The composite material must be laminated to provide the required bending stiffness. This is accomplished by designing a stack sequence or 'layup' that aligns a significant proportion of the reinforcing fibres along the long axis of the pole. There are many commercial variations of stack sequence, but a common method of pole production is to use a process called filament winding. This process is based on the winding of resin-impregnated fibre tows around a mandrel at a defined winding pitch. This angle, together with the pole cross-section, determines the bending stiffness of the pole. The filament-wound structure is often supplemented by additional layers of woven material, as shown in Fig. 1.3 (shown previously in *Interdisciplinary Science Reviews*[1]).

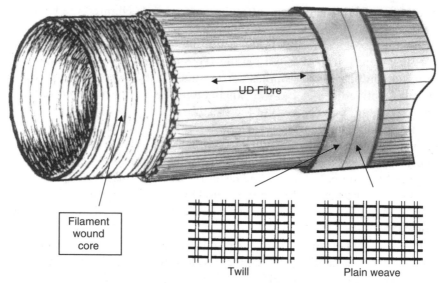

1.3 The filament wound structure.

The importance of materials processing is also evident when the athlete begins his or her descent from the bar. The energy considerations are simple: potential energy is now transformed into kinetic energy as the athlete accelerates toward the ground head first. To save the competitor from very serious injury, a large crash mat is placed in the landing area. The mat is made from low-stiffness polymer foam that absorbs the impact energy as the athlete lands. The energy absorption is a direct consequence of the foam structure in that air can be expelled from the mat and the cell walls of the foam can collapse. The cellular structure is created during the processing of the polymer, and the foam microstructure can be controlled by utilising different processing methods.

Clearly, sporting performance depends on a number of factors, both human and materials based. However, there is also an interdependence on the materials and the design of sports equipment, in that developments in materials technology can result in materials with improved properties that facilitate evolutionary or revolutionary change in design. A good example of this is in the sport of fencing.

1.3   The relationship between materials technology and design – fencing masks

Although the sport of fencing is one of the four original events of the modern Olympic Games, the International Olympic Committee recently threatened the sport with exclusion from future Games. The committee suggested that the sport should be made more accessible to television audiences by making the competitors' faces more visible. This suggestion catalysed a change in the sport that had been slowly occurring over many years. Fencing equipment manufacturers have, for many years, desired to produce a mask that enabled clear vision of the opponent. However, numerous attempts have been made to replace the metal mesh designs, but none offered any degree of consistent safety or clarity of vision.

The design of the fencing mask was recently revolutionised by the innovation of a traditional engineering thermoplastic known as Lexan (polycarbonate). This polymer is transparent and demonstrates high impact resistance. A comparison between metal and polymer masks has been made by the Italian Fencing Federation laboratory.[3] These tests showed that the polymer sections outperformed the conventional metal mesh sections. Drop tests from heights of 55 cm with a mass of 2.4 kg fastened to a steel spike (section 3 × 3 mm square with a pyramidal point angle 60°) showed that the metal mesh was penetrated, whereas the Lexan visor was only marked by the impression of the pyramidal point. Tests on complete masks also showed that the Lexan based masks did not deform, whereas the metal mesh versions did. This deformation was deemed sufficient to injure the fencer. It is clear that the use of polycarbonate has enabled a design transformation to occur; one that is a direct result of materials technology.

1.4   Overview of 'materials in sport'

Consideration of the three sports outlined above – the triple jump, pole vault and fencing – reveals several non-human factors that affect sports performance, materials selection, processing and equipment design. These elements form the main themes of the book, with each chapter focusing on a number of the above themes. Chapters 8 and 9 introduce tennis racket frames and strings, in Chapter 9 the evolution of the frame design is discussed in terms of the materials technology and a range of construction materials are reviewed. The effects of processing are addressed by consideration of the lamination process and the final production methods. In Chapter 8, string materials and the mechanical properties are introduced. Creep and friction properties are developed and player perception is discussed. In general, the materials focus is on polymers and polymer composites.

The polymer theme is also developed in Chapters 2 and 4, where the use of polymer foams in body protection and running shoe materials is detailed. In Chapter 2, body protection, cycle helmets and football shin pads are introduced and developed as case studies. Each section covers materials selection, biomechanics and current test standards. The effect of processing on the performance of polymer foams is introduced in Chapter 4. The compression behaviour and durability of the foams is discussed in conjunction with the biomechanics of running. A range of polymeric and metallic materials are covered in the chapter on mountaineering: Chapter 11. A detailed discussion of the use of these materials in a range of items including ropes and carabiners is presented. Current research in the area is also highlighted.

The sport of cycling is discussed in Chapter 10. The discussion begins with an overview of the historical development of the bicycle and the materials used. The chapter then focuses on metallic materials and selected components on the frame. The validity of current test standards is then considered. Metallic materials are also discussed in Chapter 6, within the context of golf. In this chapter, the effect of materials, processing and design on the performance of oversized drivers is discussed. Chapter 7 focuses on the effect of processing by considering a range of surface treatments. The impact on a number of sports including golf and motor sport is then considered.

The aspects of design and materials appear in Chapter 12 and the evolution of ski design is discussed in conjunction with the developments in materials technology. Design and materials are also considered in terms of aerodynamics in Chapter 5. A range of sports is introduced, from the discus to the cricket ball. The sport of cricket is considered in a broader context in Chapter 13. In this chapter, a range of natural materials is considered including wood and leather. In addition, the design of various cricket equipment is also discussed, from bats to protective helmets. The role of sports surfaces is considered in Chapter 3, where materials and test standards are discussed. A number of sports and associated surfaces are then introduced including football, hockey and tennis.

In summary, it is clear that materials technology is fundamental to the design and performance of sports equipment. The chapters that follow aim to introduce the range of materials used in a wide variety of sports and also address the themes of materials, processing and design in sports equipment.

1.5   References

1 Jenkins, M.J. (2002). *Interdisciplinary Science Reviews*, **27** (1), 61.
2 Haake, S.J. (2000). *Physics World*, **13** (9), 30.
3 Jenkins, M.J. (2000). *Materials World*, **8** (9), 9.

Part I

General uses

2
Foam protection in sport

N. J. MILLS
University of Birmingham, UK

2.1   Introduction

Three main application areas illustrate the main types of polymer foams and the factors behind their selection, specification and use. The areas are:

(a)   static products where mass is unimportant
(b)   products worn on parts of the body, which change shape
(c)   products worn on the upper part of the head, which do not change shape.

The requirements for each product will be described, and then the material choice considered. The mechanical performance of the product is usually more important than the thermal or chemical properties. However, if foam products are worn, the wearer must cope with the thermal insulation properties. In some areas, competitive products use different foams, showing there is no 'best' foam. The mechanics analysis of the product will be related to the biomechanics of the appropriate injury mechanisms. This links to product performance tests in British or European Standards.

   The sports illustrated are football (shin and ankle protection), field sports (crash mats) and cycling (helmets). A variety of other sport applications can be linked to these case studies, for example, the use of chest protection in equestrian sport, or padding of rugby goal posts. Klempner and Frisch (1991) outline the production methods and mechanical properties of polymeric foams. The author's forthcoming book (Mills, 2003) gives more detail of the mechanical properties of foams, and links these to microstructure. Chapter 4 gives details of the ethylene copolymer closed-cell foams used in running shoes.

2.2   Static foam protection products

2.2.1   Introduction

Products can be large (gym mats, rugby goal post padding) and the mass is not important. The foam must be protected against abrasive damage, for instance, from shoes, by using loose or bonded covers. These need to be flexible (easy to bend, but not easy to stretch), so tend to be cloth coated with plasticised polyvinyl chloride (PVC) or polyurethane (PU). Since foams are a fire risk, the products must usually pass fire tests; consequently, fire-retardant additives can be added to the foam polymer. The cover assists by slowing the ingress of air to the foam. For outdoor use, the products also need protection against:

(a)   rain. A waterproof cover is necessary to prevent large quantities of water being absorbed into the open cells. Hydrolytic breakdown of some PU foams is possible over a period of years if warm moist conditions prevail.

(b)   the ultraviolet component of the solar radiation. The high surface area of the foam and the ready access of oxygen make the PU vulnerable to photo-oxidation. Covers act as a barrier to UV.

Crash mats are used in a number of sporting applications, e.g. below indoor climbing walls, as the landing area in pole vaulting, and as barriers on dry ski slopes. In European Standard BS EN 12503: 2001 there are a range of mat types. The thicker mats, such as type 10 for competitive high jump, and type 11 for pole vault, are impact tested with a 30 kg mass, having a 75 mm diameter hemispherical end, that falls vertically though 1.2 metres. The peak acceleration must be less than 10 g, but for other types the peak acceleration can be higher. Neither the link between the striker shape and mass and human body dimensions, nor that between the peak acceleration and any injury mechanism, are explained. The working papers of the EN standards committee are unavailable to the public. The background to the performance tests must be established, and linked to injury mechanisms, if there is to be progress in this area.

Lyn and Mills (2001) considered the use of rugby post padding. The cylindrical shape of a rugby goal post is more difficult to protect than a flat surface. If there is a head impact with a padded post, the contact area with the head can be much smaller. Typically, a 20 J impact energy could cause concussion, if the player's head aims at the centre line of the post.

2.2.2   Material selection

For the high jump or pole vault, a thick mat reduces the fall distance, and hence the athlete's kinetic energy at impact. The application therefore determines the mat thickness. For other applications, mats could be thinner if the foam had a higher modulus. Generalising from helmet design, the ideal foam should just absorb the design impact energy without injury. With the current designs of crash mats, the headform force is usually linearly proportional to the foam deflection. As the kinetic energy of the impact increases, the thickness of the foam must also increase, and its Young's modulus must decrease to avoid injuries (see Section 2.2.7). There is a jump in properties between the response of closed-cell polymer foams, which have Young's moduli ≥100 kPa due to the air content, and open-cell flexible PU foams, which usually have moduli ≤30 kPa. Judo and other martial arts mats tend to be thinner (say 75 mm) and firmer than crash mats. Consequently, some are manufactured from closed cell cross-linked polyethylene foam (see the second case study), while others use dense PU chip foam.

The slabstock process for polyurethane flexible open-cell foam produces some scrap, which is diced into pieces of 5 to 10 mm diameter. These pieces, of differing colours and hardnesses, are mixed with more of the isocyanate and polyol chemicals and remoulded under uniaxial compression. Kay Metzeler, Ashton-under-Lyne is a major UK source of such remoulded (or 'chip') PU foam. Lyn and Mills (2001) measured the densities of mats of thickness 0.1, 0.2 and 0.4 m as 63, 76 and 72 kg m^{-3}, respectively, while the nominal density was 4 lb/ft^3 (64 kg m^{-3}). The colour mix of the 0.2 m thick foam was different from that of the other mats, and the relative compressive stress at a strain of 0.3 was 29% higher than for the other materials. Figure 2.1 shows the edge of a compressed chip in a judo mat of density 113 kg m^{-3} in which the cell edges are compressed significantly. The foam chips have variable hardness, but the mixing is effective and the particle size (<10 mm) is much less than the size of the tested block (100 mm). The variation in the mechanical properties of blocks was ±5%, and the elastic anisotropy was of the order of 10%. The compressive stress–strain relationship at low strain rates is more linear in the 0 to 20% strain range than is virgin PU foam (probably due to the pre-compression of some chips, and the composite nature of the chip foam with varying chip moduli).

2.2.3 The kinematics and biomechanics of head impact

A worse-case scenario was considered, when the head hits the mat first. If either the outstretched arms, or the legs and torso, hit the mat first, the head accelerations will be much less. Although the body mass (50 to 100 kg) is much greater than the head mass (4 to 6 kg), the flexibility of the human neck allows the effective separation, on the short timescale of the head impact, of the head and neck motion. Even in the unlikely event of the body being in line with the head, and the impact being direct on the crown of the head, the neck will buckle. Gilchrist and Mills (1996) showed, using dummies with realistically flexible necks, that the impact force, F, on the head was related to the head acceleration, a, and mass, m, by Newton's Second Law:

$$F = ma \qquad\qquad [2.1]$$

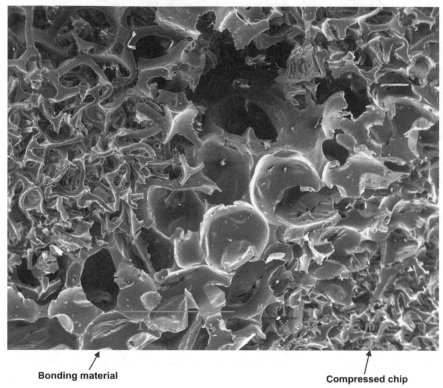

Bonding material   **Compressed chip**

2.1 SEM of unloaded PU chip foam, of density 113 kg m^{-3}, from a judo mat.

A head with velocity, V, has a kinetic energy KE given by:

$$KE = \frac{1}{2}mV^2$$ [2.2]

The maximum velocity of an out-of-control skier's head on a practice slope is 10 m/s. As climbing walls are rarely >5 metres high, the head velocity of a falling climber is unlikely to exceed 10 m/s. If V = 10 m/s, the head kinetic energy is 250 J, whereas at 5 m/s it is 62.5 J. There is a risk of concussion or minor head injury when the force on the head exceeds 10 kN – see Section 2.4.2.

2.2.4 Modelling of impacts

There are no simple approaches (for example, assuming uniform strain) for the analysis of head impacts on PU foam safety mat. The foams are non-linear materials, with rate-dependent and viscoelastic properties, so some approximations must be made. The approach was to:

(a) carry out FEA analysis of head impacts on mats, using the approximation that the foam is a non-linear *elastic* (hyperelastic) material, and validate this using experimental tests. The large changes in mat geometry, as the impact proceeds, can only be considered by using FEA.
(b) analyse whether there is a significant air flow contribution. This is potentially important in large blocks of PU foams, impacted at high velocities.
(c) leave consideration of viscoelasticity to later.

2.2.5 Finite element modelling

Hyperfoam model

The foam uniaxial compressive response, measured at moderate strain rates under conditions where the air flow contribution is negligible, is fitted with a set of adjustable parameters from Ogden's (1972) strain energy function for compressible hyperelastic solids. This is:

$$U = \sum_{i=1}^{N} \frac{2\mu_i}{\alpha_i^2} \left[(\lambda_1^{\alpha_i} + \lambda_2^{\alpha_i} + \lambda_3^{\alpha_i} - 3) + \frac{1}{\beta_i} (J^{-\alpha_i \beta_i} - 1) \right] \qquad [2.3]$$

where λ_I are the principal extension ratios, $J = \lambda_1 \lambda_2 \lambda_3$ measures the total volume, the μ_i are shear moduli, N is an integer, and α_i and β_i are curve-fitting non-integral exponents. The latter are related to Poisson's ratio v_i by:

$$\beta_i = \frac{v_i}{1 - 2v_i} \qquad [2.4]$$

Material parameters

Since polyurethanes are rate-dependent, the foam characterisation should be in impact tests lasting 0.05 s, rather than low strain rate tests lasting 60 s. The compressive stress–strain curve of rebonded PU foam was measured using a 100 mm cube sample, impacted from heights from 0.005 m to 0.65 m, by a flat anvil of mass 1.50 kg falling without friction on two guide wires. Analysis (Mills and Lyn, 2002) shows that, for such tests, air flow does not contribute to the total stress. The stress–strain curve (Fig. 2.2) shows a near linear initial portion, and a steeply rising region for strains >0.6. In the latter region the cell edges begin to contact, making further edge bending more difficult. With the 200 mm by 200 mm sample, there is a more obvious plateau in stress for strain exceeding 10%, but a similar response for strains >30%.

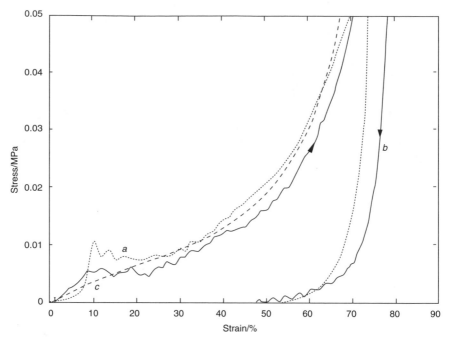

2.2 Impact compression stress–strain response for 0.6 m drop on a
.... 200 × 200 × 100 mm specimen, b → 100 mm cube of
remoulded PU foam, c — — — hyperelastic model prediction,
parameters in Table 2.1 (Lyn and Mills, 2001).

The foam Poisson's ratio was measured both in compression and tension, at moderate strain rates, using an Instron machine with transducers described by Mills and Gilchrist (1997). The Poisson's ratio in compression was small, as for virgin PU flexible foam, but in tension the value increased until it was 0.44 for tensile strains >0.1. This is evidence for a change in deformation mechanism between compression and tension.

An $N = 2$ model was used; both N terms contribute to the strong compressive hardening when the strain >0.6, but the $\alpha = -2$ term has very little effect on the tensile hardening. The parameters (Table 2.1) were altered to fit the experimental data. Figure 2.2 shows that the shape of the tensile loading curves can be matched, but not the initial 'plateau' region between 10 and 20% compressive strain. There is a reasonable match to the lateral strain response in tension.

Model geometry

The headform was modelled, as a sphere of radius 80 or 100 mm, to simulate various impact sites on the head. The foam was supported by a rigid

Table 2.1 Parameters for the hyperelastic model of rebonded PU foam

N	Shear modulus μ kPa	Exponent α	Poisson's ratio v
1	18	8	0
2	1.2	−2	0.45

flat table, but not bonded to it. A coefficient of friction of 0.75 was used between the foam and both the headform and the table. The headform was modelled as an elastic material with a Young's modulus of 70 MPa, effectively rigid compared with the foam. To reduce the computation time, an axisymmetric problem was considered with the foam being a vertical cylinder of radius 500 mm. The elements used were CAX4, 4-node bilinear axisymmetric quadrilaterals, in ABAQUS standard. The FEA mesh for the undeformed foam consisted of horizontal and vertical straight lines, with spacing biased towards the initial impact point. The contact area, between the headform and the foam, increases as the indentation increases. Figure 2.3(*a*) shows that the upper surface of the deformed foam has distorted to contact the lower half of the headform surface. The grid spacing has hardly changed at the lower and right-hand sides of the foam, indicating low strains. There are high compressive strains, ε_2, and small tensile strains, ε_1, in the foam just below the headform near the vertical axis. The distorted angles at the mesh intersections revealed shear strains in the foam at the sides of the headform. In Fig. 2.3(*b*) the contours of the compressive principal stress are closely bunched under the vertical axis of the indenting headform. The maximum compressive stress, −50 kPa, is very low compared with values of −500 kPa that occur in the foams of protective helmets (see the third case study). This is the result of the open-cell polyurethane foam being much softer than the closed-cell polystyrene foams used in helmets.

2.2.6 Falling headform impact test

The aluminium headform, of circumference 58 cm, has a mass of 4.1 kg. The headform shape, defined in BS EN 960: 1995, is longer than it is wide. Its radius of curvature at the crown impact site is 100 mm for the fore and aft section, and 75 mm for the ear-to-ear section. The radius of curvature is slightly smaller at the rear or front of the head. The headform, fitted with a linear accelerometer, aligned vertically (Fig. 2.4), was dropped from

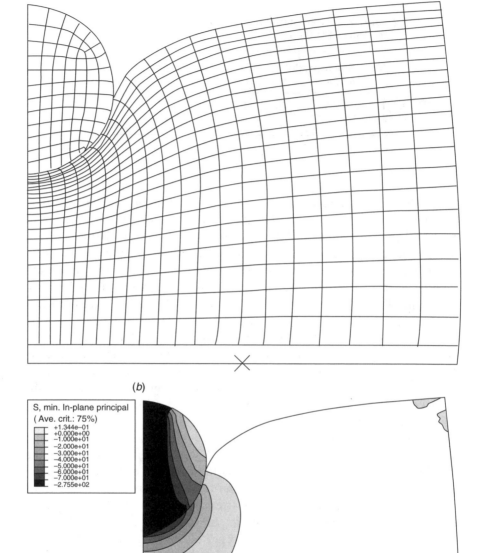

(a)

(b)

S, min. In-plane principal
(Ave. crit.: 75%)

```
+1.344e−01
+0.000e+00
−1.000e+01
−2.000e+01
−3.000e+01
−4.000e+01
−5.000e+01
−6.000e+01
−7.000e+01
−2.755e+02
```

−20 kPa

−10 kPa

2.3 (a) Predicted deformed shape of 400 mm thick foam mat after
200 mm vertical deflection by a 100 mm radius headform, (b)
Predicted contours of principal compressive stress. There is an
axis of rotational symmetry at the left of each figure (Lyn and
Mills, 2001).

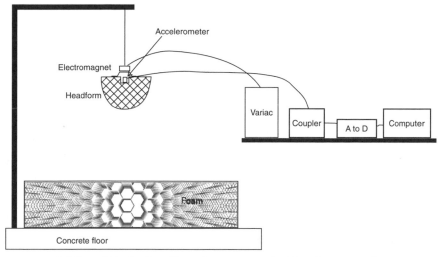

2.4 Free fall of a headform instrumented with a linear accelerometer.

heights of 0.125 m to 1.0 m. The deceleration data was converted to a 12-bit digital signal, sampled at 1 kHz. Numerical integration was used to convert this information into a graph of the force on the headform, vs. the deflection of the upper surface of the foam at the point of impact (Fig. 2.5). The loading parts of the curves are nearly linear and they almost follow a 'mastercurve' when the deflection is increasing. There are some signs of a rate-dependent response, in that the loading curves for the higher impact velocities lie slightly above those for lower velocities. When the headform velocity approaches zero, the force drops below the mastercurve, and during unloading the force is far below the loading value. The initial loading stiffnesses ($7 \, \mathrm{N \, mm^{-1}}$ for a 400 mm mat, $9 \, \mathrm{N \, mm^{-1}}$ for 100 and 200 mm mats) are almost independent of the mat thickness. However, if the strain exceeds 60%, the foam begins to bottom out, and the slope of the loading curve becomes steeper (Fig. 2.5(a)).

2.2.7 Maximum force or deceleration

The impact energy is the product of the headform mass, the drop height and the gravitational acceleration. The maximum headform deceleration a_{max} is related to the peak force F_{max} on the headform by equation [2.1]. At the maximum foam deflection x_{max}, the impacting headform has momentarily stopped, so its initial kinetic energy KE has been converted into strain energy of the foam. The area under the linear part of the force deflection curve (Fig. 2.6(a)) is given by:

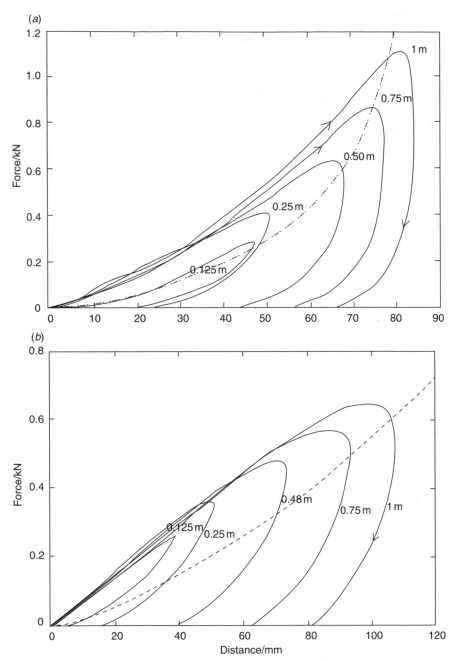

2.5 Headform impact tests, onto remoulded PU foam mats of
thickness (a) 0.1 m, (b) 0.4 m, for drop heights marked. — — —
hyperelastic model predictions, 100 mm radius headform (Lyn and
Mills, 2001).

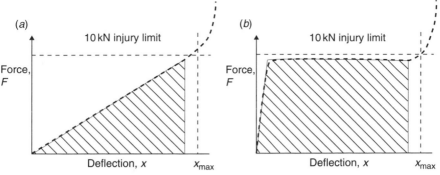

2.6 The area under a graph of headform force vs. mat deflection is energy input: (*a*) typical linear response, (*b*) response of a mat that maximises energy input without injury.

$$\text{KE} = \frac{1}{2} F_{max} x_{max} = \frac{1}{2} k x_{max}^2 \qquad [2.5]$$

where k is the foam loading stiffness. Hence the maximum deflection is given by:

$$x_{max} = \sqrt{\frac{2\text{KE}}{k}} \qquad [2.6]$$

If thicker mats are used for high kinetic energy impacts, the loading stiffness must be reduced by reducing the foam density, in order to keep the peak force lower than the injury level.

The force, energy, acceleration (FEA) model can predict the effect of high energy impacts. The energy absorbed by the foam is equal to the area under the loading graph (Fig. 2.6). If 100 g head acceleration is allowable (a force of 5 kN), a 200 mm thick mat can absorb 140 J safely, while a 400 mm thick mat can absorb a 450 J safely. Therefore, a 400 mm thick mat, containing PU chip foam of 64 kg m^{-3} density, has can prevent head injury, for impact velocities <10 m/s.

2.2.8 Comparison of FEA predictions with experimental data

The predicted headform forces (Fig. 2.5) are smaller than the experimental loading forces, at a given mat deflection, and greater than the experimental unloading forces. The prediction is elastic, with loading and unloading along the same relationship with positive curvature, whereas the experimental loading relationship is nearly linear with hysteresis on unloading. Making allowance for the higher modulus of the 0.2 m thick mat, the predicted forces at a 60 mm loading deflection were between 58% and 67% of the experimental loading forces. Possible reasons for underestimating the force are:

(a) inadequate foam modelling parameters. The FEA analysis of slow, axisymmetric indentation of virgin PU foams with a flat-based indenter (Mills and Gilchrist, 2000) predicted a relative stress 38% smaller than the experimental value for blocks of width three times the indenter diameter. This difference is slightly smaller than that for the sports mats.

(b) hysteresis due to viscoelasticity in the polymer, which was ignored in the modelling. The strain rate in the foam decreases as the headform decelerates, causing the stress to drop. It is not easy to determine the viscoelastic response of the polymer directly from foam bulk measurements; as the polymer microstructure varies with the foam processing details, and the foam cannot be remoulded into solid specimens, it is necessary to carry out micromechanics measurements on foam struts. Such work is in its infancy. Zhu, Mills and Knott (1997) suggest that the viscoelastic response of flexible PU foams may be in the non-linear range.

(c) failure to model the air flow in the foam. However, Lyn and Mills (2002) computed, from the measured air flow parameters of the foam, that the contribution of air flow to the total force was likely to be low unless the impact velocity exceeded 10 m/s.

2.2.9 The future

In the design of a minimum thickness mat, the force on the head would ideally be nearly constant (Fig. 2.6(b)) at a level just below that needed to cause concussion. However, there are currently no designs that can produce this type of response. Future work on mats will consider other injury mechanisms, and how more complex mat constructions can achieve improved performance.

2.3   Soccer shin and ankle protectors

These are an example of segmented protectors; it is not possible to play football wearing rigid calf-length boots. Separate ankle and shin protectors are mounted inside a sock, with an elasticated strap under the arch. The design of ankle protectors is linked to that of hip protectors for the elderly in falls (Mills, 1996). Both products have a shell over foam, and, in both, load can be shunted away from the protruding bone.

Franke (1977) estimated that, in Europe, 50–60% of injuries in sport were due to football. Fried and Lloyd (1992) concluded that 75–93% of the total injuries involve the lower extremities, with 17–26% involving the ankle and 17–23% the knee. Sprains, strains and contusions account for 50–88% of injuries, while fractures account for 1–10%.

The shin guard is the only protective equipment required by Fédération Internationale de Football Association (FIFA), from 1990. However, no subsequent studies have evaluated shin guard effectiveness in reducing injuries. Players may reject shin guards with ankle protection on the grounds that they inhibit performance, are bulky and uncomfortable. The size of the ankle protector must not inhibit performance or the ability to put on the boot. The foam must be sufficiently flexible to prevent discomfort or rubbing.

2.3.1 The threat and injury biomechanics

Bone fracture

In a straight-legged tackle, with the bodyweight behind the leg, the total kinetic energy of the player can be the order of 1000 J. However, due to the flexibility of the knee and ankle, the energy input to the opponent's leg is likely to be a maximum of 100 J. If the studs under the forefoot impact another player Ankrah and Mills (2003) estimate a 10 J impact energy. Cadaver tibias fracture with impact forces in the 4 to 7 kN range (Nyquist et al., 1985). If the foot is planted on the ground, and the opposing player's foot loads the tibia near its centre, the kinetic energy of the tackle is too high for the tibia to withstand. No current shin guard can prevent tibia fracture in these circumstances. Even if one existed, there would be a risk of knee damage, which is far more difficult to treat. No criterion could be found for fracture of the talus by direct impact. However, the calcaneus fails at impact forces of 3.6–11.4 kN, when impacted from below as in a vehicle crash (Seipel et al., 2001).

Bruising

The anterior border and the medial surface of the tibia, at the front of the leg, have very little soft tissue cover. Biomechanical criteria for soft tissue contusions (bruises) are not established. Crisco et al. (1996) impacted the leg muscle of rats with a 6.4 mm diameter nylon hemisphere to cause contusions. Although there were no pressure distribution measurements, the average pressure over the projected area of the hemisphere reached 9 MPa. In a review, Beiner and Jokl (2001) could not decide whether force, pressure or another mechanical variable was appropriate for a muscle contusion criterion. We assume that the criterion for bruising the soft tissue of the ankle is peak pressure, and that the minimum impact pressure to cause contusions is of the order of 1 MPa. Immediate treatment (ice, pressure) can reduce the development of bruising.

2.3.2 Product test rigs

A variety of test rig types and instrumentation have been used. Commercial shin protectors were tested by van Laack (1985), Phillipens and Wismans (1989), Bir *et al.* (1995) and Francisco *et al.* (2000), and the peak impact force was shown to be reduced by 40 to 70%. Davey *et al.* (1994) only tested flat 'shells' over flat sheets of foam, so their test geometry was inappropriate. Francisco *et al.* (1995) were the only researchers to comment on designs, making the following points:

(a) Fibre-glass shells were better than other materials in distributing the impact force.
(b) Increasing the compliance, i.e. by using air bladders, attenuated peak forces.
(c) Increased foam thickness was more important than increase protector length.

Their test rig had a rubber cover surrounding the tibia model.

The test rig mechanics should correspond with leg and foot biomechanics. Wooden legs (Philippens), with the grain running along the leg, provide too rigid a support for the sides of the protector. However, localised loading with a hemisphere will soften the wood cells and make the local contact stiffness lower; the peak force in a 'reference' impact without a protector will then be too low. Bir *et al.* (1995) used the hinged lower leg of a Hybrid III car-crash dummy. This is not biomechanically realistic; there is a central steel rod, and the PVC plastisol skin is too stiff, and too thick over the tibia. Also, the frictional conditions between foot and the ground were ill defined.

In these test rigs, the nearly-rigid striking objects were one of the following:

(a) flat or cylindrical bar across the leg, in BS EN 13061
(b) hemisphere, representing a ball, in tests related to hockey or cricket leg protectors

(c)   stud, in BS EN 13061

(d)   toe of a foot (Davey *et al.*, 1994).

None of these simulates foot flexibility, so impact loads may be excessive, due to the rigid striker and the localised load application. The kinetic energy of the striker was often adjusted to give a 'reasonable' level of peak force, rather than being related to the threat.

If just the force on part of the leg is measured, and the leg can move, the protector deformation cannot be calculated. This is needed to produce graphs of the force vs. deformation – used to identify deformation mechanisms – and to develop models of the protector performance.

2.3.3 Effective mass and effective impact energy

The test rig impact should produce an impact on the protector of similar intensity to that experienced in the real impact, in spite of the mechanics being much simpler, i.e. the striker being a single mass rather than a series of connected bony masses and soft tissues in the foot. The impact velocity, V, should be typical of the sport in case the materials properties are rate dependent. The foot mass in the test rig should be the 'effective mass M_e' of the human foot. This is defined as the mass of the rigid body with the same velocity as the foot, which produces the same peak force in an impact. Mills and Zhang's (1989) analysis of the impact between two rigid bodies with a contact stiffness, k, between them gave the initial peak force as:

$$F = \sqrt{M_e k}\, V \qquad\qquad [2.7]$$

If peak forces are measured in real-body impacts, the formula can be used to compute the effective mass of a rigid striker. This approach works if the impact is over in the order of 5 ms. However, for complex-shaped force vs. time traces that last longer, a more complex model may be required, with linked masses and springs.

It is easier to analyse a protector's performance if it is supported on a rigid anvil. However, this tends to increase the impact severity compared with a real-body impact. Hence, the striker effective kinetic energy must be calculated (Gilchrist and Mills, 1996). In a two-body impact, such as in pre-1985 motorcycle helmet standards, a striker of mass, m_1, has initial velocity V_1, whereas the initial velocity of the headform and helmet of mass, m_2, is $V_2 = 0$. Assuming that the headform moves in a straight line, momentum is conserved in the collision, so the common velocity, V_c of the masses at the moment of nearest approach is:

$$V_c = \frac{m_1 V_1 + m_2 V_2}{m_1 + m_2} \qquad [2.8]$$

The 'effective impact energy E_e' is defined as the energy input to the guard up until the time when m_1 and m_2 have a common velocity (when there is peak guard deformation). Irrespective of the coefficient of restitution of the guard:

$$E_e = \left(\frac{m_2}{m_1 + m_2} \right) \frac{m_1 V_1^2}{2} \qquad [2.9]$$

For example, if a 0.1 kg forefoot effective mass impacts a movable 0.6 kg tibia, the effective impact energy on the tibia is 86% of the forefoot's kinetic energy.

   If test rigs are complex, with several moving masses, or with deformable 'body parts', the analysis of the rig response becomes complex. It is not adequate arbitrarily to modify the impact energy in order to achieve an expected peak force level. Owing to unsuitable rig design, with unrealistic impact energies and impacting objects, some shin protector test rankings are of dubious validity.

2.3.4 Designs for shin protectors

Shin guards are worn over the front of the tibia, and ankle protectors over bony protuberances on the outside of the ankle, where soft tissue coverage is low. These protuberances have a small contact area with impacting flat or convex object, and the bone has a high Young's modulus; consequently impact forces and pressures can be high, leading to injuries.

   Shin protector shells have different bending stiffness in the horizontal cross-section and vertical directions (see Fig. 2.7(b)). If the load is high, the shell can buckle inwards. The horizontal bending stiffness should be sufficient to transfer load to the muscles at the side of the tibia, but not too great to save size/weight. There is no need for the shell stiffness to exceed 4× that of the leg muscles loaded at the sides. Its vertical bending stiffness is unlikely to approach the mean value of 180 Nm^2 of cadaver tibia in the antero-posterior direction (Heiner and Brown, 2001), so it is difficult to spread load above and below the impact site. The cushioning of the front of the tibia is important. It is surprising that protectors do not contain special foam in this area.

2.3.5 EN standard

BS EN 13061 was published in 2001 after many years in development. Its impact tests are not demanding. In its stud impact tests, a stud of diameter

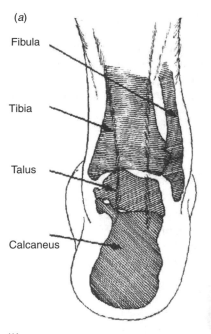

(a)

Fibula

Tibia

Talus

Calcaneus

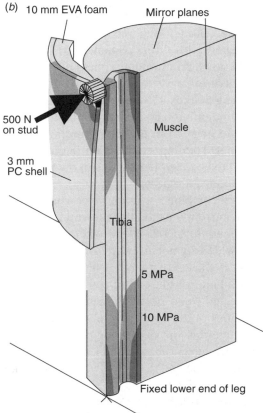

(b) 10 mm EVA foam

Mirror planes

500 N
on stud

Muscle

3 mm
PC shell

Tibia

5 MPa

10 MPa

Fixed lower end of leg

2.7 (a) Structure of the ankle (permission from Hall, 1995), (b) FEA of a shin protector impacted by a stud on player's boot, with 1.7 J energy. Contours of von Mises stress shown.

10 mm makes an almost-tangential impact at 5.4 m/s. The protector must not tear or perforate, but there is no force measurement. There is a blunt impact test with a flat horizontal bar with an extremely low 2 J kinetic energy. The peak force allowed is 2 kN. The impact energies were not justified in any way.

2.3.6 Wearability trials

Ankrah and Mills (2002) carried out trials in which volunteers undertook dribbling, sprinting and shooting tasks while wearing 5, 10, 15 cm thick EVA foam ankle discs. They were timed and asked their opinion of comfort and restriction. One-way analysis of variance (ANOVA) showed no significant change in performance with ankle disc thickness.

2.3.7 Ankle protector materials and tests

Ankrah and Mills (2002) considered the risk to players' ankles to be from impacts from projections on the sole of the opponent's boot – aluminium and nylon studs (truncated conical shapes), and thermoplastic polyurethane (TPU) blades (slightly curved, 30 mm long and 4 mm wide at the top). The worst-case scenario was tested, with direct alignment of the stud with the centre of the ankle protuberance. Aluminium replicas of three ankle bones were assembled using a polyurethane rubber adhesive. The soft tissues of the foot were simulated using Senflex 435 foam – this closed-cell semi-flexible polyolefin foam has an indentation resistance similar to that of athletes' ankles. It allows load transfer away from the talar protuberance, which was covered with 2 mm of leather. The impact site is on the lateral side of the ankle, which was orientated to be the upper surface of the test rig. The lower surface of the bone was supported rigidly.

The prototype protectors combined a foam layer and a domed shell. The foams were either the currently used ethylene vinyl acetate (EVA) of density $30 \, kg \, m^{-3}$, or a foamed 40%/60% blend of Dow ethylene–styrene interpolymer (ESI) and low-density polyethylene, of density $53 \, kg \, m^{-3}$, with improved impact absorption. Sheets of foam (10 mm thick) were thermoformed to a domed shape. 1 mm of glass-fibre composite prepreg (GRP) was vacuum-bag moulded, and 2 mm polycarbonate (PC) sheet was thermoformed into domed shapes of diameter 60 mm and radius of curvature 37 mm. The three shell materials provide a range of bending stiffness. Some commercial, 2 mm thick, injection-moulded low-density polyethylene (LDPE) discs, with the same shape and diameter, were also used. The foam was bonded to the shell inner surface.

Football studs or blades were attached to a vertically falling mass. Equivalent impact energies of 3.5 J were used, to avoid damage to the ankle

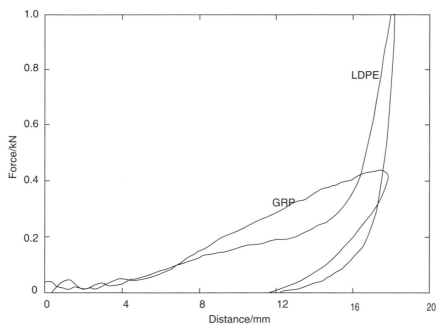

2.8 Force–deflection graphs for 3.5 J blade impacts on ankle protectors with LDPE and GRP shells and ESI foam (Ankrah and Mills, 2002).

model. The total impact force was calculated from the striker acceleration. The deflection of the ankle protector surface was computed by numerical integration of the striker acceleration.

Energy is absorbed by compression of the foam between the shell and the ankle. This increases the duration of the impact and reduces the peak impact force. However, only load shunting will reduce the impulse delivered to the ankle bone. A Tekscan flexible pressure sensing mat, placed between the protector and the ankle model, demonstrated load transfer away from the talar protuberance, when a GRP shell was used. The combination of GRP shell and ESI foam reduces the peak force to the lowest level, of around 0.5 kN (Fig. 2.8). The LDPE shell is far less effective. This suggests that the talus would experience high forces for equivalent impact energies, >3.5 J, if current commercial ankle protection were worn. Forces >5 kN on the talus are likely to cause fractures.

2.3.8 FEA of protector deformation and load distribution

A combination of FEA and high-speed photography was used to investigate the deformation mechanisms of the protectors. The ankle protector

Table 2.2 Data for the simulations shown in Fig. 2.9

Shell	Input energy (J)	Peak force (N)	Deflection (mm)	Peak pressure (kPa)
PC	0.91	328	7.0	80
LDPE	20	83	7.0	250

model (Fig. 2.9) has an axis of rotational symmetry. The stud is an aluminium tapered cylinder of height 10 mm, flat end with diameter 12 mm, and edge radius 1 mm. The protector geometry was the same as in the impact tests. The talus geometry was simplified to be a 16 mm diameter cylinder with a 8 mm radius hemispherical end. Ankle soft tissue was modelled as a nearly incompressible gel-like material, the hyperelastic model in ABAQUS with Ogden shear moduli $\mu_1 = \mu_2 = 200$ kPa, exponents $\alpha_1 = 2$ and $\alpha_2 = -2$, and inverse bulk modulus, $D = 1.0 \times 10^{-9}$ Pa^{-1}, values taken from previous modelling. EVA foam was modelled as a hyperfoam with shear modulus $\mu = 55$ kPa, Ogden exponent $\alpha = 0.5$, and Poisson's ratio = 0. The talus had a Young's modulus of 10 GPa, while the GRP, PC and LDPE shells had Young's moduli of 15, 2.4 and 0.1 GPa, respectively, and appropriate yield stresses. The ankle tissue was bonded to the ankle bone, as was the protector foam to the protector shell. No attempt was made to model the dynamics of the moving masses.

The PC shell is sufficiently stiff not to buckle (Fig. 2.9(*a*)), so the EVA foam is almost uniformly compressed between the shell and the ankle flesh. However, the LDPE shell buckles at its centre (Fig. 2.9(*b*)), reducing the volume of highly deformed foam. The predicted force vs. deflection graph was very close to that measured with 10 mm of EVA foam under a PC shell (Table 2.2).

High-speed photography showed that the LDPE shells buckled; for these shells, pressure distribution measurements showed that there was little load spreading away from the bony protuberance.

2.3.9 The future

It would be useful to relate football players' contusion patterns to the type of shin and ankle protectors worn. Further FEA modelling of the muscle and cartilage around the ankle is needed to lead to optimising the design of ankle protectors. The design of test rigs needs to be developed to be close to the biomechanics of the leg and foot so that the ranking of products is realistic. At present, materials selection is hindered by lack of knowledge of bruising criteria. If bruising occurs for pressures >2 MPa, foams must

(a)

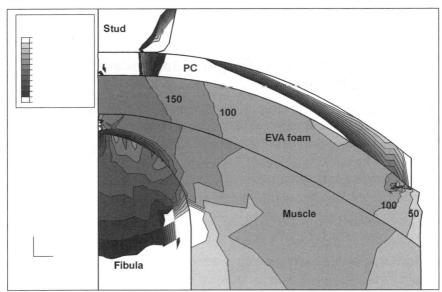

(b)

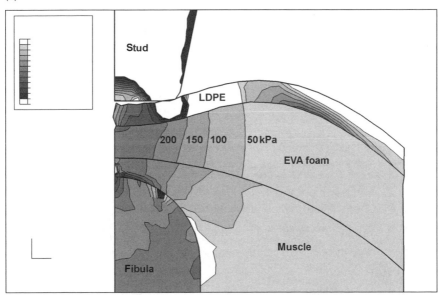

2.9 Predicted deformation of ankle protectors with (a) PC, and (b) LDPE shells, with contours of the vertical compressive stress in kPa. Radial sections are shown (Ankrah and Mills, 2002).

compress to high strains for stresses <2 MPa. The shell must be sufficiently stiff to maintain its domed shape, in order to ensure that the majority of the foam is highly deformed.

2.4   Rigid foam protection for sports wear – cycle helmets

Rigid closed-cell foams in bicycle helmets will be used to illustrate the third application of foams in sports wear. The skull is effectively rigid, so can be protected by a rigid product. However, the foam needs to absorb large amounts of kinetic energy. Figure 2.10 shows a typical cycle helmet and the components; we will concentrate on the foam liner. Most cycle helmets have a thin thermoplastic shell over the foam moulding.

2.4.1 The need for helmets

Competitive cycling involves high speeds, and, in the case of mountain biking, a risk of impacts with rocks, trees, etc. Even cycling for transportation or recreation involves higher speeds than walking. In many countries cyclists risk collision with motor vehicles, which share the road space. Head injuries are more likely to be life threatening than broken limbs, and the medical profession can rarely reverse the effects of brain damage. The use of bicycle helmets appears to reduce the number and severity of head injuries and deaths in crashes (see Section 2.4.8). One strategy is to per-

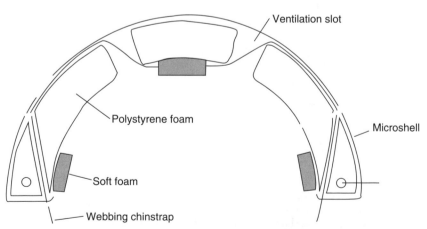

2.10 The components of a bicycle helmet, seen in cross-section. The comfort foam pads are used to fit a range of head sizes.

suade all cyclists to wear such helmets. However, helmets cannot prevent all head injuries; the aim is to minimise the social costs (of medical treatment, long-term care, loss of employment . . .) of injuries to this group of road users. Another strategy is cyclist training or enforcement of the traffic laws, in an attempt to eliminate foolhardy behaviour. In some countries, cyclists are separated from other motorists, which means that cycle helmets are less necessary.

2.4.2 Biomechanics criteria for head injuries

Serious injuries to the brain can be attributed to three causes.

Skull fractures

Skull fractures are mainly caused either by rigid convex objects penetrating the skull, or by a blow of a rigid flat object causing linear fractures without penetration. The helmet may prevent such fractures by distributing the force applied to the head.

Linear acceleration of the brain

A direct blow to the skull can cause brain swelling and bleeding below the impact point, or on the opposite side of the skull (a contra-coup injury), or distributed through a region of the brain. Large collections of blood (haematomas) can be seen in computer tomography X-ray (CT) sections through the head. Blows to the sides of the human head appear to cause more severe brain injuries than frontal blows with the same acceleration levels.

Rotational acceleration of the brain

When the heads of animals were subjected to high levels of rotational acceleration, without linear acceleration, it was possible to produce concussion, or permanent brain damage of a diffuse nature (Bandak *et al.*, 1996). Levels of rotational acceleration that cause concussion in humans are estimated to be of the order of $10\,000\,\mathrm{rad\,s^{-2}}$.

In crashes, a blow to the head rarely acts through the centre of gravity of the head, so a combination of linear and rotational acceleration occurs. On the short timescale of a typical frontal or lateral head impact, the head moves linearly while the neck 'shears'; only later is its motion constrained by the neck. The injury-causing head accelerations occur on a short timescale, whereas the neck motions occur on a longer timescale. The torso mass has little effect on the peak head acceleration because the neck reaction force is much smaller than the head contact force. Equation [2.1] relates the impact force on the head to the head linear acceleration. The neck flexibility means it is reasonable to test helmets using a free headform rather than using an anthropometric dummy.

To protect the wearer against head injury the helmet should:

(a)   not be penetrated by convex objects, nor should the local pressure on the skull exceed *c.* 2.5 MPa, or there could be local skull fracture(s)
(b)   prevent the impact force, acting towards the centre of gravity of the head, from exceeding 10 kN, i.e. the linear acceleration of a 5 kg head is <200g
(c)   prevent the force, acting tangential to the helmet surface, exceeding 2 kN. This will keep the angular head acceleration below the estimated limit for injury.

We will concentrate on the linear acceleration condition (b), because helmets are designed to meet this condition. Such designs tend to meet the skull fracture condition. At present, little is done to reduce rotational acceleration and there are no tests for rotational acceleration in the standards.

Forensic reconstructions of cycle helmet damage by Smith *et al.* (1994) and by McIntosh and Dowdell (1992) indicate that the acceleration levels to cause moderate closed-head injury range from 100 to 200g. They show that the majority of cycle helmet damage occurs for impacts equivalent to a free fall of 1.5 m or less on to a rigid flat object. Surveys show that the road surface is commonly impacted, and that the most common impact sites are on the front and sides of the head (Larsen *et al.*, 1991).

2.4.3   Cycle helmet standards

The jargon used in standards can be difficult for the lay reader to understand, and the test equipment may be difficult to visualise from the drawings. Consistent, low capital-cost, test methods simulate direct impacts with rigid flat and convex objects. The kerbstone-shaped anvil (with 105° included angle and 15 mm radius edge) simulates kerbstones or street furniture. Although some cyclists impact deformable steel panels on vehicles, and vehicle glazing, there are no tests on deformable anvils. Table 2.3 compares several national standards, as does the website www.bhsi.org/webdocs/stdcomp. The drop height, which should relate to the head height while riding of about 1.5 m, determines the kinetic energy of the impact. It is a compromise between insisting on high protection levels and product acceptability.

Differences between the tests in standards make a comparison of their protection levels difficult. The human head mass increases with the

Table 2.3 Requirements for cycle helmet impacts

Standard	Snell B95	ASTM	CPSC	BS EN 1078
Drop (m) onto flat anvil	2.2*	2.0	2.0	1.5
Drop (m) onto kerbstone (K) or hemispherical (H) anvil	1.44*(H)	1.2(H) 1.2(K)	1.2(H) 1.2(K)	1.06(K)
Permissible peak acceleration (g)	300	300	300	250

*The impact energy is specified; this is the drop height if the headform mass = 5 kg.

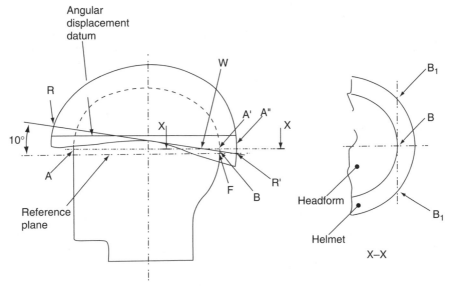

2.11 Impact test sites in BS 1078 are above the line R at the front and R' at the rear (permission from British Standards).

circumference, as do the headforms in the EN but not in the US standards. The tests are minimum requirements; actual acceleration results are occasionally published in consumer magazines. The need to meet impact test criteria at 50 °C and at −20 °C means that the foam may provide head accelerations in the range 150 to 200 g at 20 °C. Figure 2.11 shows the impact sites in BS EN 1078; these are not as low at the front and sides as some real impacts.

2.4.4 Materials and process selection

Many low-density materials can absorb energy in compression without the stress reaching harmful levels. The choice of materials is based on cost, processing and minimising the weight. Cork, used in pre-1960 motorcycle helmets, is only available in limited sizes, is expensive and denser than polymer foams of the same compressive collapse stress. Honeycomb, made from polymer, paper or aluminium foil, is optimal for energy absorption when the compression direction is along the cell axes. However, the flat sheet cannot be bent to a doubly convex shape to fit inside a helmet. It also needs protection to avoid it being split parallel to the cell axes by a wedge-shaped object. Anisotropic cellular materials, with optimal energy absorption direction in one direction, tend only to exist as flat sheets.

Polymer bead foams can be moulded into complex shapes, and the economics are such that cycle helmets are imported and sold in the USA for less than $10. The yield stress of closed-cell foams varies approximately with the 1.5th power of the relative density of the foam and with the polymer yield stress in the bulk state. Glassy PS has a higher bulk compressive yield stress ($c.$ 100 MPa) than rigid PU, so it, or PS/PPO blends, are preferable to PU. The mass of a typical cycle helmet is 0.3 to 0.5 kg. Although it is possible to wear heavier helmets, the extra mass adds to discomfort.

As the foam is a good thermal insulator, while about 30% of the body's heat loss is from the head, air-flow ventilation is necessary for comfort in hot weather. For effective ventilation, the holes at the front of the helmet must be large, and there must be fore and aft channels in the interior surface to allow air flow past the hair and scalp. Holes can be moulded in EPS liners, whereas if cross-linked foam sheets were thermoformed, holes would need to be cut in a post-moulding operation. Therefore, most cycle helmets are moulded from EPS beads, with fewer using EPP, and a few using PU foam castings. The moulds for casting rigid polyurethane foam have a low capital cost, but the process is slow, and the liner density is higher than that of EPS of the same yield stress. Although EPP liners cost several times as much as EPS to manufacture, the foam is less brittle and recovers better after an impact. It may be preferred for skateboard helmets, which suffer a large number of minor impacts. EPS recovers only a little after an impact, so the helmets must be destroyed after a crash.

The helmet must stay in place in a crash, in spite of the top of the head acting like a *ball* to the helmet *socket*. Thin cycle helmet shells provide insufficient attachment strength for webbing chin straps. Therefore, the 20 mm wide polyester webbing straps are passed through slots moulded in the foam (Fig. 2.10), allowing high crash forces (1 kN) to be taken by the foam in compression. Straps run to the front and rear of the helmet; the front straps tighten if the helmet rotates rearward, and the rear straps tighten if the helmet rotates forward. This is only effective if the strap is tightened securely under the chin, which may be slightly uncomfortable. If the helmet can be rocked back by the wearer to expose the forehead, this could drastically reduce the head protection in a crash.

2.4.5 Design: the force–deflection relationship of the helmet foam

The microshell of bicycle helmets causes little load spreading, so can be ignored for impact design. Similarly, the low shell and liner masses do not give rise to force oscillations. Although neither the human skull nor the outer surface of a helmet is exactly spherical, it is a reasonable approximation that the section where the impact occurs is spherical. Both the skull and the road surface are treated as being rigid. However, the 5 to 8 mm thick scalp is relatively soft and deformable.

Two cases are considered. The first is an oversimplification, but readily leads to a design, while the second provides the justification for the design.

A foam with constant collapse stress, and zero load spreading

The contact geometry between a flat rigid surface (the road) and a helmet of outer radius, R, is shown in Fig. 2.12. The liner crush distance, x, is much less than R (100 to 200 mm), since the liner thickness $T < 30$ mm. The foam crushes over a disc of radius, a. Applying Pythagoras' theorem to the triangle gives:

$$R^2 = (R - x)^2 + a^2$$

If the x^2 term is ignored in the expansion of the brackets, the contact area A is:

$$A = \pi a^2 = 2\pi Rx \qquad [2.10]$$

Assuming no load spreading, and that the foam has a constant yield stress σ_y while the strain is increasing, the force F transmitted by the foam is:

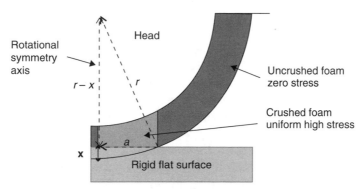

2.12 Geometry of head and helmet foam crushing, showing assumed stress contours.

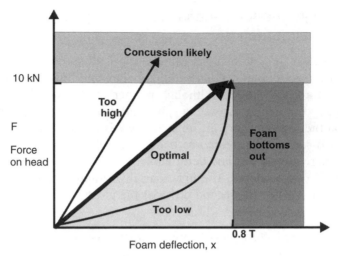

2.13 Response for a bicycle helmet hitting the road surface, for too low, optimal, and too high foam yield stress.

$$F = A\sigma_y = 2\pi R\sigma_y x \qquad [2.11]$$

The slope of this straight line is the *loading slope*, k. Substituting typical values of $R = 140\,\text{mm}$ for the front of a helmet liner, and $\sigma_y = 0.7\,\text{MPa}$, gives $k \cong 600\,\text{N/mm}$.

Helmet design

For an average-sized headform of mass $5\,\text{kg}$, undergoing a 2 metre drop, the kinetic energy at impact is $E = 100\,\text{J}$. Although the BS EN 1078 allows a $250\,\boldsymbol{g}$ headform deceleration, a margin must be allowed for material variability, and tests at high and low temperatures. Hence the design acceleration limit is $200\,\boldsymbol{g}$ at $20\,^\circ\text{C}$. This is equivalent to the condition that $F < 10\,\text{kN}$ (Fig. 2.13). The other limit in the figure is for the foam bottoming out. This occurs when the compressive strain approaches $1 - R$ (R is the foam relative density), and the stress rises rapidly. Typically, bottoming out occurs at a strain of 80%. The foam thickness, T, must be a minimum of $1.25\,x_{\text{max}}$ to avoid this.

The area under the force–deflection curve is equal to the energy input. If the force reaches $10\,\text{kN}$, as the headform comes to a momentary halt at deflection x_{max}, and equation [2.11] applies,

$$0.5 \times 10\,\text{kN} \times x_{\text{max}}\,\text{mm} = E \qquad [2.12]$$

Therefore, the optimum foam is one where the loading line meets the intersection of the two 'dangerous' areas. If foams with too low a yield stress are used, they bottom out before the load reaches 10 kN; if the foam has too high a yield stress, the force reaches 10 kN before the foam bottoms out.

The sequence of design is as follows:

(a)   Given the impact energy, E, equation [2.12] is used to find x_{max}. For $E = 100$ J, $x_{max} = 20$ mm. Allowing for bottoming out, the foam thickness, $T = 25$ mm.

(b)   Using the helmet radius at the impact site and equation [2.11], the foam yield stress is calculated to give a loading curve that passes through the point (80% of foam thickness, 10 kN) – Fig. 2.13.

(c)   The polymer and the foam density are chosen to provide the appropriate yield stress. For an impact site of radius 100 mm, the yield stress required is 0.7 MPa, hence the density of EPS should be 65 kg m^{-3} (according to the relationship between density and initial yield stress).

(d)   The loading curves are recalculated for the extremes of the test temperature range in the standard, using data for the foam at –20 °C and 50 °C.

For other design impact velocities, V, the foam thickness is proportional to V^2, and the foam yield stress to V^{-2}, to keep the head acceleration below 200g. Consequently, high speed impacts will be outside the protective capacity of helmets.

Variable collapse stress and load spreading (FEA)

There is no published FEA of the foam deformation for impacted cycle helmets. The author, using the approach of Masso-Moreu and Mills (2003), examined the stress distributions for:

(a) a headform having a radius of curvature 120 mm in the fore and aft direction, and of 80 mm in the side-to-side direction. The helmet had a uniform 30 mm thickness PS foam liner, with density 35 kg m^{-3}.

(b) an off-centre impact site on the coronal plane (the vertical plane containing both ears).

(c) impacts with either a flat rigid plane, the kerbstone of BS EN 1078, or a rigid hemisphere of 50 mm radius.

In Fig. 2.14(a) the stress contours are nearly vertical. There is a nearly constant stress in the contact area, and a rapid decrease in stress at the sides of the contact area, confirming the approximation of Fig. 2.12. For the kerbstone (Fig. 2.14(b)), there is an increase in the foam stress towards the centre of the contact area and there are non-zero compressive stresses well

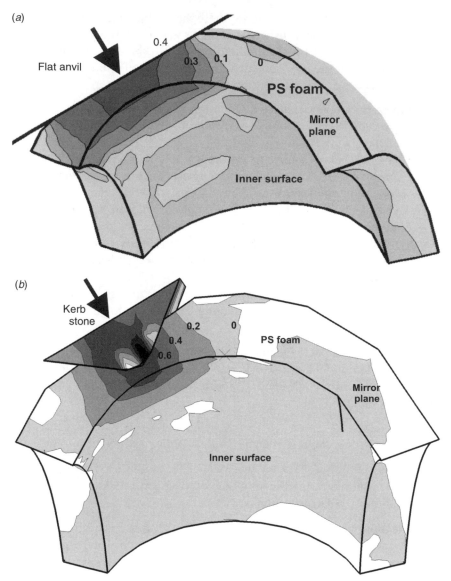

2.14 Impacts, at sites 30° from the helmet crown, onto (a) flat surface
(compressive stress contours in MPa), (b) kerbstone (seen end
on).

outside the contact area. Consequently, the stress distribution is quite different from that in Fig. 2.12.

Figure 2.15 compares the loading curves for the three types of anvil. The
impact sites are 30° from the crown for the kerbstone and flat anvils, but
60° for the hemisphere (close to the helmet lower edge). The responses are

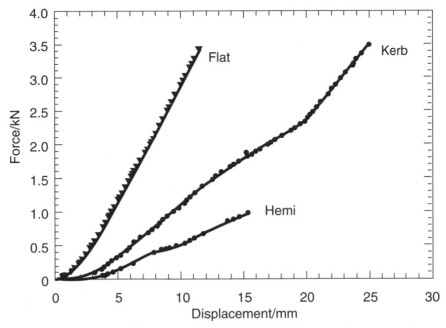

2.15 Predicted force vs. foam crush distance for a cycle helmet using XPS35 foam, on a variety of anvils.

nearly linear, confirming the simple relationship of equation [2.11]. However, the loading slopes for the flat, kerbstone and hemispherical anvils are 337, 150 and 81 N mm^{-1}, respectively, while equation [2.11] predicts a slope of 326 N mm^{-1} for a mean helmet outer radius of 130 mm and an initial yield stress of 0.4 MPa. Hence equation [2.11] underestimates the FEA loading slope by 3% for flat surface impacts.

Subsidiary analysis showed that a thermoplastic, 0.5 mm thick, shell, of Young's modulus 2 GPa, raised the loading slope by 11%. Other factors, such as the non-ideal fit of the headform to the inner surface of the helmet liner, will reduce the loading slope. Therefore, the simple analysis of equation [2.11] is reasonable for flat impacts, but cannot predict the different responses for other anvil shapes. A rule of thumb for kerbstone impacts is that the loading curve slope is 50% of that for a flat surface impact, but this ratio depends on the orientation of the kerbstone relative to the helmet. A foam that is ideal for an impact with a flat surface at a given velocity has a sub-optimal yield stress for an impact on a kerbstone at the same velocity. At the side of the helmet, where the radius of curvature is large, the foam geometry must be altered, for example, by incorporating ventilation slots, thereby reducing the contact area between the foam and the head.

However, many helmet designs have large vents in all parts of the helmet, so alternative solutions must be sought. Helmet foam optimisation depends on the impact velocity and the nature of the object struck, so it is not possible to have an optimum design for all impacts.

2.4.6 Experiments to validate the analysis

The experiments of Gilchrist and Mills (1991) were on bicycle helmets that differ from current designs. There were very few ventilation slots and the 2mm thick HDPE outer shell was much thicker than microshells. Its loading response for impacts on a flat surface was linear, in agreement with the simple theory of equation [2.11]. Mills and Gilchrist (2003) have begun to validate design methods for modern helmet designs involved in oblique impacts.

2.4.7 The need for a helmet microshell

In *c.* 1990, several designs of cycle helmets had cloth covers over the EPS mouldings. McIntosh & Dowdell (1992) showed that some of these helmets break up into pieces in crashes. In a crash, any localised crushing of EPS may initiate cracks at the sides of the indented area. If the impact is on one of the ventilation holes, the effective crack length is half of ventilation hole length plus the surface (bead boundary) crack length. The wedging force needed to propagate this crack is of the order of the 50N. As the crushing force in a cycle helmet impact is many kN, foam fractures are possible. The fracture toughness of EPS is very low, ranging from 80 to 110kN m$^{-1.5}$ (Mills and Kang, 1994). Consequently, cracks can propagate and split the helmet shell into several pieces.

The micro-shell is often a PVC thermoforming 0.5mm thick, the edges of which are taped to the foam. It provides the tensile strength on the outer surface of the helmet and prevents the foam liner from fracturing in a crash. It plays a minor part in the energy dissipation but it prevents foam damage in everyday use. A helmet with a microshell is less prone to indentation in oblique impacts on a rough road surface, so it aids 'sliding' on the surface, and reduces the force tangential to the helmet surface. Hodgson (1990) showed that there is high friction in an oblique impact between soft-shell cycle helmets and concrete. The design of ventilation slots is also critical.

2.4.8 Cycle helmet effectiveness

The effectiveness of cycle helmets has been assessed by epidemiological studies. Thompson *et al.* (1989) analysed the injury severity of cyclists admitted to hospital in 1986–7 in the USA, comparing injuries suffered by helmet wearers with those by non-wearers. Helmet use reduced the risk of head injury by 85% and the risk of brain injury by 88%. Maimaris *et al.* (1994), who studied admissions to a British hospital, found that wearing a cycle helmet reduced the risk of a head injury by a factor of 3.25 (with 95% confidence limits of 1.17 to 9.06), but head injuries still occur for some severe collisions with motor vehicles. The Snell Foundation in the USA partly commissioned the Harborview survey of 527 helmets (Rivara and Thompson 1996). They comment, 'Brain injury increased very slightly with increasing (helmet) damage score up to the point where the helmet received catastrophic damage. Then the injury rates shot up dramatically.' Hence, impacts that do not completely crush the cycle helmet foam will only cause minor head injuries.

The website www.bhsi.org lists research literature for and against the use of cycle helmets. Hillman (1993) argued that a cyclist with a helmet may take greater risks, for instance, increasing his speed until his perceived feeling of risk is the same as before. He argues that cycle helmets cannot be relied upon to protect riders from all the types of head injury – 'To avoid impairing vision or hearing, cycle helmets are designed to be worn high on the head and thus do not afford protection to parts of the head, neck and upper face which account for half of the so-called "head" injuries of cyclists. In addition they do not protect the head from rotational trauma, which can seriously damage the brain and brainstem and which is quite common when cyclists are hit a glancing blow from a motor vehicle' (McCarthy, 1992).

Spaite *et al.* (1991) showed that wearing a helmet reduces the rate of injury to parts of the body other than the head; the inference is that helmet wearers are careful riders. Australia, New Zealand, and several US states have made the wearing of cycle helmets compulsory. The change in Australia in July 1990 significantly reduced head injuries, partly attributed to a reduction in the number of cyclists on the roads. The site http://rmstuart.uthscsa.edu/fatal_brain/helmet.html has case histories and medical information. Recent New Zealand statistics show a drop in pedestrian injuries in parallel with cyclists' injuries, in spite of the former group not wearing helmets. The interpretation of injury statistics is a minefield!

2.4.9 The future

It is likely that cycle helmet design will change to provide greater protection. A helmet of thickness 50mm could provide head protection for a fall from 4m onto a road surface, or for higher speed impacts with vehicles. However, it might be unacceptable ergonomically or aesthetically. The mass would be only 25% greater, since an EPS of 63% of the density would have a compressive yield stress of 50% of the EPS used in the current 25mm thick helmets. There would be greater thermal insulation, and greater angular inertia. The field of vision of the wearer must not be compromised

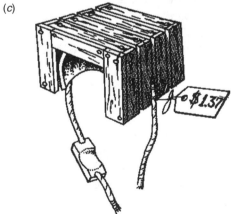

(a)

Multilayered laminated
concentric spheres

Small circular
eye ports to
minimise stress
concentrations

(b)

Chin strap
exceeds all
known stds

Anterior inferior
posterior strap
passes under
crotch and attaches
to rear of shell
in latch

(c)

$1.37

2.16 (*a*), (*b*), (*c*) Cartoons showing the perspective of the shell
designer, the retention system designer, and the manufacturer
(permission from Newman, 1978).

in spite of the need to protect the wearer's forehead. Figure 2.16 (Newman, 1975) illustrates some constraints on the design process.

There can be high rotational accelerations of a helmeted headform in an oblique impact with a road surface (Hodgson, 1990) and one research aim is to reduce this acceleration.

2.5   Further sources of information

There are no books available yet on this area. The author's forthcoming book (Mills, 2003) has chapters on sport applications, and also considers foam designs in the transport, medical and engineering fields.

Information sources can be classified.

Specialist journals

For journals see those in the reference list. There is no single journal specialising in foams for sport protection.

Trade associations

British Association of advisors and lecturers in physical education at www.baalpe.org
The Sports Industry Federation at www.sportslife.org.uk
Polyurethane Foam Association at www.pfa.org

Medical associations

Institute for Preventative Sports Medicine www.ipsm.org, which has papers on shin guards and sport goal protection.
International Research Council on Biokinetics of Impacts www.ircobi.org, which runs annual conferences on transport injury biomechanics. Its secretariat is at www.inrets.fr

2.6   Summary

The foam selection for the three case studies have been in three general areas:

(a) elastomeric open-cell foams for large static protection products
(b) closed-cell foams of semi-crystalline polymers for protection worn on moving bony projections of the body
(c) closed-cell foams from glassy polymers for head protection.

In all the case studies the foam density is kept below about $80 \, \text{kg m}^{-3}$, so that more than 90% of the volume is air. The design of all these products has been shown to be a compromise, so it is not possible to maximise the protection at the expense of the ability to perform the sport.

2.7 Acknowledgements

This chapter is based in part on the research of Stephanie Ankrah, Iona Lyn and Adam Gilchrist; I am grateful for their help.

2.8 References

Ankrah S & Mills N J (2002). Ankle protectors in football shin guards. *The Engineering of Sport 4*, ed. S Ujihashi and SJ Haake, 128–135. Blackwell, Oxford.

Ankrah S & Mills N J (2003). Performance of football shin guards. *Sport Engng*, Submitted.

Bandak F A, Eppinger R H & Omaya A K, eds. (1996). *Traumatic Head Injury*, Mary Ann Liebert Inc, Larchmont, NY.

Beiner J M & Jokl P (2001). Muscle contusion injuries: a review, *J Am Assoc Orthop Surg*, **9**, 227–237.

Bir C A, Cassata S J & Janda D H (1995). An analysis and comparison of soccer shin guards, *Clin J Sport Med*, **5**, 95–99.

BS EN 960 (1995). *Headforms for use in the testing of protective helmets*, British Standards Institution, London.

BS EN 1078 (1997). *Protective Helmets for Pedal Cyclists*, British Standards Institution, London.

BS EN 12503 (2001). Parts 1 to 7, *Sports mats*, British Standards Institution, London.

BS EN 13061 (2001). *Shin guards for association football players*, British Standards Institution, London.

Consumer Product Safety Commission (1998). 16 CRF Part 1203, Safety standard for bicycle helmets, Federal Register, **63**, 11712–11747.

Crisco J J *et al.* (1996). Maximal contraction lessens impact response in a muscle contusion model. *J Biomech*, **29**, 1291–1296.

Davey D W, Morrison C J *et al.* (1994). Body protective wear, with particular reference to football injury. *The Engineering of Sport*, ed. S Haake, 117–124.

Ekstrand J & Gillquist J (1983) Soccer injuries and their mechanisms: a prospective study. *Med and Sci in Sports and Exercise*, **15**, 267–270.

Francisco A C, Nightingale R W *et al.* (2000). Comparison of soccer shin guards in preventing tibia fracture. *Am J Sport Med*, **28**, 227–233.

Franke (1977). *Traumatologie des Sports*. VEB Verlag Volk und Gesundheit, Berlin.

Fried T & Lloyd G (1992). An overview of common soccer injuries. *Sport Med*, **14**, 269–275.

Gilchrist A & Mills N J (1991). The effectiveness of foams in bicycle and motorcycle helmets, *Accid Anal Prev*, **23**, 153.

Gilchrist A & Mills N J (1996). Protection of the side of the head, *Accid Anal Prev*, **28**, 525–535.

Hall S J (1995). *Basic Biomechanics*. Mosby – Year Book Inc, p. 234.

Hawkins R D & Fuller C W (1999). Epidemiological study of injuries in four English football clubs, *Br J Sport Med*, **33**, 196–203.

Heiner A D & Brown T D (2001). Structural properties of a new design of composite replica femurs and tibias. *J. Biomech*, **34**, 773–781.

Hillman M (1993). *Cycle Helmets: The Case For and Against*. Centre for Policy Studies, London.

Hodgson V R (1990). Impact, skid and retention tests on bicycle helmets. Report of Wayne State Univ, Dept of Neurosurgery, Detroit.

Klempner D & Frisch K C, eds. (1991). *Handbook of Polymeric Foams and Foam Technology*, Hanser, Munich.

Larsen L B, Larson C F et al. (1991). Epidemiology of bicyclist's injuries, *IRCOBI Conf*, 217–230.

Lyn G & Mills N J (2001). Design of foam crash mats for head impact protection, *Sports Eng*, **4**, 153–163.

Lyn G & Mills N J (2002). Design of foam crash mats for head impact protection. *4th International Conference on The Engineering of Sport*, ed. S. Ujihashi, 89–94.

Maimaris C, Summers C L et al. (1994). Injury patterns in cyclists attending an accident and emergency department – a comparison of helmet wearers and non-wearers. *Br Med J*, **308**, 1537–1540.

Masso-Moreu Y & Mills N J (2003). Impact compression of polystyrene foam pyramids, *Int J Impact Eng*, **28**, 653–676.

McCarthy M (1992). Do cycle helmets prevent injury? *Br Med J*, **305**, 881–882.

McIntosh A & Dowdell B (1992). A field and laboratory study of the performance of pedal cycle helmets in real accidents. *IRCOBI Conf*, 51–60.

McIntosh A S, Kalliaris D et al. (1996). An evaluation of pedal cycle performance requirements. *Stapp Car Crash Conf*, SAE, 111–119.

Mills N J (1990). Protective capability of bicycle helmets. *Br J Sports Med*, **24**, 55–60.

Mills N J (1996). The biomechanics of hip protectors. *Proc I Mech E, part H*, **210**, 259–266.

Mills N J (2003). *Polymer Foam Mechanics-Engineering Medical and Sports Applications*. Unpublished data.

Mills N J & Gilchrist A (1990). Body protectors for horse-riders. *IRCOBI Conf*, 155–166.

Mills N J & Gilchrist A (1991). The effectiveness of foams in bicycle and motorcycle helmets. *Accident Anal Prev*, **23**, 153–163.

Mills N J & Gilchrist A (1997). The effect of heat transfer and Poisson's ratio on the compressive response of closed cell polymer foams. *Cell Polym*, **16**, 87–119.

Mills N J & Gilchrist A (2000). Modelling the indentation of low density polymer foams. *Cell Polym*, **19**, 389–412.

Mills N J & Gilchrist A (2003). Reassessing bicycle helmet impact protection. *IRCOBI Conf.*

Mills N J & Kang P (1994). The effect of water immersion on the fracture toughness of polystyrene foam used in soft shell cycle helmets. *J Cell Plast*, **30**, 196–222.

Mills N J & Lyn G (2001). Design of foam padding for rugby posts. *Materials in Science and Sports*, ed. FH Froes, 105–117. TMS, Warrendale, PA.

Mills N J & Lyn G (2002). Experiments and modelling of air flow in impacted flexible polyurethane foams. *Cell Polym*, **21**, 343–367.

Mills N J & Zhang P S (1989). The effects of contact conditions on impact tests on plastics. *J Mater Sci*, **24**, 2099–2109.

Mills P (1989). Pedal cycle accidents; a hospital study. Transport and Road Research Lab research report RR220, Crowthorne.

Newman J A (1975). On the use of the HIC in protective headwear evaluation. *19th Stapp Car Crash Conf*, 615–639.

Newman J A (1978). Engineering considerations in the design of protective head-gear, *Proc Assoc Adv Auto Med*, Ann Arbor, MI.

Nyquist G W, Cheng R *et al.* (1985). Tibia bending: strength and response, *29th Stapp Car Crash Conf*, SAE Warrendale, PA.

Ogden R W (1972). Large deformation isotropic elasticity. *Proc Roy Soc Lond* A, **328**, 567–583.

Phillipens M & Wismans J (1989). Shin guard impact protection. *IRCOBI Conf*, 650–676.

Rivara F P & Thompson R S (1996). Report for the Snell foundation (on *www.smf.org*). Circumstances and severity of bicycle injuries.

Seipel R C, Pintar F A *et al.* (2001). Biomechanics of calcaneal fractures. *Clin Orthop*, **388**, 214–224.

Smith T A, Tees D *et al.* (1994). Evaluation and replication of impact damage to bicycle helmets. *Accid Anal Prev*, **26**, 795–802.

Snell Memorial Foundation (1995). Standard for protective headgear for use in bicycling, New York.

Spaite D W, Murphy M *et al.* (1991). A prospective analysis of injury severity among helmeted and non-helmeted bicyclists involved in collisions with motor vehicles, *J Trauma*, **31**, 1510–1516.

Thompson R S, Rivara F P *et al.* (1989). A case control study of the effectiveness of bicycle helmets, *N Engl J Med*, **320**, 1361–1367.

van Laack W (1985). Experimentelle Untersuchungen über die Wirksamkeit verschiedener Schienebeinschoner in Fussballsport, *Z Orthop*, **123**, 951–956.

Zhu H, Mills N J & Knott J F (1997). Analysis of the high strain compression of open-cell foams, *J Mech Phys Solids*, **45**, 1875–1904.

3

Performance of sports surfaces

COLIN WALKER

University of Strathclyde, Scotland, UK

3.1 Introduction

When sports were developed and their rules formalised, mostly in the nine-teenth century, they were played on whatever surface was readily to hand. For this reason, golf developed on the sand dunes of St Andrews. Lawn tennis derived its name from its origins on the grassy lawns of English country houses. Football and rugby developed on the natural turf of village greens or school fields. Since these early days, profound changes have taken place in the way in which sports are played, but the characteristics of the playing surface in many ways continues to define the games themselves. In this chapter, the nature of sports surfaces will be discussed, along with the ways in which their performance is assessed. The introduction of new surfaces, and the ways in which they have been integrated into the playing of games, is currently a matter of ongoing interest in many sports. We have, for instance, the use of water-based pitches in hockey, or the variety of sur-faces which are in use for tennis – carpet, clay, all-weather, as well as the variations in the grass surfaces. Despite the efforts that have been expended to assess surface performance in a way in which is repeatable and relevant, the fact remains that most surface development still proceeds on a trial-and-error basis. The existence of comprehensive performance standards does, however, ensure that, as new surfaces appear, they do much to protect the players from surface-related injury.

3.2 Why do we have a diversity of sports surfaces?

Before beginning to consider the range of surfaces and their performance, it is reasonable to ask why there should be such a range of surfaces. The real answer is that sport is now played in all kinds of climates and locations, indoors and outdoors. In addition, the pressure on facilities is such that the number of games to be played per year is a major consideration in the choice of a surface – one artificial pitch may be able to accept as much as

100 times the usage of a grass pitch, and will be playable almost regardless of the weather. There are surfaces, too, such as the clay court in tennis, which may originally have been seen as a passable substitute for grass, but which now has developed its own characteristic style of play. In all sports, too, there is a desire to maintain the spectacle, and for this reason new rugby turf pitches are usually laid down on a 'soil' that is largely sand. The free drainage that the sand provides allows play to proceed during and after rain without the pitch degenerating into the quagmire that has been common in the past. The ability to play indoors, protected from the weather, is a capability that is valued in extreme climates and so a whole range of surfaces has been developed for indoor use. In addition, since grass may not thrive in the modern, enclosed stadia, turf technologies have been developed whereby grass may be nurtured outside the stadium, ready to be installed before each game.

3.3   The measurement of surface performance

The performance of sports surfaces is measured by a battery of physical tests. These will be discussed in this section, bearing in mind that not every test is applicable or appropriate for each surface, nor can any one specification apply to every type of surface.

The aim of a test regime is twofold: firstly to ensure that the surface is safe for use, and secondly to ensure that it does comply with the standards set down. In some cases these standards will be performance related, as in the resilience of a running track; in others it will be a case of ensuring that the surface has been supplied and installed in conformance with standards[1] that aim to ensure that the surface will perform and endure over its scheduled life. As far as performance is concerned, there are two parameters that define the way that a surface will interact with an athlete – the resilience of the surface and the torsion between the foot and the surface.

3.3.1   Measurement of resilience

The resilience of a surface is measured by dropping a weight on to a measurement foot that incorporates a spring. The standard device is known as the 'Berlin Athlete' (Fig. 3.1). It uses a 20 kg weight dropping on to a 3 kg measurement foot that incorporates a spring with a spring constant of 2000 MN/m (i.e. it takes a force of 2 kN to cause a deformation of 1 mm in the spring). The mechanics of the Berlin Athlete have been analysed in detail.[2] Over the last 20 years, this has been the basis of testing, but the test does have flaws, which imply that it is a less than perfect device. Firstly, since the Berlin Athlete is of all-metal construction with low damping, there is a high degree of mechanical noise that degrades the signal. When the test was

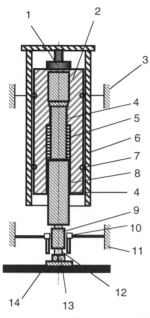

1 –	Electromagnet
2 –	Weight 50 kg
3 –	Carrier
4 –	Slide bearings
5 –	Spring 2 MN/m
6 –	Outside tube
7 –	Steel balls
8 –	Steel leg 8.3 kg
9 –	Strain gauge
10 –	Deflection device range + 3 mm or + 25 mm
11 –	Holder
12 –	Steel cylinder
13 –	Test foot diameter 69 mm or diameter 40 mm
14 –	Surface to be tested

3.1 Schematic of the Berlin Athlete as modified for surface displacement measurement.[16] The *Stuttgart Athlete* uses a similar system, with a different drop weight and spring constant.

devised, the weight and spring constant were chosen to have a resonant frequency of <50 Hz. The measurement foot and the surface form a second resonant system with a resonant frequency of 100–250 Hz. This signal, along with the mechanical noise, has to be removed from the fundamental. In fact, the two resonances are rather close together for successful filtering without degradation of the signal.

In operation, the Berlin Athlete is calibrated by dropping the weight on to a standard surface; the test surface is then defined by the force reduction observed when the Berlin Athlete is dropped on it, where

$$\text{Force reduction} = \frac{\text{Maximum force on test surface}}{\text{Maximum force on calibration surface}}$$

While this would appear at first sight to be a realistic way of measuring the surface resilience, several factors intrude on this basic simplicity. The test was originally envisaged for surfaces such as gymnasium floors, and for this purpose it may well be accurate and realistic, since the force–deflection curve of the typical floor is linear and elastic.

For surfaces with a non-linear force–deflection characteristic, or those which show viscoelastic properties (and this class includes all outdoor surfaces), the Berlin Athlete needs its performance to be qualified.

(a)   The measurement foot is 70 mm diameter. Its surface indentation characteristics are quite different to a shod heel (with a diameter of ~50 mm). This point has been discussd in the literature under the heading of 'point contact' vs. 'area contact'. The indentation of the measuring foot into the surface is certainly area contact, i.e. the compression of the area under the foot is uniform. The heel impact is mainly point contact, i.e. the compression of the surface is greatest close to the middle of the heel area, and is less at the rim of the shoe.[3]

(b)   Where the systems shows viscoelastic properties, the measured force–deflection curve is time dependent, so that the apparent surface stiffness depends upon the timescale of the test. In surface testing, the timescale of the test should approximate to that of a footfall. In this respect, the Berlin Athlete is on the high frequency side of ideal for the footfall as a whole, but close to ideal for the first phase of contact when the heel hits the surface and is decelerated.[4]

(c)   Where the force–deflection curve of the surface is non-linear, the maximum force should closely approximate a footfall. Since the dynamic modulus, i.e. the actual stiffness that is experienced by the athlete's legs, is defined by the tangent to the force–deflection curve (at the correct timescale) (see Fig. 3.2), the Berlin Athlete should be

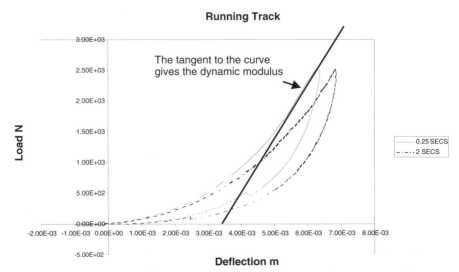

3.2 Force–deflection curve for a typical sports surface (in this case a running track). The main characterising parameter is the dynamic modulus at the peak load reached by the athlete – normally >2.5 times the athlete's body weight. The two curves indicate that the surface properties change with the timescale of testing.

dropped at varying heights until a peak force of ~2.5 kN is obtained. This may then be compared with a drop from the same height on to a standard surface. For the reason explained in (a) and (b), even then this will not really be an accurate reflection of a footfall on the surface. If the weight is simply dropped from a standard height, the resulting dynamic compliance will not reflect the behaviour of the surface when an athlete runs on it, since it is dependent upon the actual load level achieved in the test.

In fact, the Berlin Athlete should be used in the manner of a secondary standard as follows. The properties of a surface may be measured in the laboratory, using a rapid-acting servo-hydraulic tester, with a heel-shaped indenter. The 'correct' or 'desired' force–deflection curve is now known. The drop weight tester may then be operated as in (c) to yield, say, a peak load of 2.5 kN at a drop height of 65 mm. The surface, as laid outdoors, may then be compared with this figure, and suitable variance from this agreed upon between the contractors and the client, e.g. that the 2.5 kN level should be reached from a drop height of between 62 and 68 mm.

To sum up, then, the Berlin Athlete works well as a device for ensuring compliance of a surface with the standard. On its own, it is not capable of making comment on whether or not a surface is likely to be kind to the legs of those playing on it. At this point in time, there has been no definitive study relating specific surface properties to patterns of injury, so that it is scarcely surprising that the Berlin Athlete cannot do this. It is the test specified in all of the standards, and, as such, it is well known and widely used. It is of interest that the football authorities in the shape of FIFA have defined a similar, but different, test, the 'Stuttgart Athlete' as the standard test for their project of developing artificial pitches for soccer.

3.3.2 Surface torsion

The measurement of surface torsion is almost more important than resilience, since there have been many instances where surfaces have been held to cause knee injury due to the locking of the foot in the surface during a turning manoeuvre. As may be imagined, the likelihood of the foot locking is related to the speed of running prior to the turn, and so it is more likely for faster athletes and for those who play in positions that demand sprinting speed.

The test for surface torsion involves the use of a metal 'foot' with a pattern of studs similar to those used in play on the surface.[5] The 'foot' is placed on the ground, weighted down and the torque required to turn it is measured, either using a torque wrench or a pulley and weights. It has been shown that the torque measured is proportional to the load imposed, so that a full load equal to three times the athlete's weight is not required.

Once again, this is not an absolute measurement, and it is not directly related to the likely level of torque that an athlete's knee or ankle can stand. Rather, the surfaces have been developed, along with the appropriate footwear, such that foot lock injury is acceptably uncommon. With this in mind, the torque measurement does give a relative ranking for different surfaces.[6] The problem arises from the players themselves, who may choose footwear that seems to maximise the grip on the surface, unaware or choosing to ignore the fact that they may be exposing themselves to 'foot lock' and consequent knee or ankle injury.

3.3.3 Other tests

All artificial pitches which meet DIN 18365 have to pass a test that simulates athletes falling and striking their head on the surface. There are other tests of a basically geometric nature – flatness, the ability of a ball to roll on it in a straight line, drainage, the alignment of lines etc. A complete list of the tests used by FIFA is available in reference 8.

3.4   Sport-specific surfaces

3.4.1 Football

At the highest level, football is played on mature natural turf, nowadays grown in a base that is 85–90% graded sand to allow water to drain away readily. Under ideal circumstances, with sunshine, water and light usage, such turf is a delight to play on, since the turf ensures that the surface is compliant and comfortable to run on; the surface is level, and the grass is maintained at the correct length to make the ball sit up a little off the ground. A player who falls on the turf will skid on the grass, which will obligingly shear and release the liquid content of its cells, lubricating the interface between athlete and ground until he comes to a halt, largely unharmed. Likewise, a player who turns sharply will feel the turf shearing under his boot, limiting the torque applied to his leg, and reducing the likelihood of injury.

The situation was quite different for anyone playing football on a first-generation artificial turf pitch, where a polymer fibre carpet was filled with graded sand to prevent the fibres lying flat. Anyone falling on such a surface would find the skin smartly abraded from any limb that made contact. When pitches of this type were installed in the UK for league football, the players quickly adapted to the new surface by avoiding the sliding tackle. Since this is a particularly decisive defensive move, it was found that the game itself changed, the ball moved around the pitch much more, and the players ran further.[7] It was this latter point, rather than any specific property of the pitch, that led these early pitches to be described as 'hard on the legs' – the players were really weary after a game as they had run up to 50% further than they did on a natural turf pitch. (There were other factors that gave the first-generation pitches a poor reputation – the ball bounced much higher, and there was a lack of know-how as to how to maintain them.) An inquiry was set up in England under Sir Walter Winterbottom[7] and, as a result, the sand-based pitches were effectively banned from the professional game. It is only in the last few years that pitch development has caught up with football's needs, to the extent that FIFA have published specifications for artificial turf pitches that meet with their approval, and the first few installations have received a 'recommendation' under the FIFA Quality Concept Project.[8] It has been such a long trip for artificial turf from its first faltering steps to now being accepted as a training pitch for the French Football Federation and by top-class clubs such as Porto, Kaiserslautern and Nantes, that one is entitled to ask just why the football authorities would wish to turn their backs on grass. In fact, at every level, grass suffers from excessive wear especially around the goalmouth, and in the centre circle. With pressure on time, games are only cancelled when the ground is quite unplayable, and this leads to games proceeding when the pitch is sodden after heavy rain, but not unplayable. Since the turf is wet, it is easily damaged. The damage done to the turf is then difficult to repair before the next game, when the pitch will still be wet. Over a European winter, grounds deteriorate badly, to the detriment of the quality of the football. In a business which is fundamentally about entertainment, the degradation of the basic product is bad news, hence the search for a surface that plays like turf, but is weather, and wear, resistant.

Before considering the details of artificial pitches, it is important to remember that turf grass pitch development has not stood still. There has been a steady improvement in the way in which the seed bed has been prepared, and also in the types of grass used. Systems of increasing sophistication have been developed for drainage and for keeping the pitch frost-free. Turf is now available readily as rolls, which can be used to relay all or part of a pitch in a short time. As ever, though, the new turf needs time to develop its own root system before it is played on. Perhaps most interesting is the concept of growing the turf, complete with its soil, in a square box. The whole pitch is constructed from an array of interlocking boxes, suspended above the drainage and heating system. Damaged areas may then be replaced with fresh boxes of turf, which are ready to play upon with no delay due to the need to develop a new set of roots. This system has been installed in the Millennium Stadium in Cardiff (see Fig. 3.3).[9]

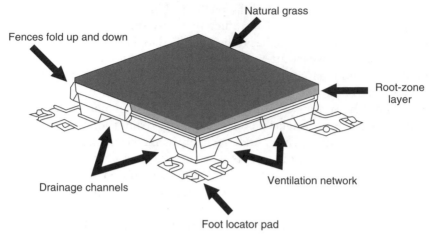

Natural grass

Fences fold up and down

Root-zone layer

Drainage channels

Ventilation network

Foot locator pad

3.3 Stadium turf grown in a box (approximately 2 m × 2 m) for ease of growth and replacement. Note the fences that fold down to ensure that they do not interfere with play. Channels below the turf allow drainage and heated ventilation if required to prevent freezing.[12]

3.4.2 Modified turf – DD Grassmaster and Astrograss systems

From the previous discussion of the problems of turf grass, it may be seen that there are two main concerns: the damage caused by play, particularly when the turf is wet, and the difficulty of repairing the turf once damaged. The Desso Company in Holland and SWI in the US looked at this from the point of view of a well-established artificial turf installer and innovator, and devised a halfway house in which the turf is reinforced with an array of vertical fibres inserted into the pitch (Figs 3.4(*a*), (*b*)). This is not a quick fix, since the project to install Grassmaster begins with the laying of a sand-based grass pitch. The vertical fibres are inserted 20 cm into the surface at 2 cm intervals, and are left 2 cm proud of the surface. In total, some 43 000 km of fibre, i.e. more than enough to go around the world, are inserted into the pitch to stabilise it. Within a few days, the inserted fibres cannot be distinguished from grass. Since the polypropylene fibres constitute only 3% of the total, the surface plays the same as a normal pitch. Sliding tackles are possible without injury, and, due to the stabilisation of the turf by the fibres, damage is greatly reduced. The vertical fibres assist the drainage, so that the pitch is playable under all but the most extreme conditions.[10,14] The Astrograss product gives a similar result, although as will be evident from the graphics, the technology is slightly different.

(a)

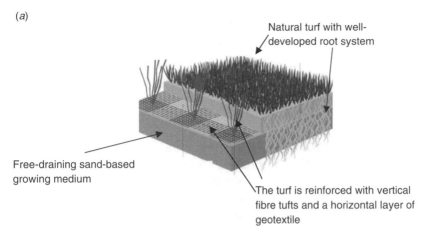

Natural turf with well-
developed root system

Free-draining sand-based
growing medium

The turf is reinforced with vertical
fibre tufts and a horizontal layer of
geotextile

3.4(a) Reinforced natural turf system with horizontal and vertical
reinforcement of the turf (Astrograss[13]).

(b)

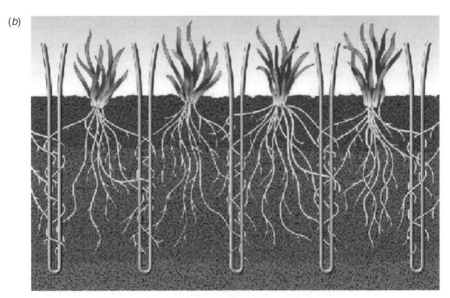

3.4(b) Reinforced natural turf system with inserted vertical
reinforcement (Grassmaster[10]).

In playing terms, then, the Grassmaster gives a surface that is ideal. In
usage terms, it will accommodate three times the number of games that
can be played on an unreinforced pitch. As a training pitch, it will cope with
900 hours of usage per season (17 games per week).

There is no doubt that this is a major innovation. The fact that the basic nature of the playing surface is unchanged makes it a highly attractive proposition, exemplified by its use by Liverpool and Monaco and Real Madrid football clubs and in the Norwegian national stadium as their primary playing surfaces, not simply for training.

3.4.3 Artificial turf

At the outset, we should distinguish between pitches that are to be used at the professional level, even if only for training, and pitches for casual use. Almost every sports centre has a number of small sand-based pitches for hire, and these may be in use for 12 or more hours a day. The norm is to play on them in flat-soled training shoes, and to wear track-suit trousers to avoid abrasions after falling. These pitches have a shockpad underneath them and meet the relevant safety standards, but otherwise are similar to the first-generation pitches. For casual use, they fill a niche, and are capable of prolonged high-intensity usage. Since the players adapt their kit to the surface, footlock tends not to be a problem.

As was mentioned earlier, FIFA, the international body that governs world soccer, have had an ongoing project to issue guidelines for artificial pitches.

A number of companies have met the standard, using their own version of a carpet system (Fig. 3.5). Most companies use a shockpad, which is a porous mat of rubber granules 10–20mm thick. The carpet lies on top of this with the tufts of fibre sticking up. Without a stabilising infill, the fibres would quickly lie flat. As one might expect, a large degree of development has gone into the fibre tufts and into the stabilising infill. In some systems, the fibres are flat and curled, the sand infill leaves the top 10mm free, and this curls over, hiding the sand, which is all but invisible. Companies have experimented with a variety of fibres – polypropylene, nylon, polyolefins and blends of polyethylene and polypropylene. These variations are aimed at achieving a fibre that is strong, resilient, non-abrasive, light-fast and resistant to weathering over a 10-year period.

The one system that is radically different dispenses with the shockpad, and integrates it into the fibre infill as a layer of rubber granules that sits on top of a root layer of graded sand. This is Fieldturf (Fig. 3.6), which was originally proposed as a surface for the NFL in North America, but has more recently been installed at a number of locations in Europe for soccer.[11] In particular, the facility at Almere, in The Netherlands has reached the standard of the FIFA Quality Concept. Fieldturf is also different from most other artificial surfaces in that it uses a long carpet fibre pile so that players can use short or long studs as they wish. The fibre stands proud of the rubber top infill, and the rubber/grass composite will shear as the player turns

Synthetic grass woven into a
permeable backing. The graded
sand infill prevents the fibres from
bedding down into a mat.

Shockpad, 10–20 mm thick laid in
roll or formed *in situ*. Rubber
crumb bound with polyurethane
resin to give a resilient, easy
draining base.

Sub-base – porous tarmac

Foundation – crushed rock to
provide a stable base

3.5 Structure of a first-generation artificial turf pitch. Subsequent
developments relate to the structure of the carpet and the
shockpad.[11]

on it, so that 'footlock' is greatly reduced. The fibre itself is a polyethylene/polypropylene blend.

The involvement of FIFA in the development of non-traditional pitches is a major step forward, and a number of other companies have built complete facilities to the FIFA specification.[8] Since these are relatively recent events (since the beginning of 2001), these are early days for there to be a body of experience deriving from controlled experiment. What one can say is that the users of these new surfaces are all enthusiastic as to their performance. One of the real benefits is the ability to train on a high-quality surface all the year round, with the implications for the improvement of individual skills and, just as important in football, the development of teamwork and set-piece moves.

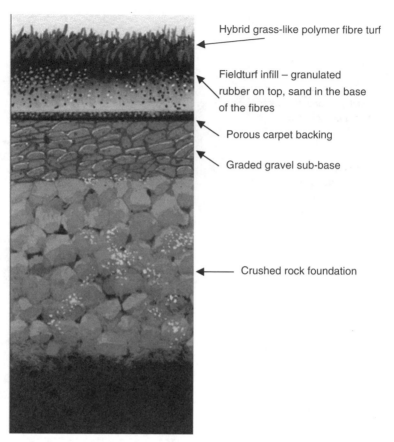

Hybrid grass-like polymer fibre turf

Fieldturf infill – granulated rubber on top, sand in the base of the fibres

Porous carpet backing

Graded gravel sub-base

Crushed rock foundation

3.6 Schematic of Fieldturf rubber/sand filled carpet system (adapted from Reference 11). Note the absence of a separate shockpad. Shock absorption is provided by the rubber crumb infill.

With the experience of first-generation pitches behind it, the industry is maturing, and is, for instance able to emphasise the need for routine maintenance. In the past, pitches were often sold as being 'low maintenance', which often translated into a state of neglect. With care and attention, the new pitches, like the older models, will last the 10-year lifespan for which they are scheduled.

3.4.4 Rugby

Rugby, of whichever code, is one of the most difficult sports for which to provide a quality playing surface. There is the need for a secure footing; the surface needs to cushion the often uncontrolled impact of large players falling on it (120 kg rugby players are now not uncommon); certain areas –

close to the touchlines and goallines – need to withstand repeated scrummages and rucks, with something close to 2 tonnes of forwards pushing against each other. Rugby is a winter sport, so that the pitch is often wet, with the added likelihood of damage. Almost everyone who ever played rugby must recall plodding from scrum to lineout in mud that made any sort of creative running or passing at best a lottery, and at worst, a gift of scores to the opposition. The situation has tended to get worse in recent years, with the move to the professional game requiring extended times of training, with the additional pitch wear that this implies.

Owing to the damage caused, most rugby training does not take place on the same pitch that is used for matches. This applies particularly to scrummaging practice, and basic fitness training. A real problem arises with areas that are retained for training purposes, as these quickly degrade into a quagmire at the beginning of winter, and do not recover until the return of drier weather at the end of winter. A version of the same problem exists in drier climates, where the worn areas do not regenerate easily, and become even harder than the grassed areas. Playing on hard pitches does cause long-term damage to the players – the syndrome known in South Africa as 'Springbok back' arises from a combination of the vigorous scrummaging that is inherent in the game there, and running about on pitches that are dry and unyielding.

Given all of these problems with natural turf pitches, it is surprising that more progress has not been made in devising improved rugby pitch concepts. For a top-class pitch to maintain its quality, one or two games a week are likely to be the norm for a traditional grass turf pitch seeded into soil. This may be improved radically for pitches which are laid with a 90% sand, 10% loam base for the turf to be seeded into, similar to the practice in soccer. It would seem that this is indeed a major step forward, but as yet one cannot assess the extent of the improvement. There is no body of experimental evidence yet available as to how these pitches perform in terms of resilience and surface torsion over a period of time, with changes in weather and the state of development of the turf. As was discussed at the beginning of this chapter, a simple test with the Berlin Athlete is not an adequate predictor of surface compliance in circumstances where the surface exhibits viscoelastic properties.

The other approach which has been adopted for rugby is the use of the Grassmaster system, which is really a further step on from the sand-based pitches, with the added stabilisation from the vertical fibres. This has been used for over eight years in the UK, and is reported to have been played on with a layer of snow; no damage resulted.[17]

The New Zealand rugby authorities have approached this in a rather different manner, in that they have installed an area of Fieldturf indoors as a training area. It is not reported whether the training extends to scrummaging on the Fieldturf.[11] One may imagine that more indoor rugby training facilities will be built. It is plain that quality training is becoming more and more important as time goes by, and the old ideas of running about in the rain do not enable the players to cultivate the precise skills that are needed to be competitive.

It is important to note, however, that simply upgrading a poorly drained pitch or constructing an entirely new one will not necessarily solve the drainage and usage problems. It is therefore essential that any pitch upgrades or new constructions of playing pitches should be designed with estimated intensities of use in mind and the availability of an appropriate maintenance programme. As yet there is no comprehensive answer to the provision of training areas that are durable and that offer a surface performance similar to turf.

3.4.5 Hockey

Hockey has been transformed by the adoption of artificial turf pitches as the standard, even down to school levels of play. When the game was played on grass, many players never had the chance to develop their skills to any extent, since anything less than a perfect grass surface injects an element of chance into the game – passes go astray, dribbling is a lottery. On artificial turf pitches, long passes can be made accurately and precise dribbling is possible, since the surface is smooth and predictable. Instead of players swinging at the ball to pass it, they use a push with the wrists, which is quicker, more accurate, and gives the opposition less chance to block the pass. The offside rule, which in essence concentrated play in the middle half of the pitch, has been abandoned, so that the game is a fast, whole-pitch activity. Just as the first football players to use artificial turf pitches found they were running greater distances, so the hockey players now find that the changes in the game imply changes in training and fitness regimes.

For hockey, the pitch of choice is water-based (Fig. 3.7), i.e. it has a carpet with fibres that are normally curled, but the whole pitch is kept wet – in fact, just short of being flooded. The main aim of this is to lubricate the athlete/pitch interface in a fall. Without the water, the players would receive painful abrasions. The carpet has a shockpad, which, along with the curled fibre, makes a comfortable surface for running. The FIH specify a high level of force reduction for the Berlin Athlete test (<65%), implying that they intend that the pitches should not cause jarring or pounding injuries to the players' legs.

Apart from water-based pitches, a great deal of hockey is played on sand-stabilised or sand-dressed pitches. The sand-stabilised pitches are similar to first-generation football pitches, but have a shockpad of 10–20 mm so that

Tufted carpet of grass-like
polymer fibre

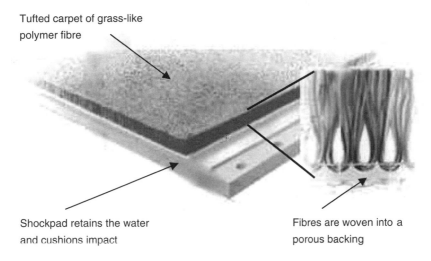

Shockpad retains the water
and cushions impact

Fibres are woven into a
porous backing

3.7 Schematic of water-based hockey pitch (adapted from Reference 11).

they are comfortable to run on. Falling on them, though, causes abrasion injury to knees, hands and elbows.

The sand-dressed pitches have a heavier fibre that is given a curl. The sand stabilises the fibres, but the fibre curls over and covers the sand. Any athlete running or falling on the surface makes first contact with the fibres, so that abrasion is greatly reduced.

The sand-based options continue to be popular since they are well understood, while the water-based pitches are more expensive to install, and require maintenance for the water-pumping system.

3.4.6 Athletics

When Roger Bannister ran his record mile time in 1954, at Oxford, he ran on a cinder track – it was a bed of crushed ashes, rolled flat, for such was the state-of-the-art in running track design. During the 1960s polymer tracks were introduced, and have been improved in the interval so that the tracks, and the spiked running shoes that have developed specifically for the polymer tracks, form a reliable system at venues anywhere in the world almost irrespective of the weather. Basically, the modern running track is a layer or layers of elastomer on a solid base – as far as the athletes are concerned, the main specification is the resilience – the extent to which it deforms under load. The question is – how resilient should the track be?

Top layer of polyurethane and rubber granules
optimised for grip and wear

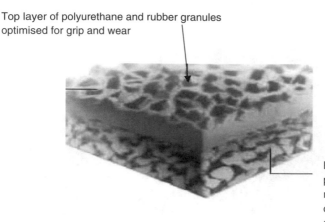

Base layer of
polyurethane-bound
rubber granules,
optimised for shock
absorption

3.8 Structure of typical polyurethane running track. The two layers
ensure that the track is comfortable to run on, with adequate grip
and wear resistance.

Too soft and the sprinters will find that their energy is being absorbed by
the track; too hard and the distance runners will suffer damage to their feet
and legs from the excessive jarring.

Some 20 years ago, McMahon and Greene developed a biomechanical
theory of running and used it to design an athlete-friendly indoor running
track. This was built and was indeed found to be comfortable to run on, and
to result in fast times being set.[15] Current developments in the design of
running tracks have tended to ignore this work, and to choose to lay down
tracks which result in fast sprint times.

Modern tracks are laid down as a layer or layers of polyurethane
elastomer, sometimes filled with polyurethane granules or rubber crumb
(Fig. 3.8). The total thickness of the polymer layer is normally 13 mm. A
typical force–deflection curve is shown in Fig. 3.2, using a hemispherical
indenter of 50 mm diameter to simulate a heel. The dynamic modulus, i.e.
the tangent to the force–deflection curve, at a load level of 2.5 kN is shown
on the Figure. This load level is that achieved by a runner of ~80 kg running
at a pace of 12 km/h, i.e. a gentle jogging pace. At higher speeds, the dynamic
modulus will be higher – in the range 1–1.2 MN/m. This figure should be
compared with the figure of <200 kN/m suggested by McMahon and Greene
as being ideal for fast times and minimal leg strain. The higher figure
complies with the IAAF specification, which is framed in terms of a
Berlin Athlete force reduction figure. In fact, current tracks are optimised
for fast sprinting times.

Distance runners, on the other hand, find that they do indeed jar their legs and experience undue levels of fatigue. There is plainly room for further experiment and development.

3.5   Future developments

While it is plain that pitches define the way in which each game or sport is played, the pitches have evolved in response to the demands that have been placed upon them. The ultimate goal of having pitch types that enhance the games and at the same time are weather, and wear, resistant is still some way off, even if major strides have been taken in that direction. The FIFA initiatives, along with experience of the modified pitches that are already in use, are likely to ensure that the playing surface for professional soccer will improve markedly over the next few years. Teams will need to adapt to these new conditions if they are to continue to prosper. The potential changes for rugby are even greater, since it is conceivable that both the pitches retained for match play will improve and the training areas are also likely to come up to a similar standard. This is likely to be of greatest benefit to teams in the wetter areas, who at present often train in sub-optimal ground conditions.

Hockey has travelled furthest along the road of pitch development, and, in the process, has revolutionised the game as a spectacle. The water-based pitches now in use are unlikely to be the end of the road, but changes are likely to be evolutionary rather than revolutionary.

As far as athletics is concerned, the performance of running tracks is already at a high level. Future developments are likely to address the matter of the interaction of the athlete and the track both in training and during competition, with a view to reducing the incidence of injury.

3.6   References

1 DIN 18035 part 6 (1978, revised). *Sports Grounds – Synthetic Surfaces – Requirements, Tests, Maintenance*. Beuth-Verlag, Berlin.
2 Kolitzus HJ (2002). Artificial Berlin Athlete: comments on function and use (updated 12/2/2002) ISSS Publications, available at www.ISSS.de
3 Misevitch KW & Cavanagh PR (1984). Material aspects of modelling foot/shoe interaction. *Sports Shoes and Playing Surfaces*, ed. EC Frederick, Human Kinetics Publishers, Champaign, Illinois, USA.
4 Miller DI (1990). Ground reaction forces in distance running. *Biomechanics of Distance Running*, ed. PR Cavanagh, Chapter 8. Human Kinetics Publishers, Champaign, Illinois, USA.
5 Canaway PM (1975). Fundamental techniques in the study of turfgrass wear. *J Sports Turf Res Inst*, **51**, 104–115.

6 Baker SW & Bell MJ (1986). The playing characteristics of natural turf and synthetic turf surfaces for association football. *J Sports Turf Res Inst*, **62**, 9–35.
7 Winterbottom W (1986). *Artificial Grass Surfaces for Association Football. Report and Recommendations.* vol. 127, p. iii. Sports Council, London.
8 www.FIFA.com
9 www.Greentechitm.com
10 www.Desso.nl
11 www.Fieldturf.com
12 www.sports-surfaces.co.uk
13 www.Astroturf.com
14 www.astrograss.net/product.htm
15 McMahon TA & Greene PR (1979). The influence of track compliance on running. *J Biomech*, **12**, 893–904.
16 Andreasson G & Olofsson B (1983). Surface and shoe deformation in sports activities and injuries. *Biomechanical Aspects of Sport Shoes and Playing Surfaces.* ed. BM Nigg & BA Kerr. University of Calgary.
17 http://www.sportsvenue-technology/com/contractors/surfaces/desso/index.html

4
Running shoe materials

N. J. MILLS
University of Birmingham, UK

4.1   Introduction

In this chapter, running biomechanics and injuries will be linked to the materials and design of shoes. The emphasis is on materials selection, using materials science to explain the performance of foams and rubbers. The biomechanics of running is a major research area. The book edited by Frederick (1984) considers modelling of heelstrike and interactions with playing surfaces. Nigg's book (1986) considers running biomechanics, while that by Segesser and Pforringer (1989) considers a range of athletic shoes and playing surfaces. The extensive research literature since 1989 has been partly summarised by Shorten (2000), but there has been no comprehensive review. The website of his company Biomechanica has an introductory bibliography www.biomechanica.com/help.html#force.

In running, the shoe is part of a larger system. Shoe properties, such as the cushioning of heel strike, must be related to the cushioning inside the heel and the leg, and to the running surfaces, to put the shoe contribution in context. The emphasis is on running shoes rather than those for other sports. In running the motion is mainly in a straight line, with some acceleration or braking. Basketball shoes need stiffer cushioning for the greater landing velocities, while tennis shoes require more support for sudden lateral motions (Segesser & Pforringer, 1989). The mechanics of heel-strike cushioning, and the attenuation of the impact forces, will be considered. Orthotic foam inserts in shoes, used to correct a variety of walking and running problems, are considered by Hunter *et al.* (1995) (an orthosis is an external device, which applies force to part of the body, to correct some problem).

4.1.1   Sources of information

The websites of major shoe manufacturers show their design innovations, for example www.Adidas (asicstiger/mizunousa/newbalance/Nikebiz/puma/

reebok/saucony).com. Some of these illustrate the biomechanics of running. The magazine *Biomechanics*, partly on the web at www.biomech.com, considers both orthotic and running shoe issues. The history of trainers, and background about running shoe materials, is at www.sneakers.pair.com. Educational case studies on shoes include www.scire.com, aimed at US high schools; its section on *Slam Dunk Science* introduces the roles of shoe components. There are sports shoes lessons at www.cookeassociates.com/seesite/SHOES/shoes_teachers.html and at www.sep.org.uk.

Sport injury medics, such as Dr Pribut at www.drpribut.com/sports/, explain running injuries and comment on shoe design and choice. International Society of Biomechanics, Technical Group on Footwear Biomechanics, organises biennial conferences on Footwear Biomechanics. Abstracts from the 1999 and 2001 meetings are available at www.staffs.ac.uk/isb-fw. www.uni-essen.de/~qpd800/FWISB/sneakers.html has videos of foot motion and dynamic pressure maps. Journals in the area are mentioned in the reference list. Patents, which may explain technology and illustrate shoe designs, can be downloaded from the US (www.uspto.gov) or UK patent (www.patent.gov.uk) offices.

4.1.2 Shoe components

Shoe mass is important, since extra mass will add to the energy consumption of the athlete. Elite athletes may use lightweight (0.1 kg) spiked shoes with no midsole for races, but heavier, more durable shoes for training. We will concentrate on trainers, which weigh about 0.3 kg. Figure 4.1 shows the component parts:

(a)   outsole of rubber
(b)   midsole of foam
(c)   innersole of foam with cloth cover
(d)   heel counter
(e)   upper, cloth or leather.

An old pair of trainers can be cut in half with a band-saw to reveal the thickness of the various components. The midsole foam moulding is wedge shaped, running the whole length of the shoe. Traditional designs of tennis shoes have no foam midsole.

4.2   Shoe construction

The construction of shoe uppers involves much non-automated cutting out and sewing. Most manufacturers have factories in the Far East, where labour costs are low (Vanderbilt, 1998). The US site of www.newbalance.com has a 'Factory Tour', showing some operations on

(a)

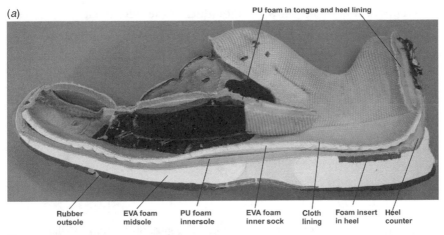

PU foam in tongue and heel lining

Rubber
outsole

EVA foam
midsole

PU foam
innersole

EVA foam
inner sock

Cloth
lining

Foam insert
in heel

Heel
counter

(b)

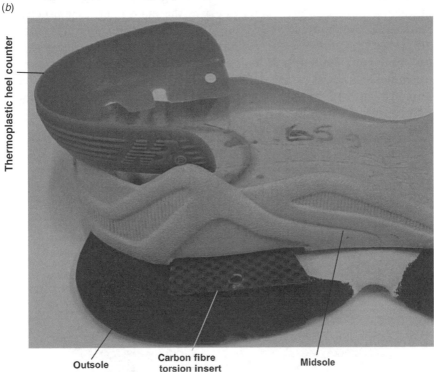

Thermoplastic heel counter

Outsole

Carbon fibre
torsion insert

Midsole

4.1 (a) Section through a used Adidas shoe, showing the main
components, (b) separate components from a New Balance shoe.

shoe uppers. It does not show the moulding of the soles in closed presses – Section 4.2.5 considers the processing technology for EVA midsoles.

4.2.1 Patented features in shoes

Sales depend on brand names, sponsorship by elite athletes, and on the external shape and colours. The distinctive external appearance and outsole tread pattern may emphasise (real or imaginary) internal components that provide cushioning, or control torsion of the foot. Features that can replace some of the midsole foam are listed below.

Nike's large bubbles of 'air'

Nike's air bubbles apparently contain SF_6 gas, in a thick-walled polyurethane rubber chamber, as an insert in the main foam (www.sneakers. pair.com/airtech.htm and US Patent 4219945). The PU wall prevents the SF_6, or other high molecular weight gas, from diffusing out at any significant rate, but air will diffuse in after the shoes have been used, to restore the initial overall gas pressure. US Patent 5625964 gives some of the background. Figure 4.2 shows that the shaped PU wall of the gas chamber forms four springs.

Asics's gel insert in the heel

The small size of Asics's gel insert suggests that it plays a minor role in the shoe's shock absorption performance. US Patent 5718063 describes the prior art in this area, but gives very little detail on the nature of the gel.

Puma's thermoplastic honeycomb in the heel region

US Patent 5084987.

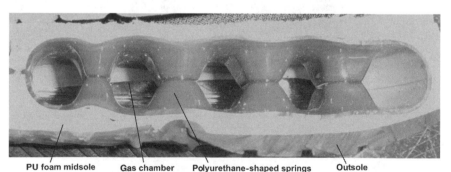

PU foam midsole   Gas chamber   Polyurethane-shaped springs   Outsole

4.2 Section through the heel of a Nike Air shoe.

Nike's polyurethane rubber springs in 'Shox' shoes

The performance of Nike's rubber springs is discussed later.

4.2.2 Outsole materials and design

The role of the outsole is to:

(a) provide grip with a range of ground surfaces
(b) prevent damage or wear to the midsole foam.

The location of the sole flexing can be controlled to some extent by the use of 'hinges' in the outsoles – an inverted U-shape running across the foot, which means that the midsoles is thinner at this point. Figure 4.1 shows how, in the well-worn shoe, there is a permanent bend at the metatarsal heads (Fig. 4.5 shows the foot anatomy).

The outsole provides 'friction' on the running surface. There has been more consideration of running surface characterisation, see the chapters in Frederick (1984) and Sesseger (1989), than of sole friction. Some extremes of sole–ground interactions are as follows:

(a) If the rubber outsole is flat, and there is a film of water on a polished hard surface (marble at a shopping mall entry), hydrodynamic lubrication leads to very low friction.
(b) Running tracks made of bonded rubber crumb are often used with spiked soles.
(c) Cross-country running requires rubber 'studs' as part of the outsole to provide grip in mud, etc.

Rubber compounds in car tyres can be selected to increase grip in the wet, at the expense of wear rates. For shoe soles, the wear conditions are less extreme. Valiant (1997) described friction measurements in which athletes performed acceleration, braking and lateral motion tasks, with their shoe soles in contact with a three-axis force plate. The greatest traction demands were in the propulsive phase, rather than in the heel strike phase. To avoid slipping, the coefficient of friction of the outsole on the running surface should exceed 0.6 for the heel and 0.7 for the forefoot. The coefficient of friction is a function both of the sole material (1.2 for 'rubber' and 1.8 for PU on dry surfaces) and whether the surface is wet or dry (the respective values reduce to 0.6 and 0.4). Dust on a smooth surface can also affect the value.

Manning *et al.* (1998), who were concerned with slipping on polished surfaces, showed that, if the surface roughness R_{tm} of the shoe outer was increased from 5 to 20 μm, the coefficient of friction on a water-lubricated glass increased from 0.08 to 0.3. R_{tm} is defined as the mean peak to trough distance over a sample length. However, Kim *et al.* (2001) show that wear often reduces the initial surface roughness of shoe surfaces.

Ahnemiller (1997) described the formulation of a polyether polyurethane for outsoles that has transparency, adhesion and good wet traction. He commented that these materials cannot compete with natural rubber or EPDM on a cost basis, so tend to be used in high-wear areas of the shoes. The outsole must be bonded to the midsole, and must resist cutting from sharp surfaces. The thickness that provides adequate midsole protection, and sufficient wear life, would have been found by experience. Thermoplastic PU, such as Desmopan 481 (www.plastics.bayer.com), is claimed to have extreme wear resistance, and it bonds well to uppers.

4.2.3   Influence of processing on materials selection

Modern shoe sole designs have complex shapes, and often use several colours. These are possible due to the process technology available for multi-step injection moulding. For instance, see www.maingroup.com, an Italian producer of multi-station injection moulding machines. These designs would not be possible without the polymer processing technology. Outsole material selection is related to product manufacturing cost, via the need for rapid cycle times. If conventional rubber is selected, it is possible to injection-mould the polymer plus cross-linking agent, but the mould must be kept hot for several minutes while the cross-linking reaction is completed. For a multi-colour outsole, the overall cycle time becomes excessive. Hence thermoplastic elastomers are selected: a two-phase structure develops on cooling, and the hard segments in the microstructure trap the ends of rubbery molecules. For athletic shoes thermoplastic polyurethanes (TPU) are preferred. See, for example, www.plastics.bayer.com, who use the Desmopan trademark. Block copolymers of styrene and butadiene have also been used, but the wear resistance is inferior. For these thermoplastic materials, the mould is cold, and the cycle time is determined by the thickness of the moulding, as heat must diffuse out of the melt before it solidifies. Overall, the cycle time is much shorter than for rubber. If non-marking outsoles are needed for sports hall use, to avoid marking wooden floors, the most common outsole is natural rubber, used without the addition of carbon black.

Midsole design is also linked to technology. If the midsole has a complex shape, or if it encapsulates an air pocket, etc., direct injection moulding of EVA or PU foam is necessary.

4.2.4   Foot rotation control features

Pronation control for rearfoot runners

The relative rotation of the heel and tibia, about an axis into the paper, in the sense shown in Fig. 4.3, is referred to as pronation. Excessive pronation

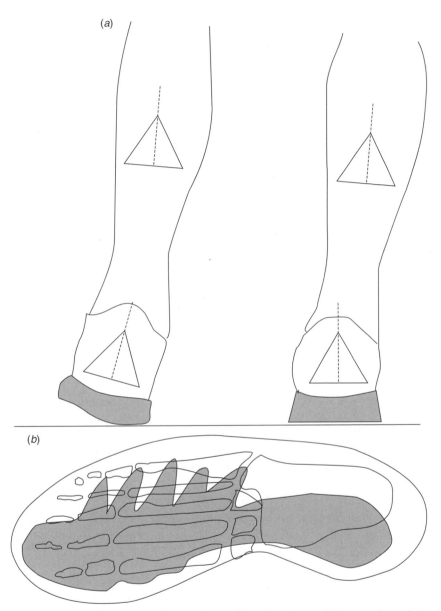

4.3 (*a*) Rear view of right foot, with talo-calcaneal joint supinated at initial heel contact, then pronated at mid-stance (adapted from Shorten, 2000), (*b*) carbon fibre sheet (shaded) used to control torsion of the foot (adapted from US patent 6477791).

can cause injuries, such as plantar fascitis (a pain near the heel, due to damage of the tendon attached to the lower surface of the calcaneus). The associated rotation of the tibia about a vertical axis, due to the complex nature of the tibia/heel joint, can lead to knee injuries. Shoe designs attempt to control pronation by softening the landing on the outer edge of the heel soft, while stiffening the inner edge of the heel with a higher modulus foam. Some runners may have less than usual heel rotation (supination), so the heel can be flared out on the medial side, to stiffen the initial impact. Selection of these 'motion control' shoes should be the result of a biomechanics analysis of the particular runner's style.

Heel counter

The heel counter, a moulded plastic insert between the leather outer and the padded liner, encourages the shoe heel to move with the calcaneus bone (Fig. 4.1(b)). However, the light construction of trainers is not as effective as heavy leather construction of boots. They are often cut away at the position of the Achilles tendon, to avoid chafing. There is cushioning inside the heel counter to avoid rubbing; when the heel foam compresses, the heel moves vertically relative to the heel counter.

Controlling foot torsion

Stacoff et al. (1991) studied foot rotations in forefoot running, and measured the torsional stiffness of shoes and the foot. For a 2 Nm torque, applied about a toe-to-heel axis, spike shoes with no midsoles rotated by $39 \pm 4°$, running shoes with midsoles by $9 \pm 3°$, while athletes' feet rotated by $25 \pm 10°$. Hence, the use of running shoes reduced foot torsion. Analysis of photographs of footstrike showed minor differences between spike shoes and running shoes, but significant differences from barefoot running. They suggest that the running shoe torsional stiffness should be reduced by reducing the foam thickness or width in the midfoot region or by using more flexible materials.

Thin carbon fibre-reinforced composite plates have been used in the heel of shoes to control torsion. A recent Adidas patent (US 6477791) describes the use of much larger plates (Fig. 4.3(b)), that increase the torsional coupling of the forefoot with the heel. The plates have some effect on the bending stiffness of the sole.

4.2.5 Midsole foams

The main loads on the midsole are:

(a) localised compressive loads from the heel, metatarsal heads and hallux (big toe). The midsole cushions the impact forces in these regions.
(b) bending loads during part of the foot motion. The bending stiffness of the sole will affect the runner.

These loads will be analysed in Section 4.4. The 'heel lift' is the difference in height between the rear and front of the shoe. It affects how much the midsole thickness tapers towards the toe. Dixon *et al.* (1999) varied the heel lift from zero to 15 mm. For experienced athletes, although the peak Achilles tendon force reduced with increased heel lift, the change was not significant.

Foam selection – EVA foams

The majority of running shoes (trainers) have midsoles that are compression or injection moulded from ethylene/vinyl acetate copolymer (EVA) foam. Vinyl acetate (VA), with structure $CH_2=CH–O–C=O–CH_3$, and ethylene monomers are randomly arranged in the copolymer. The typical 18% VA content reduces the crystallinity of the EVA to about 20%, far less than the 40% to 50% in low density polyethylene. Consequently, the Young's modulus of the EVA is very low, and the crystal melting temperature is about 70 °C.

The reason for selecting EVA foam is probably that shoe manufacturers had steam-heated presses for compression moulding rubber outsoles; these could be adapted to compression mould foamed EVA, which cross-links at similar temperatures to synthetic rubbers. Reasons for the use of a density $200 \pm 100 \, kg \, m^{-3}$ emerge during the case study. The density can be increased to $300 \, kg \, m^{-3}$ in regions intended to give extra support.

The white colour of basic material allows the incorporation of pigments, hence the multi-colour mouldings of the exterior of trainers. The polymer itself is colourless, but the air bubbles strongly scatter light, providing the white appearance. The durability of the foam is not as great as solid rubber components, but the sides of shoes suffer less wear than the soles. The shoe components must remain bonded together in use; it helps if the polymer has several mechanisms of adhesive action: cross-linking reactions in the EVA form covalent bonds to the surfaces of other components. EVA copolymers are widely used as hot-melt adhesives because they contain polar groups and the melting point is low. The initially thermoplastic EVA can flow between the fibres of leather, before it cross-links. Hence the melt viscosity of the polymer affects the bonding. As the molecular weight of the polymer increases, the melt viscosity increases rapidly, so melt flow into small gaps becomes more difficult. Adhesive bonding is much cheaper than the alternative of stitching, as a method of securing the midsoles to the shoe upper.

Processing and microstructure of EVA foam

In earlier shoe constructions, wedges were cut from EVA foam sheet and adhesively bonded to the outsole and shoe upper. EVA is often used in a

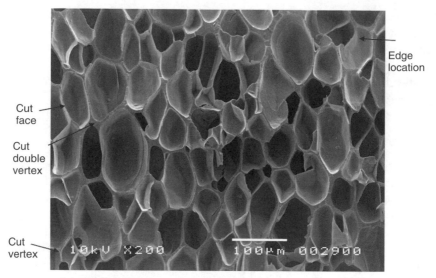

Edge
location

Cut
face

Cut
double
vertex

Cut
vertex

4.4 SEM of EVA foam from running shoes (Verdejo, unpublished). The
thickness direction of the sole is horizontal.

two-stage process (Sims and Khunniteekool, 1996). A mixture of EVA
polymer, blowing agent and cross-linking system is compounded at 100 °C,
without foaming or cross-linking. The compound is extruded into sheet
form, and the chemical blowing agent decomposes, generating a foam. The
foam sheet is cooled. In a second stage sections are compression moulded
at a higher temperature with the other shoe components and allowed to
cross-link. The product is referred to by Nike and some other manufacturers as 'Phylon'.

The microstructure of running-shoe EVA foam is shown in Fig. 4.4. This
cross-section reveals the inside of closed cells. The microstructure can be
idealised as polyhedral cells, although the cell sizes and shapes vary. These
consist of:

(a) nearly planar faces, which act as barriers between the air in two neigh-
     bouring cells.
(b) nearly straight edges, which are the intersection of three faces. In Fig.
     4.4 the face orientation affects the contrast, so lines where the con-
     trast change reveal the position of edges. If the edge is cut through, it
     has a typical tricuspid shape (the Plateau border).
(c) vertices where, in general, four edges meet. Since the relative density,
     R ($\equiv$ foam density/polymer density) of the EVA foams is considerably
     higher than typical polystyrene packaging foams (0.02), there is more

polymer in the edges and vertices. Consequently, some 'degenerate' vertices, where two or more vertices have linked up, connect more than four edges.

The closed-cell foam has a fraction of polymer in the cell faces $\phi > 0.7$. The cell faces are relatively thick, compared with low-density polyolefin foams. Near the sole surface, the cells are flattened.

Alternative foams

Promotional articles for novel polymers may convince the reader that these materials are commercially important. Thus Pillow (1997) described the use of constrained geometry catalyst technology to make narrow molecular weight distribution ethylene–octene copolymers (87:13 molar ratio), which have 13% crystallinity and start melting at 42 °C. They can be cross-linked using peroxides, and foamed. Diegritz (1998) described the use of Dupont Dow *Engage* ethylene–octene copolymer foams as replacements for EVA shoe midsoles. The material can be injection moulded, which improves the economics of manufacture. It has better thermal stability than EVA, so the processing window is wider. After loading for 10^5 cycles, the residual deflection is 56%, less than the 70% for the EVA foam, showing that less creep has taken place. The foam is claimed to have a lower hysteresis than EVA foam, so it should not heat up as much.

The main commercial alternative to EVA is PU foam. Indesteege *et al.* (1997) discuss the use of water-blown PU foam for midsoles; see Table 4.1. The PU foam density is higher but the compression set (defined in Section 4.5.5) is much lower, presumably due to the lower permeability of the polymer to air. Efforts to reduce the PU foam density, while maintaining the mechanical properties, are discussed since extra mass is a disadvantage for running shoes. PU midsoles are more common in basketball shoes,

Table 4.1 Comparison of PU and EVA foams for midsoles

Test	EVA	PU
Density (kg m^{-3})	200	320
Split tear strength (kN/m)	4.5	3.5
Compression set (60 °C/6 h)	58	5
Asker C* hardness	62	60
Pendulum rebound resiliency (%)	35	30

*This uses a 5 mm diameter hemisphere pressed with a force of a few Newtons.

where there are greater landing velocities. They also bond better to PU air bladders. Sawai *et al.* (2000) discuss the development of lower density PU foams (*c*. 250 kg m^{-3}) for use in sports shoes, and discuss how the split tear strength increases with foam density. The split tear strength measures the force required to continue a crack in a foam, after a split has been started, divided by the thickness of the foam.

Alley and Nichols (1999) extend the discussion, differentiating between injection-moulded EVA (IP) and compression-moulded EVA (Phylon). There is little difference in terms of the properties in Table 4.1, except that the IP had a slightly lower density (170 to 220 kg m^{-3}) than the Phylon (200 to 250 kg m^{-3}). However, the waste from compression moulding is >50% whereas it is <10% for IP, which makes the average cost per Phylon midsole $1.80 to $2, compared with $1.2 to $1.4 for IP. For PU foam the cost is $1.55 to $1.65, since the tooling cost is less. Over the years 1996 to 1998 EVA gained at the expense of PU, with the 1998 data being 54% Phylon, 13% IP, and 33% PU.

Foam selection for foot stability

Materials which provide heel cushioning may not provide foot stability. Plasticised PVC foam 'Airex S' is used for stability training – it is difficult to stand on a sheet of this foam without the ankles twisting. This closed-cell foam has a Young's modulus that is nearly high enough for shoe use. However, the polymeric cell walls are elastomeric, and can stretch easily with low energy absorption. Hence the foam has a high Poisson's ratio; a cube of the foam bulges out readily at the sides when it is compressed in one direction. Oscillatory (shear) motions of the foam should be damped. This is possible with EVA, a semicrystalline polymer, but foams made from rubbery polymers may have lower damping.

Foam grades

For rubbers and foams, hardness is a quality control measure; for rubbers to check the degree of cross-linking, for foams to check the compressive modulus. In the Shore A hardness test, a steel pin of diameter about 1 mm is pressed into the rubber surface by a spring; the greater the *elastic deflection* of the rubber surface the lower is the rubber hardness. The hardness is related to the Young's modulus, but the correlation is not well established. Different hardness measures (Shore, Asker in Table 4.1, etc.) can only be correlated for a particular material of a particular thickness. As there is a complex strain field under the indenter, and the foam has a non-linear response in compression, the hardness does not correlate well with the initial foam Young's modulus. Nike's US patent 4535553 (1983) comments that EVA midsoles must have a Shore A hardness >25, otherwise the foam will bottom out. Also, low hardness foams compromise lateral stability of the heel.

4.2.6 Upper materials

Uppers were traditionally made of leather, but nowadays synthetic materials are often used. The material needs: tensile strength, flexibility, and durability against rubbing and flexure. It should allow the foot to breathe, while being waterproof. The www.scire.com website discusses the design of sport shoe uppers. Some ultra-light racing shoe uppers are, in effect, Lycra socks that attach the midsole to the foot. While the fit of the upper depends on the last used to construct the shoe, the lacing system provides some additional adjustment. Foot stability depends on the heel counter, which holds the shoe to the heel. The flexibility and height of the heel counter can be altered. For running, the requirement for the upper to protect the foot from abrasion and contact injuries is modest. Its flexibility is more important, especially in the frontal region (Fig. 4.14(a)), so lightweight constructions are used.

4.3   Running

4.3.1 Pressure distribution in the shoe

Shorten's (1993) maps, of pressure distributions in running shoe midsoles, show peak pressures of the order of 400 kPa in the heel region. Hennig and Milani (1995) tested 20 types of shoes, with athletes running at a marathon pace of 3.3 m/s. For some shoes the pressure reached 1 MPa, but typically it was of the order of 500 kPa. An animation of one test is viewable at www.uni-essen.de/~qpd800/research.html. Figure 4.5(a) shows Verdejo and Mills's (2002) measurements for a runner on a treadmill, wearing new shoes with EVA midsoles. The peak pressure, 280 kPa, is somewhat lower than in earlier research, but the treadmill surface is more compliant than a road surface. Tekscan (www.tekscan.com/medical) F-scan technology was used to measure the pressure distribution.

4.3.2 Foot-strike forces

High-speed photography of runners shows that the average vertical velocity of the foot when it strikes the ground is 0.5 to 0.7 m/s. The force on the shoe vs. time, measured by a force plate, is shown in Fig. 4.6(a) for heel-strike runners. The shaded band shows the range of variation. The force, in the direction of motion, is much less. Swigart et $al.$ (1993) split the vertical force component vs. time trace into two peaks; the initial impact peak with maximum force, F_1, and the subsequent active peak, with maximum force, F_2, where leg muscles push the body from the ground.

(a)

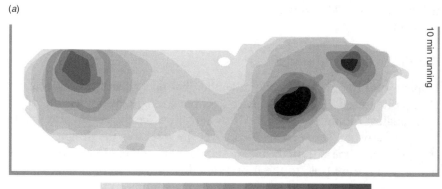

0   21   42   63   63   104   125   146   167   188   208   229   250   KPa

(b)

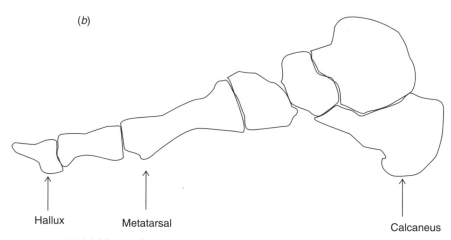

Hallux   Metatarsal   Calcaneus

4.5 (a) Maps of peak pressure at each part of the shoe during running
(heel at left) (Verdejo and Mills, 2002), (b) foot anatomy showing
the projections that cause pressure peaks.

Bobbert *et al.* (1992) used motion markers on runners' bodies, and modelled the running action as the motion of nine linked body segments. They
showed that the initial force peak, F_1, approximately 25 ms after touchdown,
is due to the support leg segments colliding with the floor; the lower leg
does not rotate (about a horizontal lateral axis) in this period and the vertical knee and ankle accelerations are the same. The upper leg rotates, limiting the vertical acceleration of the hip. For the rest of the ground contact,
between 40 and 100 ms, rotation of the lower leg limits the vertical forces
on the foot. The upper and lower leg muscles do work on both landing and
take-off, so the majority of the energy in the foot strike is consumed.
Gunther and Blickman (2002) showed that the non-linear response of the

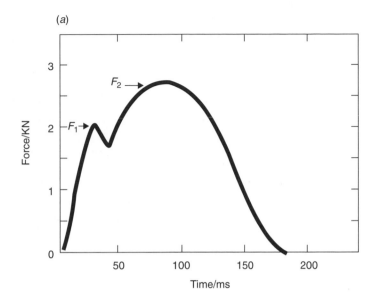

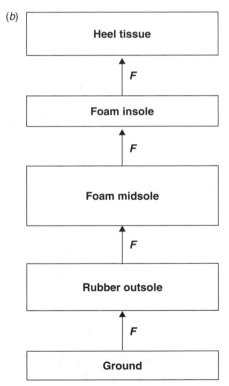

4.6 (*a*) Schematic of normal force vs. time, for heel-strike runners, (*b*) series loading of the heel and shoe components during heel strike.

ankle and knee play an important part in the determining the shape of the ground reaction force graph.

4.3.3 Biomechanics of the foot

There are three deformable bodies in a heel strike; the heel pad of the foot, the shoe and the ground surface. Concrete or tarmac hardly deforms during running. Figure 4.6(*b*) shows the force from the running surface passing in turn through the sole components and the fat pad of the heel. In this *series* arrangement, the same force F is experienced by each component, while the deflections are added. The shoe must be at least as compliant as the heel or ground to cushion the impact. There is no need for shoe cushioning for running on sand or soft grass, but the heel strike on a road surface must be cushioned. However, excessively thick, soft shoe heels may allow excessive rotation of the ankle joint and hence injury.

4.4   Shoe foam stress analysis

4.4.1 Isothermal compression of the air in the foam cells

Compressive stress is taken both by the cell gas pressure, and by the deformation of the polymer structure. In the first approximation these can be considered separately; the strength of the polymer structure is unaffected by the gas pressure, since there is an equal gas pressure in neighbouring cells, except near the foam surface. Mills and Gilchrist (1997b) used a computer model to consider the heat transfer. By comparing the predictions with experimental compressive stress strain curves of PE foams, they showed that the conditions are still nearly isothermal even at impact strain rates.

For the analysis of the air pressure contribution, each cell is assumed to act as a piston; consequently the analysis of one large piston will suffice. Figure 4.7 shows the polymer and gas in 1 m^3 of foam, separated so that the

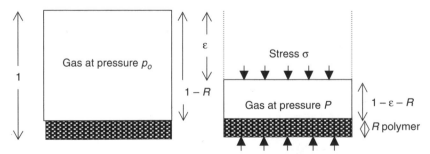

4.7 The gas volumes in a foam of zero Poisson's ratio and relative density R, before and after uniaxial compression.

relative volumes can be seen. For a zero foam Poisson's ratio, the compressive strain, ε, is also the volumetric strain. The polymer is taken to be incompressible, so the volume of gas in the compressed foam is $1 - \varepsilon - R$. The absolute pressure, p, of the isothermal gas in the compressed foam is given by:

$$p_a(1 - R) = p(1 - \varepsilon - R) \qquad [4.1]$$

The compressive stress σ is measured relative to atmospheric pressure so:

$$\sigma = p - p_a = p_a \frac{\varepsilon}{(1 - \varepsilon - R)} \qquad [4.2]$$

Gibson and Ashby (1997), assuming that the polymer contribution σ_0 was constant, obtained:

$$\sigma = \sigma_0 + \frac{p_0 \varepsilon}{1 - \varepsilon - R} \qquad [4.3]$$

For low density EVA foams, σ_0 can be evaluated by fitting the loading part of a graph of stress against $\varepsilon/(1 - \varepsilon - R)$ with a straight line, and extrapolating to zero strain. However, for typical shoe densities, the graph does not have a linear portion. Micromechanics analysis (Mills and Zhu, 1999) predicts that the polymer contribution will vary somewhat with the strain. If equation [4.3] is used, p_0 is the 'effective' gas pressure; its value can exceed 100 kPa as it represents hardening from both gas and polymer.

Uniaxial compressive response of EVA foam

Figure 4.8 shows a typical compressive force vs. deflection response, at impact strain rates, for EVA foam of density 146 kg m^{-3} from a running shoe. There is hysteresis on unloading, and the response changes slightly after 1 hour of simulated running. The foam specimen was 14 mm thick, with a loaded area of 400 mm^2.

When both Poisson's ratio and the polymer contributions are zero, the compressive stress strain response under adiabatic conditions becomes

$$\sigma = p_a \left(\frac{1 - R}{1 - \varepsilon - R} \right)^{\gamma} - p_a \qquad [4.4]$$

where γ is the ratio of specific heats under adiabatic and isothermal conditions.

Tordon (1993) measured the impact compression response of a foam described as EVA65. Inspection of the foam revealed that it is a cross-linked EVA foam, produced by the Zotefoams process, with density 65 kg m^{-3}, of lower density than midsole foams. When the isothermal response of the cell air was subtracted from the foam response (Fig. 4.9), the polymer contri-

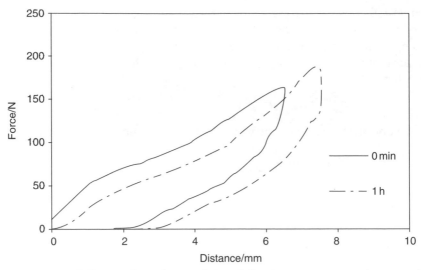

4.8 EVA shoe foam impact force–deflection response, at the start of testing and after 1 hour of impacts at 1.5 Hz (Verdejo, unpublished).

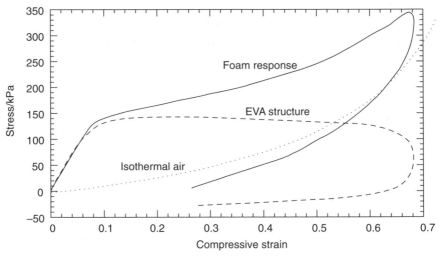

4.9 Impact compressive stress–strain curve for EVA foam of density 65 kg m^{-3}, with the contributions from the isothermally compressed air, and the EVA. (Adapted from Tordon, 1993.)

bution was almost constant at 140 kPa when the strain was increasing, and −30 kPa when it was decreasing. He concluded that the process of compression of the gas is close to isothermal, and that the polymeric contribution is significant.

The polymer contribution is responsible for the foam energy losses in compressive impacts. If a rubbery polymer matrix was used, it might reduce

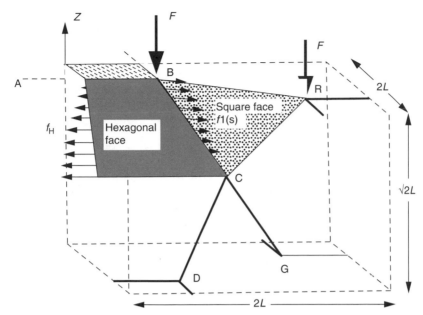

4.10 Structure cell of the Kelvin foam, for compression in the Z direction, with horizontal tensile forces f_1(s) and f_H(s) in two types of face. The external boundaries act as mirror planes.

the foam energy losses, but there would be disadvantages (less foot stability, and a higher density material needed to achieve an adequate Young's modulus). Conversely, if LDPE foam (50% crystallinity) of much lower density were used, there would be permanent cell face stretching after compressive impact. Hence the durability of the foam would be poor.

Micromechanics of the polymer contribution to the foam compression

Micromechanics modelling reveals the mechanisms that contribute to the foam-compressive response. Gibson and Ashby (1998) suggest mechanisms observed in the deformation of honeycombs (two-dimensional foams). However, their analysis of the stress–strain curve is qualitative. Mills and Zhu (1999) used the regular Kelvin foam structure (Fig. 4.10) to make quantitative predictions of the stress–strain curve. When foam is compressed in the vertical direction, the high air pressure in the cells is balanced by tensile stresses in the horizontal direction in the cell faces. As they contain oriented semicrystalline polymers of high modulus, they resist stretching, so the lateral expansion (hence Poisson's ratio) of the foam will be near zero. However, in a rubber foam, the faces will stretch and the Poisson's ratio is high. For example, Airex soft PVC foam has a Poisson's ratio of about 0.4.

The cell edges bend, and may deform axially, when the foam is deformed. The faces will bow to a convex shape if there is a pressure differential between neighbouring cells, or form parallel wrinkles in a direction perpendicular to the direction of compression.

4.4.2 Stress analysis of the foam in the midsole heel

Different levels of analysis are possible.

Uniform strain

Misevich and Cavanagh (1984) modelled the foam response, assuming that the shoe prevents any foam lateral expansion, and that the strain in the compressed foam was uniform. The air in the EVA foam cells was assumed to compress adiabatically. The cell air was taken to be an ideal gas, with the parameter $\gamma = 1.4$ in equation [4.4]. Although they stated that heat flow to the cell walls caused hysteresis energy loss, heat flow was not modelled. The polymer contribution to the stress was neglected. The impact of a 8.5 kg striker on the foam was modelled at a series of small time increments. When the initial gas pressure $p_0 = 100$ kPa, the predicted striker peak accelerations were too high. Rather than considering a polymer contribution to the compressive stress, they used $p_0 = 169$ kPa to give a better prediction. To justify this, they argued for a high gas pressure in the undeformed foam in various unconvincing ways (such as the release of gas bubbles when the foam is cut under water).

Easterling (1993) emphasised the energy return from shoe heels, tested in the laboratory at moderate strain rates; typical values of 65% energy return were quoted. Shorten (1993) computed, from the variation in plantar pressure distribution throughout a foot strike, and the viscoelastic response of the foam at compression rates up to 0.5 m/s, maps of energy dissipated and work done by the midsoles. Of a total 11.5 J of input energy, 7.9 J were returned to the foot. The majority of the energy return was from the heel, in the early stages of the foot strike. However, since energy losses in the legs are far higher, the potential benefits of reducing the energy absorption in the shoe are limited.

Considering the stress field in the heelpad and the foam midsole

Figure 4.11(a) shows a section of a child's heel. The calcaneus bone is cushioned by adipose tissue and by the fibrous outer layer. The total thickness averages 18 mm in adults. The fat is contained under pressure in a matrix of fibrous tissue. Experiments were made with a flat-faced pendulum, with a load cell on the pendulum face, striking the heel at velocities of 1 to 2 m/s (Cavanagh et al., 1984). Figure 4.11(b) shows the force vs. displacement relation is non-linear with hardening at high strains. The considerable energy loss is represented by the area inside the loading and unloading loop. Aerts

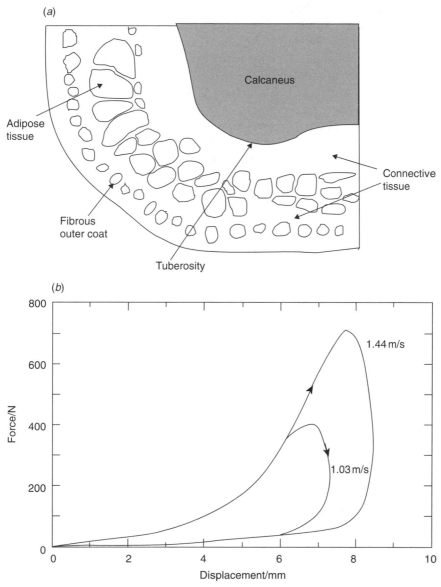

4.11 (*a*) Section of a child's heel, (*b*) force vs. displacement relation, for impact on a human heel pad. (Redrawn from Cavanagh *et al.*, 1984.)

et al. (1995) also described the force–deflection response of the human heelpad. Gefen *et al.* (2001) gave the mean pressure at the interface between the heelpad and a flat rigid surface as a function of an average heelpad thickness strain. However, they gave no heelpad stress–strain data.

Verdejo and Mills (2002) analysed the stress field under the heel, using ABAQUS FEA. The problems tackled were axisymmetric, with a vertical axis of rotational symmetry. The geometry of the calcaneus bone of the heel was grossly simplified to be a hemisphere of radius 15 mm, attached to the end of a 20 mm long vertical cylinder of radius 15 mm. The heelpad is assumed to be bonded to the heelbone surface. This allows some of the load to be transferred by shear to the cylindrical surface. The heelpad was simulated using the Ogden hyperelastic material (see Chapter 2), using shear moduli $\mu_1 = \mu_2 = 200\,kPa$, exponents $\alpha_1 = 2$ and $\alpha_2 = -2$, and inverse bulk modulus $D = 1.0 \times 10^{-9}\,Pa^{-1}$. This allowed the reproduction of the heel-strike data of Aerts. The EVA midsole foam was taken as vertical cylinder of radius 35 mm and height 22 mm, with flat end faces. The Ogden hyperfoam data for the foam were $\mu = 100\,kPa$, $\alpha = 0.5$, Poisson's ratio $= 0$. Figure 4.12 shows the vertical compressive stress σ_{22} contours, at a total deformation of 17 mm. At a load of 0.70 kN, the upper midsole surface has become concave, and the heelpad has spread laterally increasing its contact area with the foam. The maximum foam stress is 300 kPa, at the centre of the contact area on the foam upper surface. This confirms the Tekscan data for the pressure map on the upper surface of the shoe midsole, with a 300 kPa stress area of diameter approximately 20 mm. The predicted force vs. deflection relationship is shown in Fig. 4.13.

4.4.3 Load deflection response of other midsole designs

Aguinaldo *et al.* (2002) describe the load deflection response of the heels of Nike Shox shoes (with four polyurethane springs), and a related cushioning system from L&L International (with four egg-cup shaped polyester elastomer springs). Their cyclic loading of the heel portion of the shoes at 5 Hz from 0 to 1.4 kN is more severe than the loading in running, so may lead to increased hysteresis heating. In spite of the laboratory tests showing that both 'discrete spring' heel constructions were significantly softer than a shoe with an EVA midsole, trials with runners showed the peak ground reaction forces were independent of the construction, at $F_1 = 1.9 \pm 0.2\,kN$ and $F_2 = 2.5 \pm 0.3\,kN$ (see Fig. 4.6(*a*) for the definition of the peaks). They comment that further studies of rearfoot stability and durability are required for these novel shoe designs.

4.4.4 Foam flexure and heel stability

Figure 4.14(*a*) shows the flexure of a midsole. The curvature of the sole depends on the local application of forces from the metatarsal heads. If there is a deliberate hinge in the outsole, the curvature may be localised at this point. However, the local compression of the foam by the metatarsals

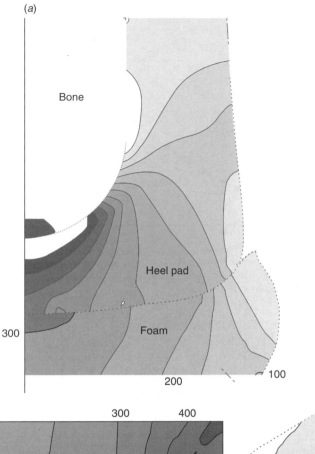

(a)

Bone

Heel pad

Foam

300

200

100

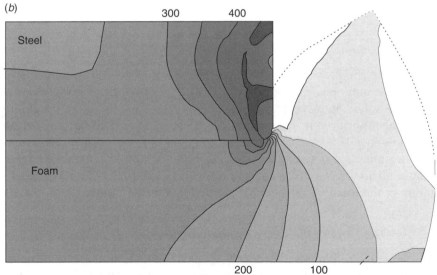

(b)

300   400

Steel

Foam

200   100

4.12 Contours of vertical compressive stress (kPa) in: (a) heelpad and
EVA foam at 17mm deflection, (b) ASTM F1614 steel heel and
EVA foam at 11mm deflection. Both supported on a flat rigid
surface; axis of rotational symmetry at left (Verdejo and Mills,
2002).

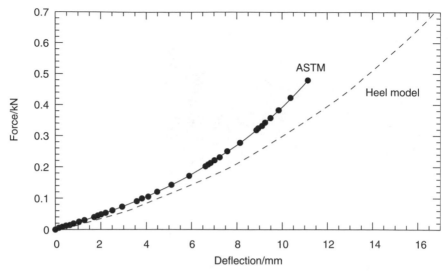

4.13 Predicted force vs. total deflection for the two models of Fig. 4.12 (Verdejo and Mills, 2002).

makes flexure easier, since the second moment of area of the foam cross-section is proportional to the thickness cubed. Consequently, measurements of the bending stiffness of the sole without any compressive load are irrelevant to the shoe mechanics. The consequences of excessive sole bending stiffness could be strains in the tendons under the foot arch – planar fasciitis (see www.drpribut.com/sports/).

Heel stability

During the compression of EVA foam there is energy loss, so only a fraction of the input energy is available to cause deformation elsewhere in the foam. However, in a system without energy loss, there might be instability of the heel position. In Fig. 4.14(*b*) a large force is exerted by the foot on the shoe heel. If the upper surface of the heel tilts, with associated pronation or supination of the talus/calcaneus joint, the runner's potential energy may fall. This would release energy for further shear of the foam. The indentation of the foam by the heel partly stabilises the situation.

The effect of the shoe on running style

The greater cushioning afforded by EVA foams allowed many distance runners to change their running style, from one where the forefoot strikes the road first, to one where the heel strikes first. Repeated heel strikes could damage joints if there were no cushioning in the shoes. Robbins and Waked

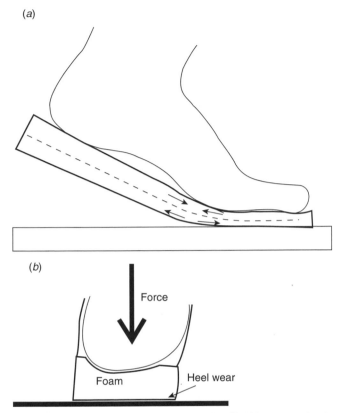

4.14 (*a*) Midsole bending during push-off, (*b*) uneven heel
compression, with the heelpad indenting the foam.

(1997a) argued that deceptive advertising (that the shoes can cushion against impact) creates a false sense of security among the users of athletic shoes, and modifies their behaviour, so that they suffer more injuries. The experiments involved volunteers stepping down by 45 mm while barefoot, onto a 25 mm thick layer of EVA foam, after being given positive, neutral or warning messages about the foam properties. They concluded that the peak force in the foot impact was affected by the message. In a related experiment (Robbins & Waked, 1997b), volunteers stepped onto EVA layers of different stiffness. When the softest foam was used, the subjects tended to land harder, possibly because they aimed to compress the soft foam to make the landing less unstable.

Shorten (2000) argued that the runner's body acclimatises to the mechanical loads during training. A sudden change to a lighter, less cushioned shoe for a race constitutes a training error, which increases the likelihood of injury.

4.5   Foam durability

4.5.1 Outsole durability

There are few papers on outsole wear. Edelmann-Nusser *et al.* (2002) describe an accelerated wear tester that produces the equivalent of a 500 km running trial (which takes more than 4 weeks) in the laboratory in 500 cycles. The shoe heel is loaded with 23 kg, while being dragged 85 mm back and forward, on a metal plate covered with 1 mm high pyramids. The 0.5 mm wear of the relatively flat thermoplastic polyurethane outsole of Adidas street-running shoes was the same as the mean wear from six runners. No comparative wear rates were given for different sole compositions.

4.5.2 Foam durability

During footstrike, the pressure distribution in the midsoles is non-uniform, with maximum values near the centre of the heel contact area and in the metatarsal area. The concern is that impacts, repeated at 0.3 s intervals, may cause fatigue damage to the foam. Most research has used laboratory fatigue tests. Cook *et al.* (1985) used a prosthetic foot to load the heel of the shoe, tilted back by 15°, from 0 to 1.5 kN at 2.5 Hz; this applies severe loads to the rear part of the heel. They only plotted the energy absorbed per cycle, and say that, after a number of cycles equivalent to 500 miles of running, shoes maintained 55 ± 10% of their initial energy absorption. They also tested shoes after 500 miles of running by volunteers and say that 70% of the initial energy absorption remained. However, such laboratory tests provide, at best, an indirect indication of performance change in a run. As no details of the EVA densities were given, and foam types have changed in the last 17 years, it is unclear if the results still apply.

The polymer contribution to the total EVA foam response is known to decrease rapidly in the first few cycles of loading, in tests that impose uniform strain (Misevich and Cavanagh, 1984). When a heel-shaped indenter loads a foam midsole, the rapid softening of foam close to the indenter leads to a progressive change in the overall strain field. A slower change in response is expected as the cycle number increases. They cycled the load, on a flat-ended cylinder of diameter 44.5 mm, between zero and 1780 N at 1 Hz, and monitored the total input energy to the EVA foam midsole in a cycle. The peak compressive stress was 1.15 MPa, higher than the peak values detected by Hennig and Milani (1995). In real life, if the shoe foam response changes with time, and the person continued to run in the same way, the peak force would increase as a run proceeded, while the impact energy remained constant. The shape of the foam force–compression relationship usually changes from being near linear to having a marked positive curvature. They found the peak input energy decreased linearly with the logarithm of the number of cycles, from 5.1 J initially to 3.0 J at 10^4 cycles. The foam thickness at zero load decreased as the test proceeded, and it was suspected that the cell gas content reduced. No comment was made about any foam temperature rise.

These experiments could give the impression that the foam response change is permanent. However, the foam recovers slowly and partially if the shoe is rested. Although SEM photographs of sections from old trainers show flattened cells near the foam lower surface, some of this is due to the initial processing; near-surface cells are flattened in the moulding process. Barlett (1995) suggested, perhaps incorrectly, that the flattening of foam cells near midsole surfaces is evidence of foam deterioration.

ASTM tests for heel performance

In the ASTM F1614 test, the runner's heel is replaced by a vertical steel cylinder, of 45 mm diameter and flat end, with a 1 mm radius edge. A high mass (8 kg) falls a small distance, imparting a 5 J impact energy. Verdejo and Mills (2002) FEA (Fig. 4.12(b)) was only stable to a load of 0.55 kN, 42% of the typical peak load in the ASTM experiments. Hence, the stresses would be higher if the simulation had reached the experimental peak load. The ASTM test does not reproduce the stress field under a runner's heel, because:

(a)   for the central 45 mm of the foam, the strain is too uniform.
(b)   for the foam at about 20 mm radial distance, there is an unrealistically high edge stress of 450 kPa, due to the presence of the indenter edge.
(c)   the peak stresses are too high, due to the lack of load being taken by the rest of the sole.

With the deformable heel model, rather than the 'rigid' ASTM heel, the deflection is 20% higher for a given compressive force (Fig. 4.13). Hence a footstrike of a given kinetic energy will produce a lower peak force than in the ASTM test with the same kinetic energy.

Thompson *et al.* (2000) made an FEA analysis of a similar situation to the ASTM test. However, they constrained the side of the foam not to move radially, without giving a reason for this unrealistic boundary condition. Tests with runners showed that, after 500 km of running, their peak plantar pressures increased slightly (Verdejo and Mills, 2002). These direct tests suggest that foam deterioration may be less rapid than predicted from laboratory cyclic tests.

4.5.3 Air loss rates and foam selection

The materials science of foam permeability involves two main areas.

Polymer permeability

The permeability of polymers is a function of the microstructure. For PE copolymers, as the crystalline phase is impermeable, the permeability decreases as the percentage crystallinity decreases. Mills (1993) gives permeability data for some polymers; the permeability is high for amorphous polymers above their glass transition temperatures, but rubbers made from different polymers differ in permeability. Data for the specific polyurethanes used in shoe foams or air bladders is not available. Mineral additives, which are impermeable to air, reduce the polymer permeability. The form and orientation of the additives may increase the effect. Talc has a platelet form; Mills and Rodriguez-Perez (2001) identify platelet-shaped particles in EVA foam cell walls, lying in the plane of the wall, that could be talc. The additive content can be 18% by weight, so the effect on the permeability is not insignificant.

Foam permeability

Low-density foams contain polyhedral closed cells (Fig. 4.4). The theory for foam permeability (Mills and Gilchrist, 1997b) starts with the equation for steady state gas flow rate Q (m³ at STP s⁻¹) through a cell face, of thickness δ and area A, between cells i and $i - 1$, between which there is a pressure differential $p_{i-1} - p_i$

$$Q = PA \frac{p_{i-1} - p_i}{\delta} \qquad [4.5]$$

where P is the polymer permeability (units: m³ at STP m⁻¹ s⁻¹ atm⁻¹). This equation was used in finite difference form to calculate the transfer of gas into a cell in a short time interval. For foams with uniform tetrakaidecahedral cells with edge length L (Fig. 4.10) the relative density R is given by:

$$R = 1.18 \frac{\delta}{L} \qquad [4.6]$$

Hence the cell face thickness is proportional to R and to the cell diameter. If the foam is treated as a continuum, it has a diffusion coefficient D_f, which is shown to be:

$$D_f = \frac{6.7 P p_a}{\phi \rho} \qquad [4.7]$$

where p_a is atmospheric pressure, and ϕ is the fraction of polymer in the cell faces. The equation states that the foam diffusivity varies inversely with its density ρ, and is independent of the cell size. For foamed rubber of $R > 0.4$, the cells are isolated spheres, changing the relationship between the foam density and its permeability.

For EVA copolymer film having an 18% vinyl acetate content, the oxygen permeability (Anon, 1995) is $435\,cm^3$ at STP mm m^{-2} day^{-1} atm^{-1} (convert these units to m^3 at STP m^{-1} s^{-1} atm^{-1} by dividing by 8.64×10^{13}). As the nitrogen permeability of LDPE is about one-third of the oxygen permeability, it is expected that the nitrogen permeability will be approximately $150\,cm^3$ at STP mm m^{-2} day^{-1} atm^{-1}. Therefore, the diffusivity of EVA foam with density of $250\,kg\,m^{-3}$, assuming $\phi = 0.5$, should be $1.4 \times 10^{-10}\,m^2\,s^{-1}$. Mills and Rodriguez-Perez (2001) used a value of $1.0 \times 10^{-10}\,m^2\,s^{-1}$ in their modelling of the gas loss from EVA foam of density $275\,kg\,m^{-3}$ under creep loading.

The parameters that affect the foam permeability also affect the mechanical properties of the foam. It is not possible to optimise PE copolymers for both gas loss and mechanical properties. If the crystallinity is increased, so is the initial yield stress of the foam. Additions of filler, or increases of crystallinity, increases the initial Young's modulus of the polymer, hence of the foam. As the foam relative density, R, increases, the rate of compression hardening increases according to equation [4.3].

Polyurethane foams have significantly different properties to EVA foams. Details of the chemistry, surfactants, etc. are given by Klempner and Frisch (2001). The PU polymers vary in mechanical properties according to the chemical mix.

Midsole gas loss

The midsole mechanical loading and size affects the gas loss via:

(a) the air diffusion distances. The thickness of the midsoles and the nature of the boundary outsole affect the total gas loss.
(b) the stress field from running. If high pressure areas are close to a 'free' surface of the foam, it is easier for the gas to escape.

No one has modelled the air loss from shoe foam during running. Such modelling should consider the interaction between air loss in the high stress regions and the stress distribution under the runner's heel. As a gas-depleted region develops near the top centre of the heel foam, the change in materials properties must be considered.

4.5.4 Modelling gas loss from EVA foam under creep loading

Gas loss modelling exists for foam under uniform compression. In Mills and Rodriguez-Perez's (2001) finite difference model, gas diffuses parallel to the

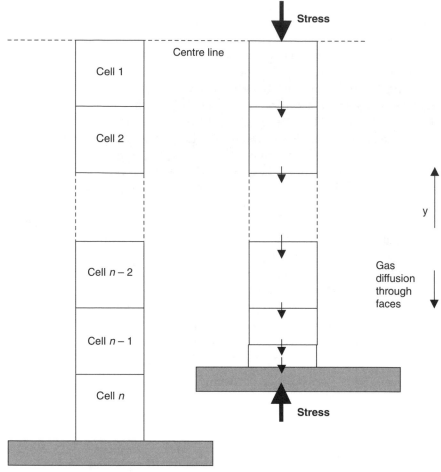

4.15 Model for air diffusion from the foam, unloaded and under constant stress.

direction of the applied stress (Fig. 4.15). The same total stress acts on each layer, but, as gas diffuses from a particular layer, the strain increases to maintain the air pressure. When the compressive strain increases in the surface layers, so does the foam diffusivity. Consequently, the surface layers collapse. For creep loading, the number, N, of cells with less than 50 kPa unloaded gas pressure was predicted to increase with time, t, according to:

$$N = 0.0457t^{0.525} \qquad [4.8]$$

This suggests a depletion rate proportional to the square root of time, a result familiar for Fickian diffusion processes in a slab.

There was no modelling of recovery after air had been expelled from cells. The polymer will recover and apply a tensile stress on the cells. However, this negative pressure is very low compared with possible pressures of 350 kPa during loading. Figure 4.9 for the lower density EVA 65 foam suggests that the polymer exerts a negative stress of 25 kPa on unloading. It is possible that adhesion between touching cell faces hinders the recovery process.

4.5.5 Compression set as a materials selection tool

Compression set is a foam selection and quality control tool. The measurement originated with rubbers, where it characterises permanent changes to cross-linked networks. Khunniteekool *et al.* (1994) showed that the compression set, measured 2 hours after a 22-hour period at 60% strain, decreases from 35% to 17% as the EVA (18% VA) foam density increased from 30 to 110 kg m^{-3}. Values are lower still for the higher densities used in midsoles. However, 'compression set' is a misleading term for foams that recover with time. Foamed ESI/LDPE blends of density 40 kg m^{-3} were held at 50% strain for 24 hours (Ankrah, Verdejo and Mills, 2002). The residual strain decreased almost linearly with log time, with almost complete recovery after 24 hours. The compression set values for 20% ESI/80% LDPE foams were 21 ± 5% at 30 min, 10 ± 2% at 24 h and 5 ± 1% at 48 h. EVA foam of density 30 kg m^{-3} recovers slowly for the first hour, but almost completely after 2 days. A compression set value, for a foam sample of a single size after a particular recovery time, is not a good indicator of the recovery process.

4.5.6 Foam temperature rise in running

The temperature sensitivity of EVA foam properties is often ignored. Kinoshita and Bates (1996) measured the temperature rise in midsole foam, for runners in Japan under a variety of conditions. There is heat transfer from the tarmac road surface and from the foot. Typically, the foam temperature rose to a plateau value after 20 minutes of running (Fig. 4.16). The heating effect of polymer hysteresis is demonstrated by the maximum midsole temperature exceeding the higher of the exterior air temperature and the road temperature by 5 to 10 degrees. The mechanical properties change of EVA foam as its melting temperature T_m of 70 °C is approached. Heat generation and heat flow should be part of a complete model of foam response in running. Kinoshita and Bates suggest that athletes select shoes with the appropriate EVA foam hardness for summer or winter use.

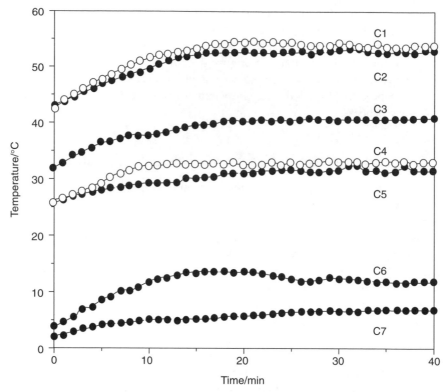

4.16 EVA midsole temperature rise during running in Japan, various runners and conditions (permission from Kinoshita and Bates, 1996).

4.6   Discussion

The analysis of foam performance in running shoes has progressed, both with the use of in-shoe pressure mapping, and with FEA analysis of the stresses in the shoe. Further work is necessary on heel tissue mechanics and FEA of the whole foot shape. The selection of EVA foam is a compromise. It is a low-cost material; its low Young's modulus allows a wedge-shaped midsole to bend relatively easily, while cushioning the impact of various parts of the foot. However, its durability to repeated impacts is not ideal. Designs where separate components cushion the heel strike are more costly, and it is not obvious that they improve the biomechanics of running. Frederick (1984b) discussed compromises in running shoe design; for instance, lighter shoes usually have less cushioning or stability, while soles that promote good traction may not be durable.

4.7   Future developments

Future developments are likely to include:

(a)   novel design features (colour, shape . . .).
(b)   attempts to increase the air flow from foot, for cooling and sweat removal (see the Adidas website for 'Climacool' and the Mizuno website for 'Intercool').
(c)   development of 'rubber mechanism' shock absorbers that provide a better balance of shock absorption and stability than EVA midsoles.
(d)   Advances in imaging techniques to view the deformation of the foot tissues during running. Dynamic measurements of the plantar soft tissue thickness were made during walking (Cavanagh, 1999), and faster data recording would make this possible during running. More widespread use of footstrike pressure distributions, or gait analysis, to allow athletes to select shoes that are best for their feet.

4.8   Acknowledgements

This chapter is based in part on the research of Raquel Verdejo; I am grateful for her help.

4.9   References

Aerts P, Ker R F et al. (1995). The mechanical properties of the human heel pad. J Biomech, **28**, 1299–1304.

Aguinaldo A, Mahar A et al. (2002). Ground reaction forces in running shoes with two types of cushioning column systems. Proc 20th Intl Sym Biomech in Sports. University of Extramedura, Spain.

Ahnemiller J (1997). Polyurethane rubber outsoles for athletic footwear. A.C.S. 151st Rubber Division meeting, paper 46.

Alley L and Nichols G (1999). Future technical challenges for the PU industry in athletic footwear production. Utech conf.

Ankrah S, Verdejo R and Mills N J (2002). The mechanical properties of ESI/LDPE foam blends and sport applications. Cell Polym, **21**, 237–264.

Anon (1995). Permeability and Other Film Properties of Plastics and Elastomers. Plastics Design Library, New York.

Barlett R (1995). Sports Biomechanics. Spon, London.

Bobbert M F, Yeadon M R and Nigg B M (1992). Mechanical analysis of the landing phase in heel–toe running. J Biomech, **25**, 223–234.

Cavanagh P R (1999). Plantar soft tissue thickness during ground contact in walking. J Biomech, **32**, 623–628.

Cavanagh P R, Valiant G A et al. (1984). Biological aspects of modeling shoe/foot interaction during running, Sport Shoes and Playing Surfaces, ed E C Frederick, 24–46. Human Kinetics Publishers, Champaign, IL.

Cook S D, Kester M A et al. (1985). Biomechanics of running shoe performance. Clin Sport Med, **4**, 619–626.

Diegritz W (1998). Well padded. *Kunststoffe-German Plastics.* **88**, 1494–1496.

Dixon S J and Kerwin D G (1999). Heel lift influence on Achilles tendon loading. *Int Soc Biomech*, Footwear Biomech Conf, Canada.

Easterling R (1993). Ch. 3, Sports shoes. *Advanced Materials for Sports Equipment*, Chapman and Hall, London.

Edelmann-Nusser J, Ganter N *et al.* (2002). Short term estimation of abrasion in running shoes. *The Engineering of Sport* **4**, Blackwell, Oxford.

Frederick E C (1984a). *Sports Shoes and Running Surfaces*, Human Kinetics Publ. Inc., Champaign, IL.

Frederick E C (1984b). The running shoe: dilemmas and dichotomies in design. *Sports Shoes and Running Surfaces*, 26–35. Human Kinetics Publ. Inc., Champaign, IL.

Gefen A, Megido-Ravid M *et al.* (2001). In vivo biomechanical behavior of the human heel pad during the stance phase of gait. *J Biomech*, **34**, 1661–1665.

Gibson L J and Ashby M F (1997). *Cellular Solids*, 2nd edn, Cambridge University Press, Cambridge.

Gunther M and Blickman R (2002). Joint stiffness of the ankle and the knee while running. *J Biomech*, **35**, 1459–1474.

Hennig E M and Milani T L (1995). In-shoe pressure distribution for running in various types of footwear. *J Appl Biomech*, **11**, 299–310.

Hunter S, Dolan M G *et al.* (1995). *Foot Orthotics in Therapy and Sport.* Human Kinetics Publ. Inc., Champaign, IL.

Indesteege J, Camargo R E and Mackay P W (1997). Considerations on the selection of PU systems for midsoles wedges in athletic footwear. *Polyurethanes World Congress*, 536–543.

Kim I J, Smith R *et al.* (2001). Microscopic observations of the progressive wear on shoe surfaces that affect the slip resistance characteristics. *Int J Indust Ergo*, **28**, 17–29.

Kinoshita H and Bates B T (1996). The effect of environmental temperature on the properties of running shoes. *J Appl Biomech*, **12**, 258–268.

Klempner D and Frisch K, eds. (2001). *Advances in Urethane Science and Technology*, RAPRA, Shawbury, UK.

Khunniteekool C *et al.* (1994). Structure/property relationships of crosslinked PE and EVA foams. *Cellular and Microcellular Materials*. ASME conf, 53–61.

Manning D P, Jones C *et al.* (1998). The surface roughness of a rubber soling material determines the coefficient of friction on water-lubricated surfaces. *J Safety Res*, **29**, 275–283.

Mills N J (1993). *Plastics*. E Arnold, London.

Mills N J and Gilchrist A (1997a). The effects of heat transfer and Poisson's ratio on the compressive response of closed cell polymer foams. *Cell Polym*, **16**, 87–119.

Mills N J and Gilchrist A (1997b). Creep and recovery of polyolefin foams – deformation mechanisms. *J Cell Plast*, **33**, 264–292.

Mills N J and Rodriguez-Perez M A (2001). Modelling the gas-loss creep mechanism in EVA foam from running shoes. *Cell Polym*, **20**, 79–100.

Mills N J and Zhu H (1999). The high strain compression of closed-cell polymer foams. *J Mech Phys Solids*, **47**, 669–695.

Misevich K W and Cavanagh P R (1984). Material aspects of modeling shoe/foot interaction. *Sports Shoes and Playing Surfaces*, ed. E C Frederick, Human Kinetics Publ. Inc., Champaign, IL.

Nigg B M, ed. (1986). *Biomechanics of Running Shoes*, Human Kinetics Publ. Inc., Champaign, IL.

Pillow J G (1997). Ethylene elastomers made using constrained geometry catalyst technology. *IRC 97 conf.*, Nuremberg, Germany.

Robbins S and Waked E (1997a). Hazard of deceptive advertising of athletic footwear. *Br J Sports Med*, **31**, 299–303.

Robbins S and Waked E (1997b). Balance and vertical impact in sports: role of shoe sole materials. *Arch Phys Med Rehabil*, **78**, 463–467.

Sawai M, Miramoto K *et al.* (2000). Super low density polyurethane systems for sports shoes. *J Cell Plast*, **36**, 286–291.

Segesser B and Pforringer W, eds. (1989). *The Shoe in Sport*, Year Book Medical Publ., Chicago.

Shorten M R (1993). The energetics of running and running shoes. *J Biomechanics*, **26**, suppl 1, 41–51.

Shorten M R (2000). Running shoe design: protection and performance, 159–169. *Marathon Medicine*, ed. D Tunstall Pedoe, Royal Society Medicine, London.

Sims G L S and Khunniteekool C (1996). Compression moulded ethylene homo- and copolymer foams. *Cell Polym*, **15**, 1–14 and 15–29.

Stacoff A, Kalin X and Stussi E (1991). The effect of shoes on the torsion and rear-foot motion in running. *Med and Sci in Sport and Med*, **23**, 482–490.

Swigart J F, Erdman A G and Cain P J (1993). An energy based method for testing cushioning durability of running shoes. *J Appl Biomech*, **9**, 26–47.

Thompson R D, Birkbeck A E *et al.* (2000). The modelling and performance of train-ing shoe cushioning systems. *Sport Eng*, **2**, 109–120.

Tordon M J (1993). Evaluation of the cushioning of closed-cell plastic foams. *Dynamic Loading in Manufacturing and Service*, Inst. Eng. Aust. Conf., Mel-bourne, 11 pp.

Valiant G A (1997). Designing proper athletic sport shoe outsole traction. *Rubb Plast News*, 1 Dec, 14–16.

Vanderbilt T (1998). *The Sneaker Book*. The New Press, New York.

Verdejo R and Mills N J (2002). Performance of EVA foams in running shoes. *The Engineering of Sport*, **4**, 580–587, Blackwell, Oxford.

5
Balls and ballistics

J. MACARI PALLIS

Cislunar Aerospace, Inc., USA

R. D. MEHTA

Sports Aerodynamics Consultant, USA

5.1   Introduction

The shape, weight and surface features of sports balls (such as cricket balls and baseballs) and ballistics (such as the javelin or boomerang) directly impact the effectiveness of the equipment and ultimately athletic performance. Collectively known as 'sports projectiles', sports balls and ballistics have a wide variety of performance objectives such as speed, distance, accurate placement, or flight duration and direction. These criteria may be the objective of the sport or a desirable characteristic that provides a competitive edge. For example, in certain boomerang competitions the winner is determined by distance or maximum time aloft, while in baseball understanding and utilising the behaviour of the stitches on the ball's cover is critical to pitching strategy.

The projectile's construction, material composition, and how the outer material is applied to the surface, all affect its performance. Balls typically have patches of materials that form the surface contour as well as protruding stitches or indentations. Subsequently, sports projectile manufacturing is interdependent with the equipment's strength, durability and performance.

Sport scientists have become more aware of the effects of surface modifications on aerodynamic flight path and ball performance. Used and scuffed balls perform differently from new balls. In certain sports, modification of the ball's surface is considered illegal ball tampering. For example, placing saliva on a baseball (known as a 'spitball') or scuffing a cricket ball modifies its aerodynamic characteristics and is illegal in that sport.

The materials and their effects on performance for nine sports projectiles that depict characteristic behaviours (cricket, baseball, tennis, golf, soccer, volleyball, boomerang, discus, and javelin) are presented. As applicable, each section includes: a brief history of the ball's or ballistic's evolution; an overview of regulations affecting the projectile's performance, shape, material composition, surface patterns or special manufacturing restrictions; the material's effects on performance; and how the materials affect athletic strategy.

5.2   Basic aerodynamic principles

Let us first consider the flight of a smooth sphere through an ideal or inviscid fluid. As the flow accelerates around the front of the sphere, the surface pressure decreases (Bernoulli equation) until a maximum velocity and minimum pressure are achieved half way around the sphere. The reverse occurs over the back part of the sphere so that the velocity decreases and the pressure increases (adverse pressure gradient). In a real viscous fluid such as air, a boundary layer, defined as a thin region of air near the surface, which the sphere carries with it, is formed around the sphere. The boundary layer cannot typically negotiate the adverse pressure gradient over the back part of the sphere and it will tend to peel away or 'separate' from the surface. The pressure becomes approximately constant once the boundary layer has separated, and the pressure difference between the front and back of the sphere results in a (pressure) drag force that slows down the sphere. The boundary layer can have two distinct states: 'laminar', with smooth tiers of air passing one on top of the other, or 'turbulent', with the air moving chaotically throughout the layer. The turbulent boundary layer has higher momentum near the wall, compared to the laminar layer, and it is continually replenished by turbulent mixing and transport. It is therefore better able to withstand the adverse pressure gradient over the back part of the sphere and, as a result, separates relatively late compared to a laminar boundary layer. This results in a smaller separated region or 'wake' behind the ball and thus less pressure drag. The 'transition' from a laminar to a turbulent boundary layer occurs when a critical sphere Reynolds number (Re_{cr}) is achieved. The Reynolds number is defined as, $Re = Ud/v$, where U is the ball velocity, d is the ball diameter and v is the air kinematic viscosity. Earlier transition of the boundary layer can be induced by 'tripping' the laminar boundary layer using a protuberance, e.g. seam on a baseball or cricket ball, or surface roughness, e.g. dimples on a golf ball or fabric nap on a tennis ball. The total drag on a bluff body, such as a sphere or sports ball, consists mainly of pressure drag described above. There is also a small contribution due to viscous or skin friction drag.

The boundary layers on a spinning sphere cannot negotiate the adverse pressure gradient on the back part of the ball either and they tend to separate, somewhere in the vicinity of the sphere apex. The extra momentum applied to the boundary layer on the retreating side of the ball allows it to negotiate a higher-pressure rise before separating and so the separation point moves downstream. The reverse occurs on the advancing side and so the separation point moves upstream, thus generating an asymmetric separation and a deflected wake. Following Newton's Third Law of Motion, a deflected wake implies a (Magnus) force acting on the ball, in a direction that opposes that of the deflected wake. On spinning (lifting) balls there is another contribution to the total drag, namely the induced drag.

5.3   Cricket

The origins of cricket are obscure and the source of much speculation, but it was probably played in England as early as the 1300s. It was certainly well established by the time of the Tudor monarchs (1485–1603). The first reference to cricket was contained in a document dated December 1478 and it referred to 'criquet' near St Omer, in what is now north-eastern France. The first recorded cricket match took place at Coxheath in Kent, England in 1646 and the first 'Test' (international) Match took place between England and Australia in Melbourne, Australia in 1877. The infamous 'Ashes' match was played in London in 1882, when an English newspaper printed a mock obituary notice after Australia had defeated England.

Aficionados know cricket as a game of infinite subtlety, not only in strategy and tactics, but also in its most basic mechanics. On each delivery, the ball can have a different trajectory, varied by changing the pace (speed), length, line or, most subtly of all, by moving or 'swinging' the ball through the air so that it drifts sideways. The actual construction of a cricket ball, and the principle by which the faster bowlers swing the ball, is unique to cricket. The outer cover of a cricket ball consists of four or two pieces of leather, which are stitched together (by hand on the better quality four-piece balls). Six rows of prominent stitching along its equator make up the 'primary' seam. On the four-piece balls each hemisphere also has a line of internal stitching forming the 'quarter' or 'secondary' seam. These primary and quarter seams play a critical role in the aerodynamics of a swinging cricket ball.

The earliest cricket 'balls' consisted of stones, pieces of wood and lumps of hide. The first 'manufactured' ball was made around 1658 by interlacing narrow strips of hide. The first 'six-seamed' ball was made in 1775 by Dukes, a family firm in Kent, England. The earlier cricket balls had a cork centre with layers of cork and wool yarn around it. The wool is wound wet under tension so that it compresses the centre as it dries. Today's cricket balls have a composite cork/rubber or a cork/rubber laminate centre (see Chapter 13 for further details on the seam and core).

Until recently the leather was always dyed red, greased and polished with a shellac topcoat. Nowadays, a white cricket ball is sometimes used in the one-day matches. For these balls, the leather is sprayed with a polyurethane white paint-like fluid and then heat treated so that it bonds to the leather

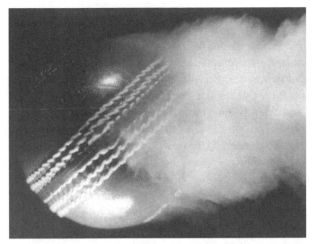

5.1 Cricket ball seam has tripped the boundary layer and delayed separation over the lower surface.

like a hard skin. As a final treatment, one coat of clear polyurethane-based topcoat is applied to protect the white surface further so that it does not get dirty easily. The weight of an approved (official) ball is between 0.15 kg (5.5 oz) and 0.16 kg (5.75 oz) and the circumference is between 22 cm (8.81 inches) and 23 cm (9 inches).

Fast bowlers in cricket make the ball swing by a judicious use of the primary seam. The ball is released with the seam at an angle to the initial line of flight. Over a certain bowling speed or Reynolds number range, the seam trips the laminar boundary layer into turbulence on one side of the ball whereas that on the other (non-seam) side remains laminar. By virtue of its increased energy, the turbulent boundary layer separates later compared to the laminar layer and so a pressure differential, which results in a side force, is generated on the ball. In Fig. 5.1, the seam has tripped the boundary layer on the lower surface into turbulence, evidenced by the chaotic nature of the smoke edge just downstream of the separation point. On the upper surface, a smooth, clean edge confirms that the separating boundary layer is in a laminar state. Note how the laminar boundary layer on the upper surface has separated relatively early compared to the turbulent layer on the lower surface. The asymmetric separation of the boundary layers is confirmed further by the upward deflected wake, which implies that a downward force is acting on the ball.

Bowlers in cricket make the ball swing into a batsman by pointing the seam towards fine leg (inswinger) and towards the slips to swing it away (outswinger). Deflections of over 0.8 m have been measured for professional swing bowlers (Mehta, 1985).

It is interesting to note that the laws of cricket allow bowlers to 'polish' the surface of the ball, but only using natural 'lubricants' such as sweat or saliva. This, of course, is a useful practice for a swing bowler since a laminar separation is required on the non-seam side to produce the necessary asymmetric boundary layer separation. If the ball surface on the non-seam side is allowed to roughen during the course of play, this will tend to trigger transition, thus reducing the asymmetry.

Note that a cricket ball is not normally changed until it has been used for over 300 deliveries. However, a used ball with a rough surface can be made to 'reverse swing' at relatively high bowling speeds (>36 m/s or 80 mph); although the maximum measured bowling speed is 44.7 m/s or 100 mph (achieved very recently), there are very few bowlers who can bowl at such high speeds. For reverse swing, transition occurs relatively early (before the seam location) and symmetrically due to the rough surface. The seam acts as a fence, thickening and weakening the turbulent boundary layer on that (seam) side, which separates early compared to the turbulent boundary layer on the non-seam side. Therefore, the whole asymmetry is reversed and the ball swings *towards* the non-seam side (Mehta, 2000). There has been a lot of controversy in cricket over the last 15 years or so with regard to reverse swing. It turns out that many bowlers resorted to the (illegal) practice of roughening the ball surface deliberately in order to try and produce reverse swing at more reasonable bowling speeds. The ideal ball would be one with a completely smooth surface on one side and a very rough surface on the other, thereby allowing the bowler to produce swing and reverse swing at will.

During the 1999 cricket World Cup there was a lot of discussion about the swing properties of the white ball used in the tournament. The white ball was introduced since it was apparently easier to see, both for the players on the field and for television viewers. The main contention was that the white ball swung significantly more than the conventional red ball. The only difference between the two balls was in the coating, as discussed above. With the conventional red ball, the final polish disappears very quickly during play and it is the grease in the leather that produces the shine when polished by the bowler. With the white ball, the additional topcoat covers up the quarter seam and the effective roughness due to it is therefore reduced. As a consequence, a new white ball should swing more, especially at the higher bowling speeds, since a laminar boundary layer is more readily maintained on the smoother surface. Also, the harder surface stays smooth for a longer time; it does not scuff up easily like the red ball and so conventional swing can be obtained for a longer playing time. Another consequence of this is that reverse swing will occur at higher bowling speeds with a new white ball and later in the innings at more reasonable bowling speeds.

5.4   Baseball

About the time of the American Civil War, a New England youngster named Arthur 'Candy' Cummings became fascinated by a new game called 'base ball'. On the beaches near his home he endlessly pitched clamshells. He found he could make the shells curve by holding and throwing the shells in a specific way. In 1867, 18-year-old Candy Cummings, pitcher for the Brooklyn Excelsiors baseball team, tried out the pitch he had been perfecting in secret for years. Over and over, throughout the game, the ball made an arch and swept past the lunging batter into the catcher's glove. Today, in the Baseball Hall of Fame in Cooperstown, New York, there is a plaque that reads: 'Candy Cummings, inventor of the curveball'.

However, for more than 100 years after Cummings introduced this new pitch, people still questioned its flight: does a baseball really curve, or is it an optical illusion? Several attempts were made throughout the years to settle the question. In 1941 both *Life* and *Look* magazines used stop-action photography to determine if curveballs really curve. *Life* concluded that they do not; *Look* claimed the opposite. As late as 1982, *Science* magazine commissioned an imaging study to measure the flight of a curveball and confirm the extent of the ball movement.

Although all sports ball aerodynamics are a balancing act between competing forces, the stitches on the baseball and the orientation of the stitches as it meets the oncoming air are key to the vast variety of pitches in baseball.

The construction of a baseball starts with the 'pill', a small sphere of cork and rubber enclosed in a rubber shell. The pill is tightly wound with three layers of wool yarn and finished with a winding of cotton/polyester yarn. This core is coated with a latex adhesive over which a leather cover is sewn. The cover consists of two 'figure of eight' pieces of alum-tanned leather which are hand stitched using a red cotton thread in a 104-stitch pattern. The rules of baseball require that the ball shall weigh not less than 0.14 kg (5 oz) nor more than 0.15 kg (5 1/4 oz) and measure not less than 22.9 cm (9 inches) nor more than 23.5 cm (9 1/4 inches) in circumference.

Combining with the effect of air pressure (due to asymmetric boundary layer separation), resulting in lift or side force, is gravity. A ball will eventually fall to earth as its velocity is reduced. Since gravity forces objects to move faster over time, its effect on the ball is more pronounced as it reaches home plate. A pitch that drops half a foot in the first half of its flight falls another 2 feet in the second half. On a classic curveball, there is an additional downward force due to the applied topspin and this, together with gravity, makes the ball drop even faster (Watts & Bahill, 2000). The force on a spinning baseball is due to asymmetric boundary layer separation which results in the Magnus effect. In Fig. 5.2 the asymmetric separation

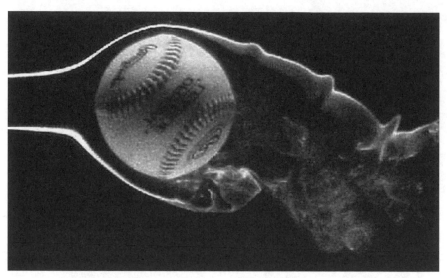

5.2 Asymmetric boundary layer separation over a baseball implies an upward Magnus force. (Photo by James J. Pallis, courtesy of Cislunar Aerospace, Inc. and NASA Ames Research Center.)

and a downward deflected wake flow are clearly evident, thus implying an upward Magnus force.

A good pitcher develops a variety of curving pitches and each with a descriptive name. The 'round house' curve ball is thrown by a right-handed pitcher to a right-handed batter with both a topspin and a twist of the wrist as if turning a door handle. This creates a high pressure area not only above but to the right of the ball, causing movement both down and to the left. The ball arrives lower and farther away than it would have if it had been thrown in a straight path. Turn in the opposite direction and the ball curves toward the batter rather than away – the 'screwball'. A baseball thrown with more force and spin about a near vertical axis drops more slowly and seems to slide to the right or left on the same plane as it was thrown; hence the name, 'slider'. A ball thrown with topspin but with less force than the slider, and without the turning motion, doesn't move left or right but simply sinks as it arrives – the 'sinker'. A fastball is thrown with considerable initial velocity (almost 44.4 m/s or 100 mph) and backspin, which generates a lift force opposing gravity.

The 'knuckleball' is held and released with low velocity and zero or very little spin; the aerodynamics of this pitch is left entirely to random effects of air pressure. Some believe that a knuckleball thrown without any spin will be at the mercy of any passing breeze. Thus, the ball dances through the air in an unpredictable fashion. However, the real reason for the 'dance' of a knuckleball is the effect of the seam on boundary layer transition and separation. Researchers have learned that, depending on the ball velocity and seam orientation, the seam can induce boundary layer transition or separation over a part of the ball, thus creating a side force. With a baseball rotating very slowly during flight, not only does the magnitude of the force change, but the direction also changes (Watts & Sawyer, 1975). This is why the ball appears to dance. It is important to note that, even if the pitcher throws the ball with no rotation, the flow asymmetry will cause the ball to rotate. The flow asymmetry is developed by the unique stitch pattern on a baseball.

5.5   Tennis

The game of tennis originated in France some time in the twelfth century and was referred to as 'jes de paume', the game of the palm played with the bare hand. As early as the twelfth century, a glove was used to protect the hand. Starting in the sixteenth century and continuing until the middle of the eighteenth century, rackets of various shapes and sizes were introduced. Around 1750, the present configuration of a lopsided head, thick gut and a longer handle emerged. The original game known as Real Tennis, was played on a stone surface surrounded by four high walls and covered by a sloping roof. The shape of the new racket enabled players to scoop balls out of the corners and also to put 'cut' or 'spin' on the ball. The rackets were usually made from hickory or ash and heavy sheep gut was used for the strings. The old way of stringing a racket consisted of looping the side strings round the main strings. This produced a rough and smooth effect in the strings and hence came the practice of calling 'rough' or 'smooth' to win the toss at the start of a tennis match. Only Royalty and the very wealthy played the game and the oldest surviving Real Tennis court, located at Hampton Court Palace, was built by King Henry VIII in about 1530.

Balls used in the early days of Real Tennis were made of leather stuffed with wool or hair. They were hard enough to cause injury and even death. Starting in the eighteenth century, strips of wool were wound tightly around a nucleus made by rolling a number of strips into a small ball. Then string was tied in many directions around the ball and a white cloth covering sewn around the ball. The present-day game of Lawn Tennis was derived from Real Tennis in 1873 by a Welsh army officer, Major Walter Wingfield. The original Lawn Tennis ball was made out of India rubber, made from a vulcanisation process invented by Charles Goodyear in the 1850s.

Balls approved for play under the Rules of Tennis must comply with International Tennis Federation (ITF, the world governing body for tennis) regulations covering size, bounce, deformation and colour. Ball performance characteristics are based on varying dynamic and aerodynamic properties. Tennis balls are classified as Type 1 (fast speed), Type 2 (medium speed), Type 3 (slow speed) and high altitude. Type 1 balls are intended for slow-pace court surfaces such as clay. Type 2 balls, the traditional standard tennis ball, is meant for medium-paced courts such as a hard court. Type 3 balls are intended for fast courts such as grass. High altitude balls are designed for play above 1219 metres (4000 feet).

Tennis balls may be pressurised or pressureless. Today's pressurised ball design consists of a hollow rubber-compound core, with a slightly pressurised gas within and covered with a felt fabric cover. The hourglass 'seam' on the ball is a result of the adhesive drying during the curing process. Once removed from its pressurised container, the gases within the pressurised balls begin to leak through the core and fabric. The balls eventually lose bounce.

Pressureless balls may have thicker cores or may be filled with microcellular material. Subsequently, pressureless balls wear from play, but do not lose bounce through gas leakage. As a cost-saving measure, pressureless balls are often recommended for individuals who play infrequently.

The tennis ball must have a uniform outer surface consisting of a fabric cover and be white or yellow in colour. Any ball seams must be stitchless.

All balls must weigh more than 56.0 grams (1.975 ounces) and less than 59.4 grams (2.095 ounces). Type 1 and Type 2 ball diameters must be between 6.541 cm (2.575 inches) and 6.858 cm (2.700 inches); Type 3 balls must be between 6.985 cm (2.750 inches) and 7.302 cm (2.875 inches) in diameter.

Type 1, 2 and 3 balls and pressureless high altitude balls must bounce more than 134.62 cm (53 inches) and less than 147.32 cm (58 inches) when dropped 254 cm (100 inches) upon a flat, rigid surface such as concrete. Pressurised high altitude balls must have a bound of more than 121.92 cm (48 inches) and less than 134.62 cm (53 inches).

Type 1 balls must have a forward deformation of more than 0.495 cm (0.195 inches) and less than 0.597 cm (0.235 inches) and return deformation of more than 0.673 cm (0.265 inches) and less than 0.914 cm (0.360 inches) when a 8.165 kg (18 lb) load is applied. Ball Types 2 and 3 must have a forward deformation of more than 0.559 cm (0.220 inches) and less than 0.737 cm (0.290 inches) and return deformation of more than 0.800 cm (0.315 inches) and less than 1.080 cm (0.425 inches) under the same conditions.

Most of the recent research work on tennis ball aerodynamics was inspired by the introduction of a slightly larger 'oversized' tennis ball (roughly 6.5% larger diameter). This decision was instigated by a concern that the serving speed in men's professional tennis had increased to the point where the serve dominates the game. The main evidence for the domination of the serve in men's professional tennis has been the increase in the number of sets decided by tie breaks at the major tournaments (Haake

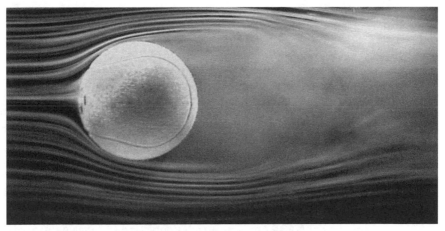

5.3 Wind tunnel smoke visualisation test over a non-spinning tennis ball. (Photo courtesy of NASA Ames Research Center and Cislunar Aerospace, Inc.)

et al., 2000). At the recreational level, the oversized ball's slower speed can make the game easier to learn and more enjoyable to play, encouraging tennis participation.

Some recent experimental studies of tennis ball aerodynamics have revealed the very important role that the felt cover plays (Mehta & Pallis, 2001b). Figure 5.3 shows a photograph of the smoke flow visualisation over a tennis ball model that is held stationary (not spinning) in a wind tunnel. The first observation is that the boundary layer over the top and bottom of the ball separates relatively early, around the ball apex, thus suggesting a laminar boundary layer separation. However, since the flow field did not change with *Re* (up to *Re* = 284 000), it was presumed that transition had already occurred and that a (fixed) turbulent boundary layer separation was obtained over the whole *Re* range tested. Although the felt cover was expected to affect the critical *Re* at which transition occurs, it seemed as though the felt was a more effective boundary layer trip than had been anticipated. The fact that the boundary layer separation over the top and bottom of the non-spinning ball was symmetric leading to a horizontal wake was, of course, anticipated since a side force (upward or downward) is not expected in this (non-spinning) case.

In Fig. 5.4, the ball is spun in a counter-clockwise direction to simulate a ball with topspin. The boundary layer separates earlier over the top of the ball compared to the bottom. As discussed above, this results in an upward deflection of the wake behind the ball and a downward (Magnus) force acting on it, which would make it drop faster than a non-spinning ball. By imparting spin to the ball, tennis players use this effect to make the ball

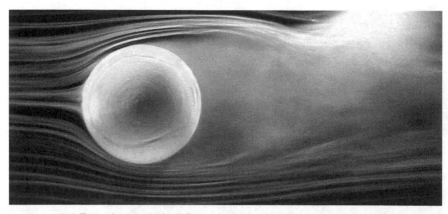

5.4 Topspin tennis ball flow shows a downward Magnus force. (Photo courtesy of NASA Ames Research Center and Cislunar Aerospace, Inc.)

curve; the direction and amount of movement is determined by the spin axis and the amount of spin. Spin about a near-vertical axis is imparted to gain sideways movement, whereas topspin and underspin (or backspin) are used to control the trajectory length (shorter for topspin, longer for underspin).

Drag measurements on non-spinning tennis balls (simulating a perfectly flat serve) revealed that the drag coefficient for tennis balls is higher than that for any other sports ball (Mehta & Pallis, 2001b). On a tennis ball, apart from providing a rough surface, the felt cover is also a porous (drag-bearing) coating since the 'fuzz' elements themselves experience pressure drag. This additional contribution was termed: 'fuzz drag' and this is responsible for the relatively high drag coefficient of tennis balls. Since the fuzz elements come off as the ball surface becomes worn, the ball drag should also decrease, and that is precisely what has been observed (Mehta & Pallis, 2001b). The recently approved oversized tennis ball drag coefficient has been found to be comparable to that for the standard-sized balls. However, the drag on the oversized balls is higher by virtue of the larger cross-sectional area and so the desired effect of 'slowing down the game' (increased ball flight time) will be achieved.

5.6   Golf

The early golf ball in the 1400s, known as a 'featherie,' was simply a leather pouch filled with goose feathers. In order to obtain a hard ball, the pouch was filled while wet with wet goose feathers. Since people believed a smooth sphere would result in less drag (and thus fly farther), the pouch was stitched inside out. Once the pouch was filled, it was stitched shut, thus leaving a few stitches on the outside of the ball. The ball was then dried, oiled, and painted white. The typical drive with this type of ball was about 150 to 175 yards. Once this ball became wet, it was totally useless.

In 1845, the gutta-percha ball was introduced. This ball was made from the gum of the Malaysian Sapodilla tree. This gum was heated and moulded into a sphere. This resulted in a very smooth surface. The typical drive with the gutta-percha ball was shorter than that obtained with the featherie. However, according to golf legend, a professor at Saint Andrews University in Scotland soon discovered that the ball flew farther if the surface was scored or marked.

This led to a variety of surface designs which were chosen more or less by intuition. By 1930, the golf ball with round dimples in regular rows was accepted as the standard design. The modern golf ball generally consists of two, three or even four pieces or layers. The multi-piece ball inner core is generally made of solid or liquid-filled rubber. The inner core is wrapped with rubber thread on wound balls or nowadays usually replaced by a solid polybutadiene layer. The outer dimpled coating or skin is generally made out of an ionomer (Surlyn), natural or synthetic Balata or polyurethane. The dimples come in various shapes, sizes and patterns. The number of dimples also varies, but most balls have between 300 and 500 dimples covering over 75% of the ball's surface. The typical drive with a modern golf ball is about 180 to 250 yards.

The rules of golf state that:

> Foreign material must not be applied to a ball for the purpose of changing its playing characteristics. A ball is unfit for play if it is visibly cut, cracked or out of shape. A ball is not unfit for play solely because mud or other materials adhere to it, its surface is scratched or scraped or its paint is damaged or discolored. The weight of the ball shall not be greater than 45.93 gm avoirdupois (1.620 ounces). The diameter of the ball shall not be less than 42.67 mm (1.680 inches). This specification will be satisfied if, under its own weight, a ball falls through a 42.67 mm diameter ring gauge in fewer than 25 out of 100 randomly selected positions, the test being carried out at a temperature of $23 \pm 1\,°C$. The ball must not be designed, manufactured or intentionally modified to have properties which differ from those of a spherically symmetrical ball.

Perhaps the most popular question in golf science is: 'Why does a golf ball have dimples?' The answer to this question can be found by looking at the aerodynamic drag on a sphere. As discussed above, most of the total drag on a bluff body such as a golf ball is due to pressure drag with a very small contribution from the skin friction drag. The pressure drag is minimised when the boundary layer on the ball is turbulent and separation is delayed, thus leading to a smaller wake and less drag. Transition of the laminar boundary layer on a ball can be achieved at relatively low Re by introducing surface roughness. This is why the professor in Scotland experienced a longer drive with the marked ball.

So, why dimples? The critical Reynolds number, Re_{cr}, holds the answer to this question. Re_{cr} is the Reynolds number at which the flow transitions from a laminar to a turbulent state. For a smooth sphere, Re_{cr} is much larger than the average Reynolds number experienced by a golf ball. For a sand-roughened golf ball or a one with a bramble surface, the reduction in drag at Re_{cr} is greater than that of the dimpled golf ball. However, as the Reynolds number continues to increase beyond the critical value, the drag increases. The dimpled ball, on the other hand, has a lower Re_{cr}, and the drag is fairly constant for Reynolds numbers greater than Re_{cr} (Bearman & Harvey, 1976).

Therefore, the dimples cause Re_{cr} to decrease, which implies that the flow becomes turbulent at a lower velocity than on a smooth sphere. This, in turn, causes the flow to remain attached longer on a dimpled golf ball, which implies a reduction in drag. As the speed of the dimpled golf ball is increased, the drag does not change significantly. This is a good property in a sport like golf where the main goal is to maintain the ball in this post-critical regime throughout its flight.

Although round dimples were accepted as the standard, a variety of other shapes were experimented with as well. Among these were squares, rectangles and hexagons. The hexagons actually result in a lower drag than the round dimples and a version of this design has recently been marketed.

Given the proper spin, a golf ball can produce lift. Originally, golfers thought that all spin was detrimental. However, in 1877, British scientist P. G. Tait learned that a ball, driven with backspin about a horizontal axis, produces a lifting force.

As discussed above, the backspin results in a delayed separation over the top of the ball while enhancing that over the bottom part. This results in a lower pressure over the top compared to the bottom, a downwards deflected wake and hence an upward lift force or Magnus force.

The dimples also help in the generation of lift. By keeping the ball in a post-critical regime, the dimples help promote an asymmetry of the flow in the wake. This asymmetry can be seen in Fig. 5.5. In this figure, the dye shows the flow pattern about a spinning golf ball. The flow is moving from left to right and the ball is spinning in a clockwise direction. The wake is being deflected downwards. This downward deflection of the wake implies that a lifting force is being applied to the golf ball.

A hook or a slice can be explained in the same way. If the golf ball is given a spin about its vertical axis, the ball will be deflected to the right for

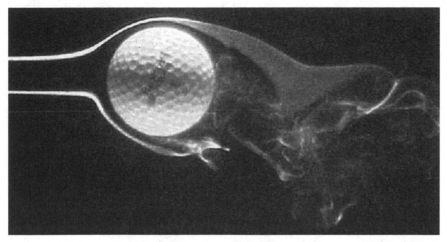

5.5 Downward deflection of the golf ball's wake implies an upwards lifting force. (Photo courtesy of NASA Ames Research Center and Cislunar Aerospace, Inc.)

a clockwise rotation and to the left for a counter-clockwise rotation. These deflections are also produced due to the Magnus effect.

5.7   Soccer/volleyball

The earliest evidence of soccer (or football as it is widely known) dates back to about 200 BC in China, where a form of the game was played that emphasised the ability of players to dribble a leather ball. The Greeks and Romans also participated in a variation of soccer that permitted ball carrying. The modern-day outgrowth of soccer is known to have started in England and the first 'ball' reportedly was the head of a dead Danish brigand. The earliest organised games were massive confrontations between teams consisting of two or three parishes each and goals 3 to 4 miles apart. By 1801 the game had been refined, requiring a limited and equal number of players on each side and confining the field length to about 80 to 100 yards, with a goal at each end. The goal was made of two sticks a few feet apart and the crossbars were lengths of tape stretched across the top. The first soccer club was formed in Sheffield, England in 1857. The London Football Association issued its first set of rules in 1863 and some order was brought to the sport. The word 'soccer' was derived from 'association'. The current 11-player teams were formally established in 1870 and the goalkeeper was established in the 1880s.

The Soccer Laws mandate that the ball must be spherical, made of leather (or another suitable material); size 5 balls have a circumference of 68.6 to

71.1 cm (27 to 28 ins) and size 4 balls have a circumference of 63.5–66.0 cm (25 to 26 ins). The ball cannot be more than 450 g (16 oz) in weight and not less than 410 g (14 oz) at the start of the match.

The earlier balls were made out of leather with a rubber bladder. The main problem with this was that the ball absorbed water in wet conditions, which made it a lot heavier and harder to control. In the last few decades, various synthetic (waterproof) materials have been introduced. A typical modern-day soccer ball features a natural latex bladder with a polyurethane cover and a multi-layer, reinforced backing. Some of the materials and designs have led players to complain, as evidenced at the 2002 World Cup where the ball was apparently 'lighter' and harder to control. Some novel patch designs have also emerged with the latest version displaying a surface covered with 'dimples'.

Volleyball was invented by two Massachusetts natives in 1892. After reaching Japan and Asia by 1900, the rules of the game were set in place over the next 20 years. The 'set' and 'spike' were created in the Philippines in 1916 and six-a-side play was introduced two years later. By 1920, the rules mandating three hits per side and back-row attacks were instituted. Japan, Russia and the United States each started national volleyball associations during the 1920s.

The volleyball rules dictate that the ball shall be spherical, made of a flexible leather or leather-like case, with an interior bladder made of rubber or a similar material. 'It shall be uniform and light in colour or a combination of colours, one of which must be light. The ball circumference must be between 65 to 67 cm (25.5 to 27 inches) and weigh between 260 to 280 grams (9 to 10 ounces) with an internal pressure of 0.30 to 0.325 kg/sq. cm (4.3 to 4.6 lb/sq. inch).'

In recent years, there has been an increased interest in the aerodynamics of soccer balls and volleyballs. In soccer, the ball is almost always kicked with spin imparted to it, generally backspin or spin about a near-vertical axis, which makes the ball curve sideways (Carré et al., 2002). The latter effect is often employed during free kicks from around the penalty box. The defending team puts up a 'human wall' to try and protect a part of the goal, the rest being covered by the goalkeeper. However, the goalkeeper is often left helpless if the ball can be curved around the wall. A spectacular example of this type of kick was in a game between Brazil and France in 1997 (Asai et al., 1998). The ball initially appeared to be heading far right of the goal, but soon started to curve due to the Magnus effect and wound up 'in the back of the net'. A 'toe-kick' is also sometimes used in the free kick situations to try and get the 'knuckling' effect.

In volleyball, two main types of serves are employed: a relatively fast spinning serve (generally with topspin), which results in a downward Magnus force adding to the gravitational force, or the so-called 'floater'

5.6 Volleyball trajectory for topspin serve. (Photo courtesy of Dr Tom Cairns, University of Tulsa.)

which is served at a slower pace, but with the palm of the hand so that very little or no spin is imparted to it. An example of a serve with topspin is shown in Fig. 5.6. The measured flight path implies that the downward force (gravity plus Magnus) probably does not change significantly, thus resulting in a near-parabolic flight path. The floater (with little or no spin) has an unpredictable flight path, which makes it harder for the returning team to set up effectively.

For both these balls, the surface is relatively smooth with small indentations where the 'patches' come together, so the critical Re would be expected to be less than that for a smooth sphere, but higher than that for a golf ball. The typical serving speeds in volleyball range from about 10 m/s to 30 m/s (22.4 to 67.1 mph) and Re_{cr} is about 200 000 which corresponds to $U = 14.5$ m/s (32.5 mph). So, it is quite possible to serve at a speed just above the critical (with turbulent boundary layer separation) and as the ball slows through the critical range, to get side forces generated as a non-uniform transition starts to occur, depending on the locations of the patch-seams. Thus, a serve that starts off on a straight flight path (in the vertical plane), may suddenly develop a sideways motion towards the end of the flight. Even in the supercritical regime, wind tunnel measurements have shown that side force fluctuations of the same order of magnitude as the mean drag can be developed on non-spinning volleyballs, which can cause the 'knuckling' effect (Wei *et al.*, 1988).

5.8   Boomerang

The boomerang is a bent or angular throwing club with the characteristics of a multi-winged airfoil. When properly launched, the boomerang returns to the thrower. Although the boomerang is often thought of as a weapon, the device has primarily served as a recreational toy. The real weapon used by the Australian Aborigines was the killer-stick. The killer-stick shares

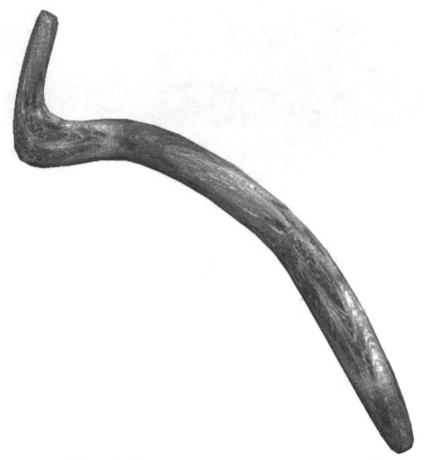

5.7 Killer-stick. (Graphic by Chadwick Okamoto, courtesy of Cislunar
       Aerospace, Inc.)

many properties with the boomerang; however, the killer-stick does not
return.

The killer-stick (Fig. 5.7) is a wooden stick shaped to have a cross-section
similar to a modern airfoil (wing-shape). Given a rotation at launch, which
stabilises the flight (similar to the today's discus and flying discs), the killer-
stick could be thrown very far, with great accuracy and at high speeds.

The boomerang was most likely derived from the killer-stick. The
boomerang is smaller and lighter than a killer-stick with a more pronounced
elbow between the 'wings'. Although the boomerang was not used to kill
game, it was used to hunt birds. As a flock of birds approached, an Aborigi-
nal hunter would imitate the call of a hawk. The hunter would throw the

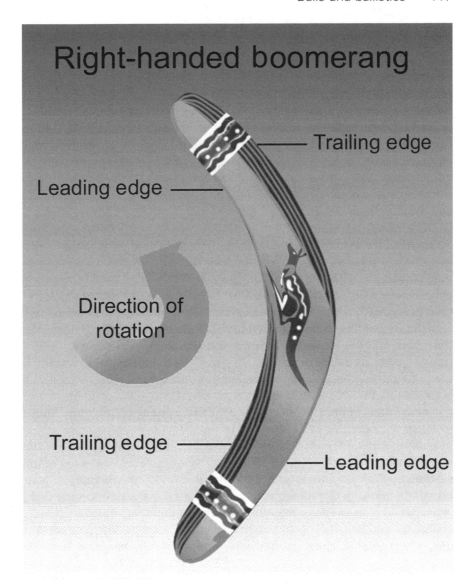

5.8 Right-handed boomerang. (Graphic by Chadwick Okamoto, courtesy of Cislunar Aerospace, Inc.)

boomerang above the birds. To evade the predator, the birds would swoop down into the hunter's positioned nets (Mason, 1974).

As can be seen in Fig. 5.8, the boomerang consists of a leading wing and a trailing wing connected at the elbow. Each wing has the typical cross-section of an airfoil. Therefore, each wing has a leading and trailing edge arranged so the leading edge strikes the air first as the boomerang rotates. Due to this configuration, there are right-handed and left-handed boomerangs. Figure 5.8 is a right-handed boomerang. A left-handed boomerang is simply a mirror image of the right-handed boomerang. The typical angle between the wings is 105 degrees to 110 degrees.

As the boomerang flies through the air, each wing produces lift. Due to the shape of the boomerang, a pressure differential exists between the lower and upper surface (on each wing), which creates aerodynamic lift. A boomerang is thrown with a spin that has two effects on the boomerang as it travels through the air: a stabilising force known as gyroscopic stability and the development of a curved flight path.

The turning force imposed on the boomerang comes from the unequal air speed of the spinning wings. For a stationary, spinning boomerang, both wings would produce the same amount of lift. Now apply the same spinning boomerang with a forward velocity and the speed of the air travelling over the wings differs. Thus, the forward moving wing experiences more lift than the retreating wing. The net result is a force that turns the boomerang.

Many shapes have been explored in the last 50 years. The angle between the wings may be altered to change the characteristics. For example, a sharper angle would decrease the tip speed, thus making the boomerang easier to catch. A modern boomerang often has several wings joined at a common juncture. Novelty designs have been developed, shaped like a bird and even letters of the alphabet. All use the same principles discussed above to return it to the thrower at the end of its flight.

With the diversity of materials available, the composition of the boomerang has evolved. Today, a vast variety of materials are used in the construction of boomerangs from tropical woods to computer circuit boards. In addition to professional designers, who often develop the flight characteristics as well as manufacture the equipment, individuals may choose to construct their own boomerangs. Boomerang construction plans are often available. Thus, material selection is not only based on the type of boomerang, its range, intended use and expense of the material, but also on the tooling and health and safety risks involved in the usage of the materials.

Performance characteristics of the boomerang are varied. Included among the classifications of today's boomerangs are LD (long distance); MTA (maximum time aloft), as well as boomerangs best suited for accuracy and ability to return. Additionally, certain materials are best suited for children or a beginner due to weight and safety factors.

Desirable characteristics in the materials include: stiffness, strength, density, ease in shaping, resistance to humidity, durability (toughness), low cost, and some flexibility. Material stiffness assures that the aerodynamic shape is maintained in wind conditions. Strength and durability are essential to fracture resistance and breakage due to impact. Materials that absorb moisture deform or warp the boomerang's shape. Flexibility allows the thrower to tune (bend) the boomerang for specific conditions. Some boomerang materials can be temperature sensitive making them more flexible during warm summer months and more brittle during cold winter months.

Originally made from wood, laminates or plywoods such as ash, oak, birch, cherry, maple, walnut, beech, and mahogany are still commonly used. Early boomerangs were fabricated from wood with a natural elbow. The wood grain followed the contour of the blades and often contained natural defects such as worm holes or small knots.

Today, common materials used in long-distance boomerangs include: wood, phenolic paper (paper impregnated with phenolic–formaldehyde resin), glass fibre and epoxy composites (GFEC), carbon fibre and epoxy composites, circuit board material (glass fibre/epoxy composite with copper sheets) and metal alloys, although the latter is prohibited in competition for safety reasons. For medium-range boomerangs, unsaturated polyester (UP) resin has also been used. Circuit board materials have been used successfully in MTA and accuracy boomerangs.

Foamed polyvinylchloride (PVC) has been used for inexpensive beginner boomerangs since it is light and easy to throw and catch. The material itself is very flexible and breaks easily. Polyamide (PA) (Nylon) has also been used. Its characteristics are similar to PVC, but PA is even more flexible and absorbs moisture in humid weather, thus deforming the boomerang. Acrylonitrile–butadiene–styrene–copolymer (ABS), an impact-resistant polystyrene, has recently been used for short-range boomerangs. The material is stiff, easily machined, humidity resistant and retains shape in warm weather.

It is common to see the same shape or design fabricated in different materials. For example, a more expensive phenolin paper or composite model might also be made in a less expensive birch plywood model. Designers continue to explore new materials in their quest for the ultimate boomerang.

5.9   Discus

The discus, like the javelin, has a long tradition in field sports, appearing in the ancient games in 708 BC. The ancient Greeks used stone, iron, bronze or lead in their construction. The discus size varied then, as it does today. In 1896, the discus became an event in the first modern Olympic Games and was standardised in 1907.

Today, the men's discus weighs 2 kg (4.4 lb) and measures 22 cm (8.66 inches) in diameter. The women's discus weighs 1 kg (2.2 lb) and has a diameter of 18.2 cm (7.2 inches). The discus for boys weighs 1.6 kg (3.5 lb) and has a diameter of 21.1 cm (8.3 inches).

The discus is greatly influenced by aerodynamic forces, and greater distances can be achieved by throwing the discus into a moderate headwind. This is due to the importance of the aerodynamic lift produced by the discus in flight.

The discus is symmetric, that is, both the upper and lower surface have the same shape. Therefore, the discus cross-section can be considered a symmetric airfoil. Like a symmetric airfoil when presented to the air at a small angle of attack, the discus will produce lift. At a positive angle of attack, the stagnation point (a point of zero velocity) will move from the centre-line of the discus to the lower surface. This signifies a higher pressure on the lower surface than on the upper surface, which produces lift. However, as with any airfoil, when the angle of attack is too large, the flow will separate, which results in a sudden loss of lift. For a discus this occurs at approximately 26 degrees angle of attack.

In a moderate headwind, the speed of the air travelling over the discus is increased. As velocity increases, lift increases. The increased lift translates to longer flight time and hence greater distance. The discus throwers must be more precise in their throwing technique to take advantage of the head-wind. However, an experienced athlete, throwing into a headwind of 10 m/s, can increase the distance thrown by 5 metres or more.

There are discuses designed for competition as well as those used for training. Very high spin, high spin, low spin, and centre-weighted discuses have been developed. In general, these projectiles would be used by elite, above average, intermediate and beginner throwers, respectively.

The main body of the discus (consisting of the upper and lower surfaces) is referred to as the sides and the outer perimeter is called the rim. Some discuses have a centre ring as well. The weight of the discus' rim is varied, depending on the intended use.

For example, a very high spin discus may have a rim weight between 80 and 90% of its total weight. This generates a high spin and turning moment and would be used by an elite athlete who can generate a great deal of speed and spin at the release. Materials that have been used for the sides of very high spin discuses include aluminium, fibre glass-reinforced plastic, reinforced carbon and ABS (acrylonitrile–butadiene–styrene–copolymer, impact-resistant polystyrene). Rims have been constructed from steel, bronze alloy and brass alloy.

In the high spin version, the rim contributes 75–80% of the total weight of the discus. Sides may be made of aluminium, reinforced fibre glass and ABS while the rims have been made from steel. Low spin discuses for use by intermediate throwers have had sides constructed from aluminium, ABS or wood with steel rims.

Beginners may use 'centre-weighted' discuses (less weight at the outer rim). Again, these are composed from ABS plastic or wood with steel rims. Some beginner discuses are made completely of rubber.

One currently available training discus has been constructed from wood and is intentionally overweighted to add resistance to build strength and improve technique. Another training discus has a removable centre ring, which can be replaced with different weights to make the discus overweight or underweight to address specific training needs.

Light-coloured rubber discuses have been fabricated especially for indoor usage. The light colour prevents dark marks on walls and floors, while the rubber provides a tough yet safe indoor projectile.

5.10   Javelin

The discus and javelin first appeared in competition in the ancient games in 708 BC. Javelin events included both target throwing and distance throwing using a sling. By 1780, the javelin was adopted as an event by the Scandinavians and the current one-handed throwing style while running was employed at this time.

In 1953, a hollow javelin was developed by Franklin Held in the United States. Since the javelin had a standard weight, the surface area was increased, which augmented the javelin's flight capability. Another characteristic of this javelin was that it landed horizontally. In 1966, the javelin was thrown over 100 metres (328 feet) by an athlete using a discus-style turn before the throw. This throwing style was judged unsafe and thereafter banned by the International Amateur Athletic Federation (IAAF). The 100-metre mark again was broken in 1984. Subsequently, the IAAF adopted new rules to ensure shorter flight times for sport safety. Originally made of wood, today, javelins are made from steel as well as combinations of steel, aluminium and/or aluminium alloy.

Limiting the javelin's time of flight has been accomplished through aerodynamics. As air travels around the javelin's shaft the air flow begins to separate on the upper surface. Separation increases the drag force. However, towards the latter part of the flight, the direction of the force is opposite the gravitational force. Therefore, the separation of the flow from the upper surface of the javelin actually increases the flight time.

The modern javelin is designed with the centre of pressure located behind the centre of gravity as shown in Fig. 5.9. This generates a nose-down pitching moment, which reduces the javelin's flight time. Although the centre of pressure's location varies during the javelin's flight, it always remains behind the centre of gravity. As a result, the nose-down pitching moment lasts throughout the flight. This nose-down pitching moment also guarantees the javelin lands point first. A point first landing ensures a safer event.

Centre of pressure

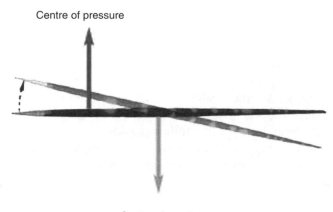

Centre of gravity

5.9 The centre of pressure is located behind the centre of gravity in the modern javelin, ensuring a 'nose-down' landing. (Graphic by James J. Pallis and Chadwick Okamoto, courtesy of Cislunar Aerospace, Inc.)

In addition, since the javelin can no longer slide across the ground, the distance the javelin was thrown can be measured with accuracy.

During flight, the javelin also encounters a spin about its longitudinal axis. As high as 25 revolutions per second, the spin stabilises the javelin in flight. In addition, the javelin also experiences oscillation down its length during flight. Oscillation is detrimental to the flight and needs to be minimised by the thrower or by the materials used in the javelin.

Modern javelins are made from steel, aluminium or aluminium alloy. Today, as in other throwing sports, the javelin is available in different weights (400, 500, 600, 700 and 800 grams), which are rated for different throw distances.

Beginner throwers use a less stiff javelin (which is more forgiving in training). As skill is developed, a stiffer javelin is used (straighter flight). The steel javelins are the stiffest and have less vibration than the aluminium variations, while aluminium offers flexibility and may be easier to throw.

Since the javelin acts like a spear, rubber tips are placed on the point of the javelin for safety (and sometimes required) in pre-college events and training. In addition to safety, soft rubber tips are added for indoor use in order to avoid marks and damage to floors and walls. Hard tips are available and are used for grass fields.

Although the javelin's materials do not vary significantly, the materials in javelin-like training instruments do. These devices have been developed to train beginner javelin throwers safely in proper throwing fundamentals and mechanics. Such aids are also utilised by seasoned athletes to improve their own techniques and accuracy. For safety and to provide a low-cost training tool, these devices are made out of plastics such as polyethylene, with soft or hard elastimer tips. One training aid even has cone-shaped 'wiffleball-like' materials located in several positions along the length of the javelin-like device. Called 'flight resistors' the plastic cones increase the drag – the faster the release, the more air resistance is developed.

5.11   Future trends

There has been extensive research into the performance of ball covers in sports such as golf, baseball and cricket. Researchers in other sports are beginning to follow suit, as noted by the surge in aerodynamic wind tunnel testing and trajectory papers recently published yet devoid until the late 1990s. For some sports, it is the intent of the governing bodies to retain particular performance characteristics of the game and not simply to attain higher speeds or farther distances. In some cases, such as the introduction of the white cricket ball in the 1-day games, the additional outer coating led to an unexpected difference in the aerodynamic properties. Thus, one emerging future trend will be National Governing Boards (NGBs) taking more proactive roles in regulating ball and ballistic performance.

Subsequently, some NGBs have launched their own research programmes to understand the effects of materials on ball and ballistic performance. The introduction of the larger 'oversized' tennis ball and the accompanying research is a good example of this. Allowable materials for ball and ballistic surface covers may be regulated more closely by performance rules (on speed, lift and drag characteristics and athlete manipulation of the surfaces such as scuffing), as well as the material traits of equipment striking the ball.

Likewise, NGBs and Olympic committees are recognising that greater knowledge of ball and ballistic performance characteristics allows a greater competitive edge. For example, the International Tennis Federation (ITF, world governing body for tennis) has supported research in the areas of ball feel, ball spin generation and ball contact time on rackets, court pace (speed), and dynamic ball-testing (conditions more realistic to actual ball-playing conditions).

Designers in sports like golf and the boomerang have exhibited a willingness to investigate and utilise a wide variety of materials and designs. This is clearly a trend that will continue.

In many sports, the study of performance evaluation has reached an almost microscopic level. Altitude and seam placement can have significant effects on ball flight and control. Athlete and equipment motion can be so fast that it is non-perceivable. Often, subtle aspects of elite athletic technique cannot be observed and evaluated without the aid of high-speed data capture equipment and motion analysis software. Invisible to the human eye, the air resistance of the athlete or equipment and the flow patterns created can enhance performance or produce undesirable effects. Along with other high technology tools, performance enhancement through the study of aerodynamics is becoming integrated with other aspects of sport science.

In addition to experimental methods such as wind tunnels, aerodynamic computational numerical methods will continue to be employed and developed for sports applications. As in aviation, numerical methods can be powerful tools used in both performance evaluation of existing equipment and in the development and prototyping of new equipment. Aerodynamic characteristic models are beginning to be embedded with both structural and biomechanic models to analyse a total performance picture.

5.12   References

Asai T, Akatsuka T, and Haake S (1998). The physics of football. *Physics World*, **11**(6), 25–27.

Bearman P W and Harvey J K (1976). Golf ball aerodynamics. *Aeronautical Quarterly*, **27**, 112–122.

Carré M J, Asai T, Akatsuka T, and Haake S J (2002). The curve kick of a football II: flight through the air. *Sports Engineering*, **5**(4), 193–200.

Flatow I (1988). *Rainbows, Curve Balls and Other Wonders of the Natural World Explained*. New York, William Morrow and Company.

Frohlich C (1981). Aerodynamic effects on discus flight. *American Journal of Physics*, **49**, 1125–1132.

Ganslen R V (1964). Aerodynamic and mechanical forces in discus flight. *The Athletic Journal*, **44**.

Haake S J, Chadwick S G, Dignall R J, Goodwill S, and Rose P (2000). Engineering tennis – slowing the game down. *Sports Engineering*, **3**(2), 131–143.

Hess F (1968). Aerodynamics of boomerangs. *Scientific American*, **219**, 123–136.

Koppett L (1991). *The New Thinking Fan's Guide to Baseball*. New York, Simon and Schuster.

Mason B S (1974). *Boomerangs: How to Make and Throw Them*. New York, Dover Publications, Inc.

Mehta R D (1985). Aerodynamics of sports balls. *Annual Review of Fluid Mechanics*, **17**, 151–189.

Mehta R D (2000). Cricket ball aerodynamics: myth versus science, in Subic A J and Haake S J, *The Engineering of Sport. Research, Development and Innovation*. Oxford, Blackwell Science, 153–167.

Mehta R D and Pallis J M (2001a). Sports ball aerodynamics: effects of velocity, spin and surface roughness, in Froes F H and Haake S J, *Materials and Science in Sports*, Warrendale, Pennsylvania, The Minerals, Metals and Materials Society [TMS], 185–197.

Mehta R D and Pallis J M (2001b). The aerodynamics of a tennis ball. *Sports Engineering*, **4**(4), 177–189.

Ruhe B and Darnell E (1985). *Boomerangs: How to Throw, Catch, and Make it.* New York, Workman Publishing.

Schrier E W and Allman W F (1984). *Newton at the Bat: The Science in Sports.* New York, Charles Scribner's Sons.

Soong T C (1976). The dynamics of discus throw. *Journal of Applied Mechanics*, **43**, 531–536.

Walker J (1979). Boomerangs! How to make them and also how they fly. *Scientific American*, **240**, 130–135.

Watts R G and Bahill A T (2000). *Keep Your Eye on the Ball: Curve Balls, Knuckleballs, and Fallacies of Baseball.* New York, W. H. Freeman.

Watts R G and Sawyer E (1975). Aerodynamics of a knuckleball. *American Journal of Physics*, **43**, 960–963.

Wei Q, Lin R, and Liu Z (1988). Vortex-induced dynamic loads on a non-spinning volleyball. *Fluid Dynamics Research*, **3**, 231–237.

Part II
Particular sports

6
Materials in golf

M. STRANGWOOD

University of Birmingham, UK

6.1   Introduction

Outwardly simple – the projection of a ball into a hole – the complexities of the human/environment/equipment interactions continue to make golf a fascinating and challenging sport. The challenge of the game has also spurred on considerable study and innovation in many areas from course preparation to player psychology including the design and manufacture of golf equipment. This has led to the development of a range of clubs – drivers, irons, wedges and putters – as well as shafts – firm, regular, whippy – and balls – high spin and distance designed to optimise a player's ability. The industry needed to support the range of golf players has thus developed into a multi-billion dollar concern with annual sales of $2.8 billion in the US alone (Moxson & Froes, 2001). The wide proliferation of equipment designs and innovations then begs the question 'is it the player or the equipment that is achieving the performance?' Innovation has always been part of sport, but this must not lead to the detriment of the sport, e.g. becoming uninteresting to spectators or so expensive/specialised as to be only accessible for a few professionals. As for soccer, where an amateur can use the same equipment (and similar pitches) as David Beckham, golfers can judge themselves against current stars, such as Tiger Woods, using the same equipment and on the same courses such as St Andrews and Pebble Beach. Hence, there is a need for some regulation of the sport in order to prevent it becoming as specialised as Formula 1, but not to stifle innovation that maintains or expands the interest of players with a range of budgets and abilities. In order to make sensible laws then, some understanding of the physics of golf is needed to assess whether innovation in materials or design is likely to be detrimental or not.

6.2   Oversized golf drivers

One specific example of innovation in golf equipment design has been in the materials and construction of drivers, which are the subject of this study. The traditional driver material was persimmon with a major modification being the incorporation of a plastic face insert and the use of metallic weights to alter the position of the centre of gravity (CG), Fig. 6.1. The balance of head speed and mass in optimising kinetic energy at impact results in driver head masses mostly in the range 185–205 g (Cochran and Stobbs, 1968), which, for a mainly solid wood head, limits the size (volume) and hence requires greater accuracy in order to strike the sweet spot. The same head mass can be accommodated in a larger volume by use of hollow metal construction so that the head has a larger sweet spot, which should be easier to use by an amateur. Any improvement in a professional's game should be small on this basis as their accuracy in striking the ball is better. The initial metallic alloys used (in 1975) were high strength precipitation hardened steels, such as 17-4 PH, Table 6.1 and Fig. 6.2. The introduction of larger, metallic drivers has resulted in a proliferation of new designs and materials encompassing other steels, aluminium-based alloys, titanium-based alloys, amorphous Zr-based alloy inserts and composites, Table 6.1. Whilst allowing less accurate players to achieve greater distances, tour drive

6.1 Traditional wooden driver (top left); 2001 titanium-based alloy driver (top right) and section through traditional wooden driver (bottom) showing plastic insert in face.

distances were also appearing to increase (Fig. 6.3; Cochran, 2002). The data in Fig. 6.3 are the drive distances recorded from two holes on each US PGA tour event for all players in those events. Thus each year consists of some 32 000 data points. The trends shown in Fig. 6.3 are a gradual increase in drive distances from 1980 until 1993, which could be ascribed to greater player physique, improved training and ball improvements (these are limited as balls were, and still are, more heavily regulated). However, post-1993, drive distances on the tour increased at a much greater rate year-on-

Table 6.1 Compositions (wt%) of selected metallic alloys used in golf club drivers

Alloy	Composition
17-4 PH	Fe-(15.0–17.5)Cr-(3.0–5.0)Ni-(3.0–5.0)Cu-(<0.07)C-(<1.0)Mn
Maraging	Fe-(7.0–8.0) Co-(18.0–19.0)Ni-(4.0–5.0) Mo-(<0.03)C-0.4Ti-0.1
250 Grade	Al-(<0.1)Mn-(<0.1)Si
AA 2025	Al-(3.9–5.0)Cu-(0.4–1.2)Mu-(0.5–1.2)Si-(<1.0)Fe-(<0.05)Mg
Ti-6-4	Ti-6Al-4V-(<0.25)Fe
Ti-15-5	Ti-15Cr-6Zr
Ti-15-3	Ti-15V-3Cr-3Al-3Sn
Vitraloy	Zr-13.2Cu-11.0Ti-9.8Ni-3.4Be

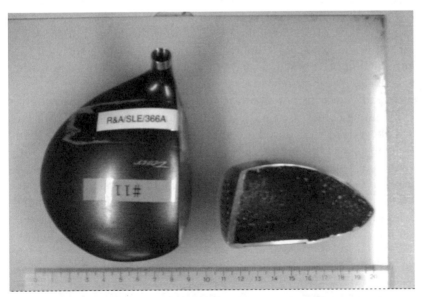

6.2 17-4 PH driver head (left) and section through hollow steel golf driver head (expanded polystyrene foam filling removed).

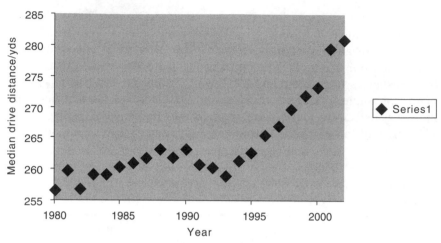

6.3 US PGA tour drive distance measurements for 1982 until July 2002.

year – the blip in 2000/2001 may be due to the replacement of the most commonly used Titleist wound ball (used by 55% of the Tour players) by a solid construction ball. 1993 corresponds to the introduction of oversized Ti-based alloy drivers and this equipment seems to be having an effect on the top players (~1000 players). Comparable statistics for amateurs, who make up the bulk of the 50 million golfers world-wide, do not show any increase in drive distance. The increase in drive distances associated with the introduction of Ti-based alloy drivers has led to concerns that further large increases in driving distances could be achieved by use of different materials and/or designs, potentially reducing the challenge of many courses and culminating in increases in the length of holes at the 2002 US Masters. In 1998, the USGA introduced a conformance criterion for drivers based on its 'coefficient of restitution' (CoR), which is based on the velocity ratio determined in a specified test (USGA, 1999). In summary, the test involves removing the shaft from the driver and adding an equivalent mass to the head (at the hozel) before inverting the head and mounting it (free standing) with the face vertical. A single ball type – the Pinnacle Gold – is then fired at the club face with an incoming (horizontal) velocity of $48.768 \, \mathrm{m\,s^{-1}}$ ($160 \, \mathrm{ft\,s^{-1}}$) spinning at 2 revolutions per second (to improve accuracy). The incoming velocity is determined using light screens, which also measure the horizontal component of the returning (post-impact) ball velocity. The ratio of incoming (V_{in}) and returning velocity (V_{out}) components is then used to define the CoR (e) using equation [6.1].

$$e = \frac{\left(\dfrac{V_{\text{out}}(M+m)}{V_{\text{in}}} + m\right)}{M}$$

[6.1]

where M = head mass; m = ball mass.

This test provides a ranking by which clubs can be compared, but a number of its features restrict its usefulness. Principally, the use of just the horizontal velocity component does not take into account any spin or trajectory parameters so that a slower ball on a more horizontal return trajectory could have the same CoR as a much faster ball with a greater vertical component spinning at a different rate. Also the CoR test requires 50 valid impacts within a circle of diameter 30 mm from a pre-determined 'sweet spot', which is time consuming, and, more importantly, little is revealed about the mechanisms operating during ball/club interaction so that the likely performance of different material/design features or balls cannot be determined. Thus aspects such as 'impedance matching' (Yamaguchi and Iwatsubo, 1999) of clubs to specific ball types cannot be assessed using the CoR test.

The CoR test does, however, indicate the general superiority of Ti-based alloy heads compared with steel, aluminium-based or persimmon heads and the relatively poor performance of hollow metallic heads with inserts. Hence, a range of heads with differing CoR values, materials and construction have been investigated in order to establish the mechanism of ball/driver interaction, although the effects of spin and loft will still need to be addressed.

6.3   Role of the face

The interaction of a club with a solid (rubber-cored) ball involves quite severe and rapid deformation of the ball (a strain of up to 0.25 in 0.5 ms), which is mostly accomplished by the core(s) as the ball covers make up a small proportion of the overall ball, Table 6.2. The elastic deformation of the ball and its recovery results in kinetic energy transfer from the club to the ball. In the case of driving from the tee, then maximum distance is generally needed so that the kinetic energy transfer from the club to the ball, ball speed from the face, and distance need to be maximised. Although a certain amount of core deformation is needed to accomplish energy transfer, the actual deformation experienced during a drive exceeds this minimum value so that energy losses occur as the rubber core deforms viscoelastically (Fig. 6.4). Hence, the possibility exists for greater distance, through reducing overall core deformation and its associated energy losses.

The impact of a ball with a solid (traditional) wooden head will result in a balance between deformation of the ball (mostly core) and head. The

Table 6.2 Dimensions and materials in typical commercial multipiece solid golf balls

Ball	Core diameter (mm)	Mantle thickness (mm)	Cover thickness (mm)
Callaway CB1 red	38.9 (PBD)	–	1.82 (ionomer)
Callaway CB1 blue	39.0 (PBD)	–	1.78 (ionomer)
Maxfli Revolution	39.8 (PBD)	–	1.43 (PU)
Nike Tour Accuracy	36.4 (PBD)	1.65 (ionomer)	1.41 (PU)
Titleist Pro V1	39.3 (PBD)	0.90 (ionomer)	0.69 (PU)

PBD = polybutadiene; PU = polyurethane.

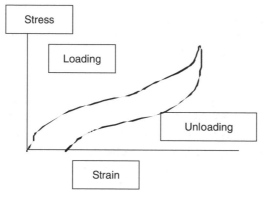

6.4 Schematic diagram of viscoelastic loading and unloading – the area between the loading and unloading curves represents the energy lost in deformation.

compressive stiffness of a solid wooden head is high (Table 6.3), so that deformation of the head is limited, resulting in large deformation of the ball and so reduced kinetic energy on release from the face (Fig. 6.5). Modification of the head to a hollow design would reduce the rigidity of the face (based on material properties, face area and face thickness). As the ball mass is limited at 46 g (along with its diameter) and the balance of mass and velocity for the head is optimised in the range 185–205 g for drivers, then the overall forces acting on impact are comparable regardless of club construction, if the clubhead speed is maintained. Thus the effect of reduced rigidity of a hollow driver would be to give increased club deformation and reduced ball deformation compared with a solid head resulting in smaller energy losses and greater ball speed/distance (Fig. 6.6). Thus, increased distances would be expected for drivers with large, thin faces that show low

Table 6.3 Compressive elastic modulus (GPa) values
for selected drive head materials

Material	Modulus
Persimmon	15–100
AA 2025	70
Ti-6-4	115
Maraging steel 250 grade	190
Ti-15-5	105
Ti-15-3	100
17-4 PH	195–200

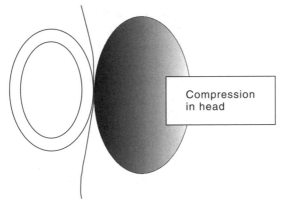

6.5 Schematic interaction of solid construction ball with solid driver
head.

6.6 Schematic interaction of solid construction ball with thin face of
hollow driver head.

rigidity – the so-called 'spring-like' or trampoline effects, which agrees with the distance increases seen as head volume has increased from 250 to 520 cm³. This effect should increase as the area increases and thickness decreases, subject to maintaining sufficient strength to resist plastic deformation or fracture on impact (repeated). Hence, the material used for the face needs high strength (yield) as well as low density. The strength of many metallic alloys is dependent on processing, so that the construction and manufacture of the head, i.e. casting or forging and welding, have a significant effect on strength and hence the degree of thinning/area increase that can be achieved. In addition to the geometry trends above, the use of materials with lower linear elastic (Young's) modulus values would, for the same overall stress system, result in greater face deformation and, hence, reduced ball deformation.

As noted above, CoR testing can reveal differences in head performance but is time consuming and so a faster test is needed. Accepting that club deformation is responsible for reduced ball deformation and greater distance, then a measurement of head stiffness should relate to CoR and performance. Compression testing of the face, e.g. using a servohydraulic or servoelectric tester, would be possible, but the strain rate that can be applied, 10^{-3} s^{-1}, is orders of magnitude below that experienced during the CoR test (0.3 s^{-1}) and so the mechanical properties revealed during testing are not those active during impact. The metallic alloys used in hollow, oversized driver heads behave linear elastically and so should show the same parameters, e.g. modulus, at low and high strains, although not at low and high strain rates. This is not the case for viscoelastic materials, such as the rubbers used in the ball core, but for the metallic alloys a low strain applied at a high strain rate strain should be able to stimulate similar behaviour to that in an impact, which is the approach adopted for this study.

6.4   Frequency spectrum testing A

Changes in the stiffness of a body should be reflected in its response to impact and its resonant vibration frequency, London's Millennium Bridge being a good example of low stiffness leading to the wrong resonant frequency. By measuring the acceleration response of a club face to a rapid sweet-spot impact, fast Fourier transform (FFT) techniques can reveal the vibration modes which should be related to stiffness of the head.

Initially, six Brüel and Kjær (B & K) 4393 accelerometers were attached to superglued studs on the driver face (Fig. 6.7). These had mass 2.4 g and so will lead to some modification of resonant frequencies, but this should be consistent for the heads studied. A total of 32 heads (Table 6.4) were studied, whose CoR had been determined by the USGA, which were subjected to an impact from a B & K 4801 mini-shaker at the sweet spot

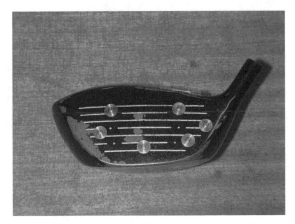

6.7 Studs cemented to driver face showing location of accelerometers around sweet spot in 'A' test.

(Fig. 6.8). The USGA testing had identified a 30 mm diameter sweet-spot zone, which had been marked on the face, and the centre of this was used for impact; testing established that impact up to 20 mm from the centre excited the same modes, i.e. frequencies were unchanged, although their amplitudes varied. By using a constant shaker/head separation and a 1 ms top-hat voltage pulse, a 0.25 ms half sine-wave force was applied to the head. Neither different duration voltage pulses (0.5–6 ms) nor different pulse shapes (e.g. Gaussian, Hanning) resulted in any change in the head vibration frequencies, although the amplitudes of the various frequencies varied. Figure 6.9 shows a typical frequency spectrum from a hollow over-sized driver head. The frequency spectra are plotted as acceleration, which does enhance the higher frequency peaks compared with a displacement/frequency plot, but does avoid drift errors associated with double integration of the accelerometer output. The spectrum for the head shows a strong resonant frequency in the range 4000–6500 Hz with lesser intensity peaks around 8000 Hz and 10000 Hz.

6.5   Test variables

Before establishing any relationships between CoR and frequency, the sensitivity of the frequency spectra to test variables needs to be established. Some of these, in terms of impact characteristics and location, have been noted above. The accelerometers are placed around the sweet spot and represent a range of distances from impact and to the face edge. Comparison of the spectra from all six accelerometers (Fig. 6.10) reveals the same

Table 6.4 Average dimensions, CoR values and construction of heads studied

Head	CoR	Face area (mm²)	Thickness of face (mm)	Crown area (mm²)	Crown thickness (mm)	Construction	Frequency (Hz)
331	0.761	3333	1.24	6300	2.98	2 piece, A1 with Ti face insert, cast F+	10340
345	0.7641	3932	6.28	6859	2.40	2 piece, A1, cast F + S, cast C	6560
330	0.7679	3410	3.22	6201	1.97	3 piece, A1 with Ti insert, cast C and S, forged F	6436
333	0.7687	2590	4.19	5279	1.17	3 piece, MS, forged F, C and S	6051
343	0.7738	3217	2.42	5881	1.37	3 piece, MS, forged F, C and S	4982
337	0.7770	3178	3.10	5822	1.08	2 piece, Ti, cast F + C, cast S	5389
335	0.7777	3626	3.49	5988	1.47	2 piece, Ti, cast F + C, cast S	5418
329	0.7781	3412	2.66	5681	1.37	3 piece, MS, forged F, C and S	5236
348	0.7781	2760	3.24	7097	1.00	2 piece, Ti, cast F + C, cast S	5891
332	0.7812	3173	3.39	6100	1.35	2 piece, Ti, cast F + C, cast S	5418
334	0.7826	3377	3.26	6187	1.32	3 piece, Ti, Forged F, cast S, C	5091
Foundry 1–3 heads	0.785	2762–2892	See Table 6.5	See Table 6.5	See Table 6.5	2 piece, Ti, cast F + S, cast C	See Table 6.5
346	0.787	3553	3.10	6185	1.22	3 piece, MS forged F, C and S	4822
341	0.7889	3892	2.66	7429	1.08	2 piece, Ti, cast F + C, cast S	5018
347	0.7894	3978	3.84	7958	1.94	2 piece, Ti, cast F + C, cast S	4582

326	0.7925	3123	2.09	5322	1.49	2 piece, Ti, cast C + S, cast F	5200
328	0.7941	3776	2.80	6340	1.73	3 piece, Ti, forged F, C and S	4982
BBT1	0.795	3376	2.40	5898	0.76	2 piece, Ti, cast F + C, cast S	5120
342	0.8045	3376	2.40	5898	0.76	3 piece, Ti, cast F, C, S	4458
325	0.8079	3170	3.44	4937	1.09	2 piece, Ti, cast F + C, cast S	4800
336	0.8119	3392	2.50	6664	1.23	3 piece, Ti, forged F, C and S	4422
K129B	0.812	3250	2.40	5800	0.66	3 piece, MS, forged F, C and S	4192
BBT2	0.814	3605	3.30	6270	1.12	3 piece, Ti, forged F, C and S	4736
340	0.8173	3929	2.70	5988	1.47	3 piece, Ti, forged F, C and S	4218
D215B	0.825	4171	2.76	7250	1.26	3 piece, Ti, forged F, C and S	4256
UBT9	0.8329	3643	2.18	7958	1.11	3 piece, Ti, forged F, C and S	5376
UBT10	0.840	4171	2.76	7250	1.26	3 piece, Ti, forged F, C and S	3964
D215A	0.841	4171	2.76	7250	1.26	3 piece, Ti, forged F, C and S	4502
338	0.8466	3776	2.90	6082	1.14	3 piece, Ti, forged F, C and S	4291
M105B	0.847	3392	2.50	6664	1.23	3 piece, Ti, forged F, C and S	4400
CNC2	0.8562	3441	2.58	7175	1.08	3 piece, Ti, forged F, C and S	3776
KSLC	0.8574	3594	2.56	7123	0.89	3 piece, Ti, forged F, C and S	3760

6.8 Location of shaker below sweet spot during 'A' test –
accelerometers not shown for clarity.

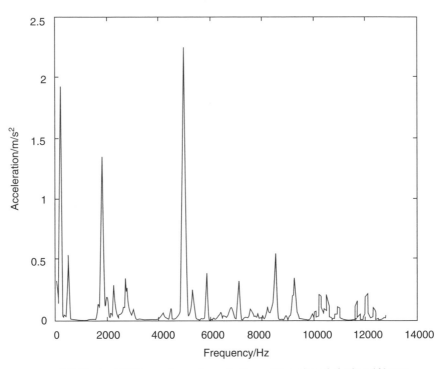

6.9 Typical FFT spectrum from hollow driver head during 'A' test.

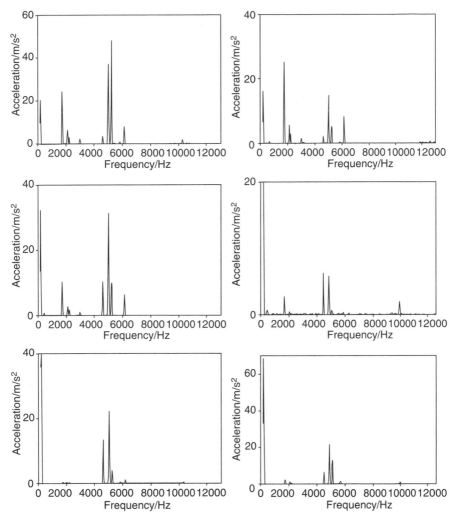

6.10 FFT spectra for all six accelerometers from one head during 'A' test.

frequencies but varying amplitudes with the pair either side of the impact spot towards the top of the face (numbers 2 and 7, where impact is 1) giving the strongest signals whilst those near the face edge, e.g. signal number 3, give very weak signals.

As all accelerometers give similar frequencies, then they are not all necessary and so the mass loading of the face could be reduced. The effect of this change in test procedure was assessed by repeated impacts but with only one accelerometer, which was moved through locations 2 to 7. These

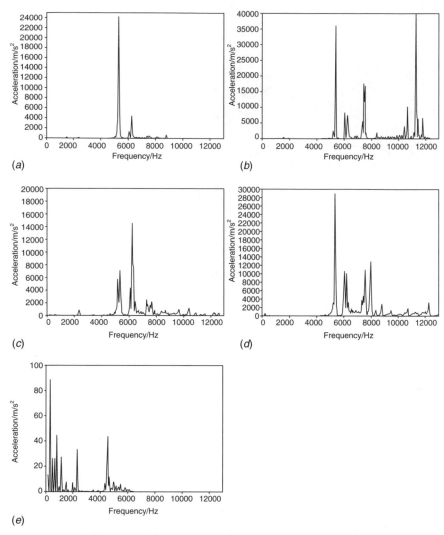

6.11 FFT spectra from simulated impact test on club used for spectra in Fig. 6.10, but with a single accelerometer in positions (a) 2, (b) 3, (c) 4, (d) 6 and (e) 7; impact at position 1 and no detectable signal from position 5.

revealed (Fig. 6.11) that there was a slight shift in frequency, but that the change in frequency was less than 500 Hz; there was also a reduction in damping and so slightly stronger, higher frequency signals were noted.

As in the CoR test, the frequency test is carried out on a head without a shaft, although the short time of ball contact with the head during driving

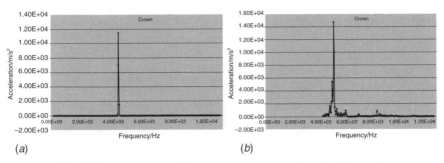

6.12 FFT spectra from 'A' test carried out on head with shaft clamped at (*a*) shaft grip and (*b*) at hozel.

(0.3–0.5 ms) compared with the low (100 Hz and less) fundamental frequency of shafts means that the shaft should play little role in the impact phenomenon. Frequency spectra were recorded for a head alone and then the same head attached to a firm and a regular flex shaft. The impact was carried out with the grip end clamped, and this resulted in bending of the shaft during impact and reduction in amplitude, but the same frequencies were recorded (Fig. 6.12), indicating that the impact test is determining the characteristics of the head unaffected by the nature of any shaft attached.

6.6   CoR–frequency relationship

The absolute amplitude values varied significantly between heads despite the use of similar impact conditions largely because of the curvature of the face and so variation in the precise impact force. Hence, the frequency spectra were plotted as amplitude normalised against the most intense peak acceleration. From these, the major vibration frequencies excited during impact were determined and compared with CoR values (Fig. 6.13). This shows a general trend of increasing CoR with decreasing major frequency, although there is still considerable scatter with such a small sample size. The major peaks mostly fall in the range 3750–6500 Hz, which is well below the range expected for an elliptical disc with corresponding dimensions and material properties as those of the face. Modal analysis (e.g. Nakai *et al.*, 2002) would predict a fundamental frequency in the range 8000–10 000 Hz, which is seen for some of the secondary peaks. Thus the major vibrational mode does not correspond to the face's fundamental mode but must arise from some other deformation mode of the head. This was investigated by using an impact test with one accelerometer each on the face, crown, sole, heel and toe, test B (Fig. 6.14). The reduction in face loading had already been shown to increase the major frequency by 500 Hz, but this test now

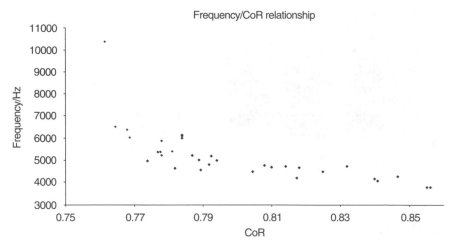

6.13 Relationship between dominant frequency and CoR for hollow oversized drivers in this study.

6.14 Cemented studs showing location of accelerometers on face (three studs but only one used), crown, sole, heel and toe for 'B' test.

allowed the vibrational responses of the different components during impact of the face to be assessed. Figure 6.15 shows a comparison of the frequency spectra for test A (6 accelerometers on the face) with the frequency spectra recorded for face, crown and sole during test B; the signals for heel and toe were too small to be useful. Comparison of tests A and B for all heads examined showed that, in all but one case, the amplitude of the crown was greater than that of the face and that the amplitude of the major frequency (from test A) was greater than those of other frequencies

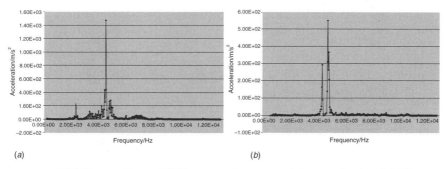

6.15 Comparison of FFT spectra for same club tested using (*a*) 'A' test with six face accelerometers and (*b*) 'B' test with one face accelerometer and accelerometers on crown, sole, heel and toe.

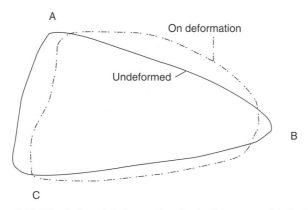

6.16 Whole head deformation for ball impact with hollow driver head.

by a larger factor than for the other components (face and sole). Hence, the vibration mode is dominated by the crown and deformation of the head during impact is associated with all components not just the face (Fig. 6.16). Accepting the errors inherent in the USGA CoR test due to the use of one head (Fig. 6.13) could be used to predict CoR based on head (mainly crown) frequency, but to what accuracy?

6.7 Variability within a single club type

Test B was carried out on 24 examples of the same club head, which comprised eight examples from each of the three foundries producing this head. These heads allowed a measure of manufacturing variables between different examples of the same head to be determined and so set a limit to the accuracy that the frequency could be used to predict CoR for commercially available heads. The spectra for the eight examples of each foundry are presented in Fig. 6.17(a)–(c); heads would be characterised by a single USGA CoR value (of 0.785). Figure 6.17 shows that there is a spread of some 500 Hz about the dominant frequency of 6000 Hz for the eight examples of each head and that each foundry produces a head with a significantly different dominant frequency. This is seen for face and crown frequencies; the more complex and variable sole geometry (including internal mass loading to control centre of gravity) gives results that are less reproducible. The frequency spectra for crown and face also show a greater dominance of the lower frequency modes by the crown.

Examination of the physical characteristics of the examples from the different foundries by image analysis, sectioning and ultrasonic methods reveals little variation in the area, length or breadth of the components (face, crown and sole) but significant and systematic variations in component thickness (Table 6.5). Decrease in thickness of the components results in a decrease in dominant frequency for that component. As these heads were cast in Ti-6 Al-4 V (wt%) as a combined face + crown with a separate cast sole welded in place, then thickness variations could have caused a change in the as-cast microstructure (Fig. 6.18). Image analysis did not reveal any significant differences in phase balance or grain size between heads from the different foundries so that metal properties were consistent. There was a variation in welding technique between the foundries with two using tungsten inert gas (TIG) welding and the other electron beam welding (EBW), which resulted in slight differences in weld bead dimensions and structures (Fig. 6.19).

6.8   Head design criteria

The previous study has revealed a link between dominant vibration frequency and CoR with higher CoR values corresponding to lower frequencies, which are dominated by vibration modes in the crown. The manufacturing variations between nominally identical heads indicates that the frequency is only characteristic within a 500 Hz band (corresponding to 0.02 in CoR value) and that frequency decreases are brought about (at least in part) by component thickness reduction. Thus it is possible to define a number of criteria necessary for a high CoR driver head in terms of material and construction as below:

(a)   Components should be as thin as possible and hence the material should have high yield stress in order to minimise the risk of plastic deformation during use.

(b)   Forged components are generally preferred to cast as they achieve higher strength in thinner sections with fewer defects (Fig. 6.20).

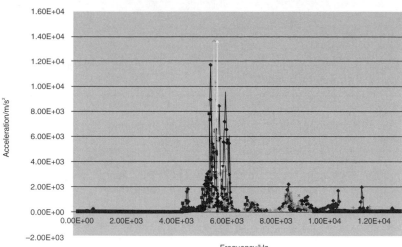

(a)

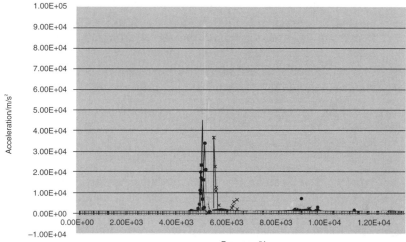

(b)

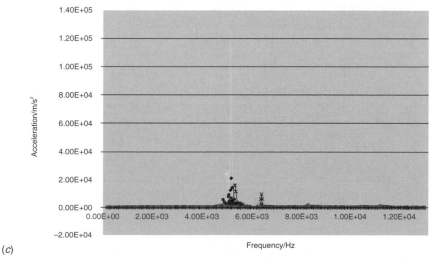

(c)

6.17 FFT spectra for 24 examples of a single head type; 8 samples
from each of foundries (*a*) 1, (*b*) 2 and (*c*) 3.

Table 6.5 Thickness (mm) and frequency (Hz) variations for 24 heads of same design from three foundries

Foundry	Face frequency	Average face thickness	Face thickness range	Crown frequency	Average crown thickness	Crown thickness range
1	5984–6048	3.81 ± 0.15	3.57–4.00	5376–5696	1.39 ± 0.07	1.28–1.50
2	6080–6368	3.57 ± 0.07	3.42–3.64	4960–5472	1.06 ± 0.03	1.02–1.14
3	6112–6336	3.54 ± 0.11	3.40–3.73	5056–5440	1.15 ± 0.04	1.08–1.20

(c)   Construction of the head, i.e. any joints between face, crown and sole, should be matched to the strength of the head components or fall between the strengths of adjacent components if the latter differ significantly. However, the joints should not provide excessive constraint to prevent deformation being transferred from face to the rest of the head.

(d)   The components should maximise linear elastic deformation, i.e. the materials should have low elastic modulus values.

Criteria (a) and (d) can be combined into a material performance index of:

$$\frac{\sigma_y}{E} \qquad [6.2]$$

where σ_y = yield stress and E = Young's modulus. Approximating yield stress to Vicker's hardness and determining E for each component from measurements of density and speed of sound from the component, the materials used in the commercial oversized driver heads studied can be plotted as in Fig. 6.21. The alloy nearest to top left (highest $\frac{\sigma_y}{E}$ ratio) is the amorphous Zr-based alloy followed by metastable β-Ti alloys, $(\alpha + \beta)$-Ti alloys, maraging steels and finally age-hardening Al-based alloys.

As well as its constituent material's properties, the stiffness of a component is geometry dependent so that the $\frac{\sigma_y}{E}$ ratio should be modified by a shape factor. For circular or elliptical discs, the maximum normal central deflection, y_{max}, is given by Young *et al.* (2001):

$$y_{max} \: \alpha \: \frac{area}{Et^3}(1-v^3)\left(\frac{1}{1+v}\right) \qquad [6.3]$$

where area is that of the disc, t is the thickness and v is the Poisson's ratio. The maximum central deflection (inverse of stiffness) corresponds to lower frequency, which is minimised for large component area, small thickness

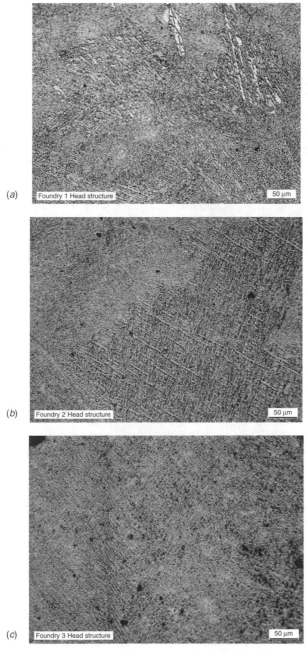

(a) Foundry 1 Head structure   50 μm

(b) Foundry 2 Head structure   50 μm

(c) Foundry 3 Head structure   50 μm

6.18 Micrographs from cast crown of an example head from foundry
(*a*) 1, (*b*) 2 and (*c*) 3, showing similar but not identical
Widmanstätten α laths in a fine transformed β matrix.

6.19 Examples of TIG (left) and EBW (right) welds in heads featured in Figs 6.17 and 6.18.

6.20 Solidification porosity observed in centre of face for cast Ti-6 Al-4 V (wt%) head.

and low modulus, subject to high yield strength. The variation in Poisson's ratio between metallic alloys is likely to be small and can be treated as constant. Correlating maximum central deflection (i.e. low stiffness) with increased CoR and incorporating strength through H_v (material hardness) then the following relationship would hold:

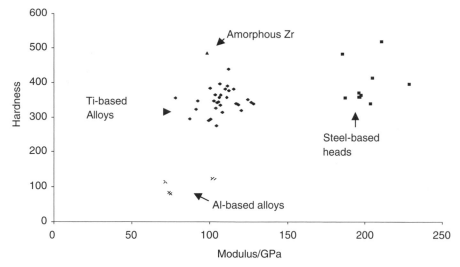

6.21 Plot of hardness vs. *E* for materials in heads sectioned in this study.

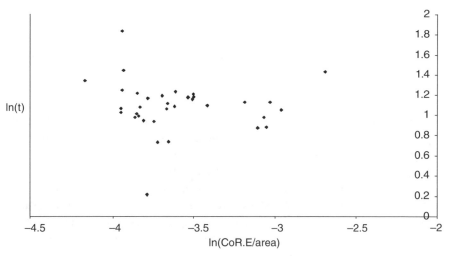

6.22 Plot of ln (parameter based on face geometries and CoR) vs. ln(t) for faces of clubs in this study.

$$\ln\left(\frac{CoR.E}{area.H_v}\right) = const. - \frac{1}{3}\ln(t) \qquad [6.4]$$

Figures 6.22 and 6.23 present these plots for the face and crown, which indicate that the faces do not follow this behaviour, whilst the crown plot shows the expected negative slope, although of value 1/2.4 rather than the

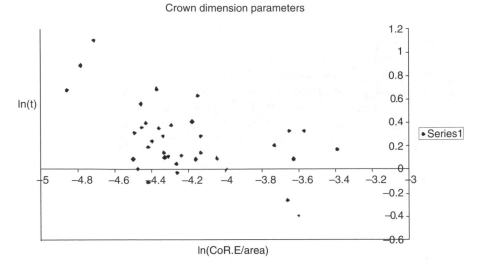

6.23 Plot of ln (parameter based on crown geometries and head CoR) vs. ln(t) for crowns of clubs in this study.

predicted 1/3. This reinforces the previous, frequency, results indicating the significant role of the crown in deformation and CoR. Figure 6.23 still shows a wide scatterband, which can be related to the construction of the head. Equation [6.4] treats the components separately and does not consider deformation transfer across the joints between components as would occur on impact of a complete head. Analysis of a large number (>500) of heads is underway in order to reduce the scatterband. Dimensional (non-destructive) characterisation has been completed but construction (Table 6.4), property and microstructural analysis are still needed for these heads. These data are needed because the degree of deformation transfer across a joint, e.g. from face to crown, is dependent on the material properties and the dimensions of the joint so that transfer for face and crown with the same component dimensions will differ depending on their properties and construction. As summarised in Table 6.4, the construction of the heads examined fell into a number of categories, which were:

- Cast two piece; the face was cast as a single piece with either the sole (F + S) or crown (F + C) with the remaining component – crown or sole, respectively – welded in place to complete the hollow head.
- Cast three piece; all three components – face, crown and sole – were cast separately and welded together to form a hollow head.
- Forged three piece; the three components were forged separately to shape and welded together giving joint lines generally around the edge of the face and at the crown–sole interface.

- Inserts: examples where different face materials were inserted into cast two piece bodies were seen, although the insertion of a face into a one piece cast/forged or two piece forged body is equally possible.

6.9   Construction effects

As noted in Table 6.4, there are a number of methods of construction noted in oversized hollow drivers and these have been seen to influence the vibrational frequencies observed on impact and, hence, CoR. As a limited number of aluminium-based alloy and steel-based heads are included in Figs 6.13 and 6.23, this will mostly discuss the titanium-based alloy heads clustered in the centre of Fig. 6.23.

6.9.1 Inserts

Only one of the heads in Fig. 6.13 shows a dominant frequency above 10 kHz, which corresponds to the natural fundamental frequency of the face. This head consisted of a titanium alloy face in a cast aluminium alloy body, which, on sectioning, was separated by a glass fibre-filled epoxy layer. The effect of the epoxy layer was to separate the face from the body by a (stiff) viscoelastic damping layer so that deformation was contained almost entirely within the face. The isolation of the face means that it vibrates as if it were freely suspended, transferring little deformation to the surrounding head. This limits the deformation of the head (face + body) during impact and so, on impact, the ball deforms more leading to a lower CoR value. This effect is also seen for the amorphous Zr-based alloy face (5856 Hz, 0.78 CoR). From Fig. 6.21, this material should be superior to Ti-based alloys, but it derives its strength from an amorphous structure with only a few small dendritic crystalline regions (Fig. 6.24). Welding would cause crystallisation of this structure, resulting in reduction in strength and an inducement of brittleness. The use of amorphous Zr-based alloys is therefore restricted to face inserts that have to be adhesively bonded into the body (Ti-based alloy). The support for the adhesive joint is a thicker section of the cast Ti-based alloy (Fig. 6.25), which provides considerable constraint around the edges of the face. The effect of this constraint is to prevent deformation transfer to the rest of the head so that this acts as an isolated face with higher frequency and lower CoR.

6.9.2 Cast components

The relatively small production runs and high tolerances needed for cast driver heads mean that investment casting is the appropriate casting route. The use of a ceramic mould material means that, as well as the presence of

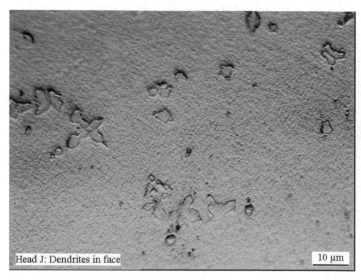

6.24 Optical micrograph of amorphous Zr-based alloy face, showing isolated dendrites in a general featureless amorphous matrix.

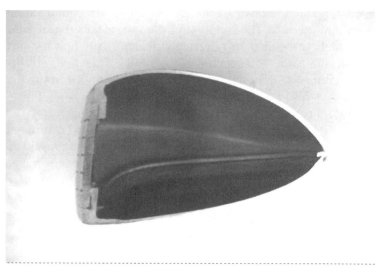

6.25 Section through head featuring amorphous Zr-based alloy face.

casting defects (Fig. 6.20), the as-cast structures are relatively coarse so that the strength of these alloys is less than that of forged or rolled components. The castability of the alloys means that fusion welding is a suitable joining method for both two and three piece heads, as the as-cast and as-welded

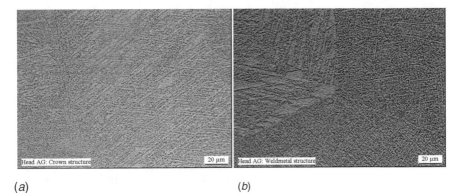

(a)   *(b)*

6.26 Base *(a)* and weldmetal *(b)* structures for cast Ti-6 Al-4 V (wt%) head.

structures will be similar with comparable properties (Fig. 6.26). In the case of cast heads, the transition between face and, e.g. crown will tend to be thicker if these are cast together as a single piece in order to allow fluid flow and reduce the effect of solidification porosity. This thickening will provide more constraint around the edge of the face so that deformation is concentrated in the face rather than the crown, raising the vibrational frequency. The use of a welded joint between the components, particularly for welding processes with smaller weld beads, e.g. electron beam welding (EBW) rather than tungsten inert gas (TIG), would provide less constraint spreading more deformation to the crown and resulting in a lower frequency. The need for fluid flow and the low strength, however, mean that the components of the head have to be thicker resulting in increased vibration frequencies and lower CoR values. This is reflected in equation [6.4] and in Fig. 6.23 through incorporation of the H_v term, as the thickness value used is the average for each component, although the greater variability of material properties for cast components contributes to the greater scatter seen for heads of this construction type.

The type of weld also influences weldmetal properties as the degree of atmospheric contamination varies; this is particularly important in the case of Ti-based alloys. The gas shielding used in TIG is less effective at preventing uptake of gaseous species such as nitrogen and oxygen by the molten weld pool than the vacuum needed for EBW. Thus the weldmetal formed on solidification after TIG welding is likely to contain higher levels of trapped O and N. Both of these elements are α-stabilisers (Polmear, 1995) and so the weldmetal will have a higher proportion of α in its structure than the basemetal, resulting in an increase in both Young's modulus (typically 80 GPa for 100% β rising to 125 GPa for 100% α) and yield stress.

The increase in α for atmospheric contamination may just offset the effect of faster weldmetal cooling rates, but often the welds exhibit differing properties to the base materials (face, crown and/or sole). Thus, a TIG weld with atmospheric contamination for a face/crown weld would give an increased section (due to weld bead dimensions exceeding those of the base components) with increased strength and stiffness between the face and the crown. This represents a region of increased constraint (in a similar manner as for the edges of inserts and cast components in the previous paragraph) and so restricts the transfer of deformation from face to crown. This effect would be less noticeable for smaller, less contaminated EBWs, and this variation in weldmetal–basemetal properties would contribute to the scatter seen in Fig. 6.23, whilst some of the differences noted in Fig. 6.19 would arise from the different welding processes utilised.

6.9.3 Forged components

The mechanical properties, e.g. yield and tensile strength, of forged (or rolled) components are generally superior to those of cast components (as noted above) due, in part, to the finer structures that can be generated (Fig. 6.27). The forged components observed in this study were in more highly alloyed systems, such as maraging steels and metastable β-Ti alloys, and the increased alloy levels, coupled with the ageing heat treatments for the systems cited, result in fine, hard precipitates and further strength increases. The higher strength and more highly alloyed nature of these systems, however, present problems on fusion welding as the forged structure is replaced by cast weldmetal (Fig. 6.27), surrounded by a heat-affected zone (HAZ) where the aged structure of the basemetal has received an extra heat treatment (in the solid state) and so suffers a loss in strength. Thus the dimensions of the weldmetal and surrounding HAZ (basemetal thickness) have to be increased to provide sufficient overall strength given the lower (local) material strength. The resulting increase in thickness again provides more constraint at the joint and restricts deformation transfer to crown and sole. In the case of maraging steels, this requires an extensive weld around the face that does not allow the full $\dfrac{\sigma_y}{E}$ ratio to be utilised as this could be up to 0.018 – comparable with the best β-Ti value, whereas Fig. 6.21 shows that the latter have a superior ratio, as the former's full strength levels are not utilised. This would be partly due to atmospheric contamination increasing the strength of the weldment in the case of β-Ti alloys. Post-weld heat treatment (PWHT) could also be used for β-Ti (and $(\alpha + \beta)$-Ti) alloys to recover properties by redissolving any precipitated α before quenching and ageing to give a fine distribution. This would be more effective for EBWd heads where atmospheric contamination would not be involved. The scale

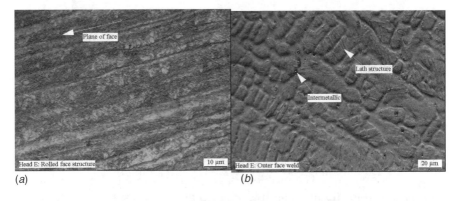

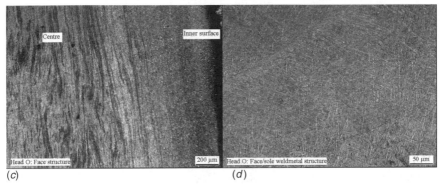

6.27 Base and weldmetal structures for forged maraging steel (*a*) and (*b*) and β-Ti alloy (*c*) and (*d*) heads.

and stability of the intermetallics in as-welded maraging steels makes this a much less effective process within a production environment.

6.10   Conclusions, further work and design trends

The need to consider an oversized golf driver as a whole head and not just the face has been established. In particular, the incorporation of thin components with large area has been shown to reduce the characteristic vibration frequencies of the head by making it less rigid. Increased head deformation reduces ball deformation and its associated energy losses, leading to higher CoR and greater distance off the face (neglecting spin and launch angle effects). Thinner, larger area components are easier to achieve for the crown than face or sole (accidental impact with the ground means that this is generally somewhat thicker), but are only effective if deformation can be transferred from the face, i.e. constraint must be minimised consistent with a durable design. The selection criteria established have indicated that the optimum alloys for commercial production remain forged β-Ti alloys with electron beam welding being used for assembly. The increase in CoR that can be achieved with currently available materials and manufacturing techniques is unlikely to significantly increase drive distances in the same way as the introduction of Ti-based alloy drivers did in the mid-1990s, based on ball speed off the face.

Since the commencement of this study's frequency testing in January 1998 and the determination of the mechanism of increased CoR, a number of high CoR drivers have been released. These mostly bear out the principles established in this study and are welded from forged thin β-Ti components with large area. The problem of constraint is reduced by moving the weld away from the meeting line of face and crown/sole onto the latter components so that the face becomes a 'cup' shape. In addition, the edges of the faces are being thinned compared with the centre of the face consistent with reduced constraint and easier deformation transfer.

As noted above, this study has only dealt with normal velocity off the face (as measured in the CoR test) and, although greater understanding has been obtained into the deformation mechanisms, the effects of head deformation on spin rate and launch angle and with other ball types still need to be elucidated. This further work is being pursued by fully modelling the deformation of hollow 3-dimensional bodies as analytically as possible by building up models and physical heads from components with known (and modelled) vibrational characteristics with increasing complexity.

6.11   Acknowledgements

The author would like to thank the Royal and Ancient Golf Club of St Andrews for their support of this work and Professor I R Harris for provision of laboratory facilities. The contributions of Professor Blake, Drs Otto and Muirhead, and E M Faulkner, S A Monk, E F Bird and M H Hussien at Birmingham during this study are gratefully acknowledged, as are the many stimulating discussions with a range of golf equipment manufacturers.

6.12   References

Cochran, A J (2002). The impact of science and technology on golf – a personal view. *The Engineering of Sport* **4**. *Proceedings of the 4th International Conference on Sports Engineering*, Kyoto, S Ujihashi and S J Haake (eds), pp. 3–16. Blackwell Science Ltd, Oxford.

Cochran, A J and Stobbs, J (1968). *Search for the Perfect Swing*, pp. 205–207. Triumph Books, Chicago.

Moxson, V S and Froes, F H (2001). Production of sports equipment components via powder metallurgy. *Materials and Science in Sport*. TMS, San Diego.

Nakai, K, Wu, Z, Sogabe, Y and Arimitsu, Y (2002). Shape Optimisation of Golf Club Heads. *The Engineering of Sport* **4**, *Proceedings of the 4th International Conference on Sports Engineering*, Kyoto. S Ujihashi and S J Haake (eds), pp. 369–375. Blackwell Science Ltd, Oxford.

Polmear, I J (1995). *Light Alloys: Metallurgy of the Light Metals*, 3rd edn. Arnold, London.

USGA (1999). *Procedure for Measuring Velocity Ratio of a Club Head for Conformance to Rule 4-1e, Appendix II, Revision 2*. Far Hills, NJ, 8 February.

Yamaguchi, T and Iwatsubo, T (1999). Optimum design of golf clubs considering the mechanical impedance matching. *Science and Golf III*, M R Farrally and A J Cochran (eds), pp. 500–509. Human Kinetics, Champaign, IL.

Young, W C, Budynas, R G and Roark, R J (2001). *Roark's Formulas for Stress and Strain*, 7th edn. McGraw-Hill International, New York.

7

Surface engineering in sport

H. DONG

University of Birmingham, UK

7.1   Introduction

The adoption of new materials and processing methods have provided for many advances in sports equipment, thus improving sports performance and enjoyment. Indeed, the ever-increasing demand for enhanced sports performance and for improved sports equipment have been the main drivers for interfacing modern materials technologies with sports engineering. Man performs better with improved sports materials and equipment design. In addition, advanced sports materials and novel designs can give new dimensions to the quality of life of people with disabilities.[1]

Some bulk materials used for sport cannot meet the ever-increasing requirements for diverse combined properties arising from the challenging design requirements of advanced sports products. For example, titanium and titanium alloys are very attractive to the racing car industry owing to their high strength–weight ratio. As a result, there has long been interest in substituting steel components with titanium ones in racing cars and engines. However, the poor tribological properties of titanium alloys have retarded such replacement, and the full potential of titanium alloys in racing car was not recognised until some novel surface engineering technologies were developed.[2]

To increase the awareness of the importance of surface engineering design and novel surface engineering technologies in sports, this paper first briefly discusses the role of such surface properties as friction and wear in sport, followed by the principles and applications of surface engineering technologies in sports equipment. Successful applications of advanced surface engineering technologies in improving the quality of sport products and enhancing sport performance are given in two major case studies, one concerned with friction control of gulf clubs and the other with surface engineered titanium components for motor sport. Finally, the prospects for the further application of surface engineering in sport is briefly discussed.

7.2   Surface properties and surface engineering

Most, if not all, sports activities involve relative movement and therefore there are many different types of interacting surfaces in motion. Accordingly, the nature and the properties of sports surfaces can influence the sport itself. It is not intended to discuss the surfaces on which sports are performed in detail since this subject is well covered in Chapter 3 of this book. However, it seems logical to examine some properties of real surfaces.

7.2.1 The role of friction

Friction is the resisting force tangential to the common boundary between two bodies when, under the action of an external force, one body moves or tends to move relative to the surface of the other.[3] Friction is one of the most important properties of surfaces and has played a very important role in sport. In ball games (such as in tennis, table tennis, cricket, snooker, squash and golf), the spin of balls can be controlled by positively tailoring the friction between the hitting surface and the ball. Meanwhile, friction also influences the flight of a ball in air. When friction between air and ball increases, the flow becomes turbulent. In turbulent flow, the air is distributed and 'drag' occurs.[4] For example, the friction between a golf club face and a golf ball can, to a large extent, affect the spin, and eventually the driving distance and accurate control, of the golf ball.

Friction is a doubly important force in motor sport. In the engine, and other moving parts, it is an enemy, which is fought against largely by lubrication. On the auto sport track, as the wheels spin, just the right amount of friction is needed for the tyres to grip the track. Too little, and the car slips and slides, as if on ice. Too much, and the car slows down.

Nowhere is friction more critical than in winter sports and water sports. Surface or skin friction accounts for 70–80% of the total water resistance. It is found that the friction between a pair of surfaces depends mainly on the chemical interaction or adhesion between the contacting surfaces, and the mechanical interaction of the surface finish (roughness and texture). For instance, in canoeing the smoother the surface, the fewer water molecules become lodged between the small asperities in the surface or bounce off into the surrounding layers of water which slows down as they pass from the front to the back of the canoe as it moves forward.[5]

In summary, friction is a surface property, which can be altered by changing the surface chemical composition and/or surface finish. For example, the skin friction of a ski can be effective reduced by engineering the surface of the ski with wax; drag of a canoe hull surface can be slightly reduced if it is engineered to contain fine grooves running in the flow direction.[6] Clearly, surface engineering has great potential in improving sports performance via controlling the friction of sports surfaces.

7.2.2 Wear of sports surfaces

Wear is the removal of material from interacting surfaces in relative motion,[7] and is one of the most common types of surface degradation of sports equipment. In some cases, wear will lead to failure of the sports equipment and/or sports injury. Severe wear of mountaineering equipment, e.g. fraying of climbing ropes, may eventually lead to failure and cause an accident; excessive wear of racing car tyres will lead to the loss of grip to the track and loss of control; wear of footwear and synthetic playing surfaces or running tracks would lead to unsuitable friction between them. While too much friction makes movement difficult, too little friction could makes slight change of speed or direction dangerous.

Equally, excessive wear of sports surfaces will impair sporting performance significantly. As discussed in the preceding section, an appropriate level of friction is essential to ensure desirable spin of balls in ball games. For example, some hitting faces of golf clubs are surface engineered using sand blasting or surface texturing to achieve suitable and thus desirable spin of the ball. However, wear will lead to the change of the surface finish and the friction between the hitting face and the golf ball, thus limiting effective performance.

A new ski normally has a flat base with sharp metal edges; however, after a couple of days' skiing the ski base will become concave or convex owing to the abrasive action of snow crystals and fine abrasive materials (such as sand and dirt) embedded in the snow. Meanwhile, severe abrasive wear occurs to the edges, thus making them rounded because of the high forces at their edges resulting from constant turning in downhill skiing. As a result, skiing performance would be significantly impaired without proper tuning. This is because skis with concave bases have erratic control characteristics and are very difficult to turn effectively. A convex base is difficult to control since only a very small portion of the base is in contact with the snow. Meanwhile, if edges are not bevelled, the ski will have a tendency to catch the snow, thus making turning very rough.[8] To improve the performance of the ski, tuning is frequently required to flatten the base and sharpen and bevel the edges. Similarly, owing to the wear caused by ice, skate blades need regular sharpening to restore their performance.

Clearly, wear of sports surfaces not only will significantly impair sporting performance but also may cause injury or even accidents in some cases. Therefore, how to modify the wear resistance of sports surfaces is a challenge from both a technological and sport science point-of-view. Advanced surface engineering technologies developed during the past two decades have great promise in meeting the above challenges and developing high-quality sporting goods with resultant enhancement of sports performance.

7.2.3 Surface engineering

Notwithstanding the fact that surface engineering has been practised in various crude forms for several thousand years and it has a fascinating history, surface engineering, a term originally coined by Bell, c. 1983,[9] is now seen as a new interdisciplinary subject in its own right, which has became widely recognised since the 1980s. The generally accepted definition of surface engineering as given by Bell is:

> the application of traditional and innovative surface technologies to engineering components and materials in order to produce a composite material with properties unattainable in either the base or surface materials. Frequently, the various surface technologies are applied to existing designs of engineering components but, ideally, surface engineering involves the design of the component with a knowledge of the surface treatment to be employed.[10]

The rapid development of the interdisciplinary subject of surface engineering has been driven and stimulated by the increasing recognition that most engineering components can fail catastrophically from surface-initiated effects, which can be mitigated against by the growing commercial maturity of a variety of electrically based modern surface technologies.[11]

Surface engineering is a wide-ranging technology, covering traditional painting, electroplating, weld surfacing and thermochemical treatments, through to modern techniques including thermal spraying, physical vapour deposition (PVD), chemical vapour deposition (CVD), ion-implantation, and energy beam surface modification, together with many recently developed hybrid technologies (such as plasma immersion ion implantation) and duplex treatments. The thickness and hardness of a surface-engineered layer can span wide ranges, as are summarised in Fig. 7.1[11] and Fig. 7.2,[12] respectively.

Moreover, surface engineering, an enabling technology, is applicable to all classes of sports materials (metallic alloys, polymers and composites) and capable of greatly enhancing a range of surface properties of sport equipment:

- tailored frictional characteristics
- enhanced resistance to surface degradation (wear, corrosion and aging)
- improved cosmetic appearance.

Only in recent years have efforts been made in the development and application of surface engineering technologies for sports materials. Surface

Thickness of coatings and depth of diffusion treatments

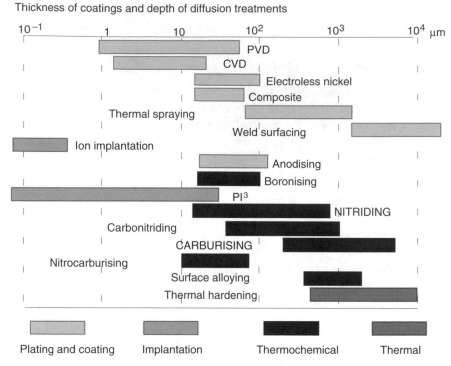

7.1 The typical thickness of surface-engineered layers.

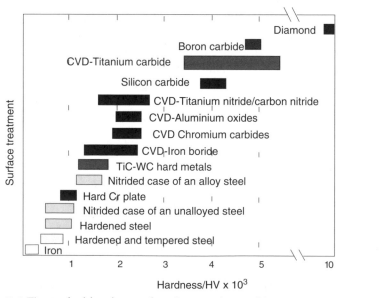

7.2 The typical hardness of surface-engineered layers.

engineering techniques which may be applied with varying degrees of success to sport equipment can be divided into three broad categories as follows: (a) surface coatings, in which a layer of material with a different composition and microstructure is added onto a surface; (b) surface modifications in which the composition and/or microstructure of surface is altered or modified; (c) duplex or multiple surface treatments, which combine two or more techniques from the first and second categories to obtain properties which are unattainable from the individual techniques. The principles and the applications of these surface engineering technologies will be discussed in further detail in Sections 7.3 and 7.4.

7.3   Surface coatings

7.3.1   Painting/organic coating

Painting is a generic term for the application normally of an organic coating to the surface of a material mainly for decorative (colour and gloss) and protective (corrosion and weathering) purposes. Most organic coatings or paints are based on a film former, which is dissolved or dispersed in a solvent or water. This film-forming material constitutes the vehicle in which pigments are dispersed to give desirable colour, opacity and other properties to the dried coating. A wide variety of film-forming materials are available, including oils, varnishes, synthetic resins and polymers (such as cellulose, vinyl, epoxy and polyester).[13]

A paint serves its function only when applied to a substrate. The equipment available to apply paints and coatings is diverse, varying from a simple paint brush to electrostatic spray robots. Most sporting and fitness equipment are painted mainly by such modern painting methods as electrostatic spraying, electrodeposition and powder coating. From an environmental and sustainable development point-of-view, powder coating is particularly attractive owing to the fact that no solvent is used and material usage can approach 100% in conjunction with a high quality product. Powder coating involves the application of 100% solid pigmented powder to the articles to be painted. The powder coating is 'fused' to a continuous film in an oven curing cycle or applied to a preheated article. Powder coating may be applied by thermal (flame and plasma) spray, fluidised bed and electrostatic spray.[14,15]

Surfaces of some wooden sports equipment (such as cricket bats and baseball bats) normally need to be protected from blistering through wet or dampness. This can be done traditionally via regular oiling, but in recent years some bat surfaces are coated with varnish or other polymer coatings.[16] Surfaces of other sports equipment, such as rackets, bicycles and fitness goods, are coated with many different types of painting materials to protect them from corrosion and to impart aesthetic appearance. For example, IF bicycle frames are painted using a spray method to protect them from corrosion and to produce an aesthetic appearance. The frames are painted first with grey epoxy primer, then with sealer, followed by decals and finished with clear coat. Finally, the frames are placed in oven and allow to cure and dry.[17]

Polymer coatings are also applied to the surface of racket strings, which is made of bundle strands of high-strength nylon fibres immersed in a matrix of a polymer elastomers, to protect them from moisture and wear.[6]

7.3.2 Anodising

'Anodising' is the generic term applied to electrolytic methods of converting the surface of non-ferrous alloys (such as aluminium, titanium and magnesium alloys) to form coherent porous oxide coatings for purposes of decoration and protection. The process derives its name from the fact that the part to be treated becomes the anode in an electrolytic cell. Although both painting and anodising can produce coatings for similar purposes, much higher adherence to the substrate can be achieved by anodic coatings than by painting since the converted coating is an integral part of the substrate material.

When a current is passed oxygen is generated at the surface of the aluminium anode and combines with the aluminium to form a layer of porous aluminium oxide.[18] Hydrogen is liberated at the cathode. The amount of aluminium oxide formed is directly proportional to the current density and time. The three principal types of anodising process and their characteristics are given in Table 7.1.

Table 7.1 Characteristics of three anodising processes

Process	Electrolyte	Thickness	Appearance	Remark
Chromic	3–10% CrO_3 (~40 °C)	4–7 μm	Opaque grey	Good chemical resistance
Sulphuric	12–20% H_2SO_4 (~20 °C)	8–25 μm	Colourless	Good corrosion protection
Hard	Sulphuric acid (–5 ~ +5 °C)	25–100 μm	Grey or brown	Excellent wear resistance

Anodising is typically associated with aluminium and its alloys, though it is also applicable to titanium and magnesium alloys. Some reasons for anodising aluminium are outlined below.

Improving abrasive resistance

The anodic oxide coatings, especially the hard anodic oxide coatings, are much harder than the aluminium metal, so the resistance to abrasive wear is greatly increased by anodising.

Imparting aesthetic appearance

All anodic oxide coatings are lustrous, the degree of which depends on the condition of the substrate metal before anodising. Many different colours can be produced by dyeing aluminium oxide on the anodised aluminium or by controlling the thickness of the titanium oxide on the anodised titanium.

Enhancing corrosion resistance

Sealed anodic coatings of aluminium oxides are barriers to atmosphere and sea-water attack.

Increasing painting adhesion

The adhering porous anodic coating confers a chemically active surface for most paints.

For some applications, post-treatments (noticeably sealing and colouring or dyeing) are sometimes necessary to confer some special properties. Although all anodic films improve the corrosion resistance, the improvement is limited by its porous nature. Significant enhancement in corrosion resistance can be achieved by blocking or sealing these surface pores by heat treatment in slightly acidified hot water or hydration sealing (i.e. partial converting of the alumina to a monohydroxide). In decorative anodising colour and aesthetic finish are major design features. The desirable colour can be produced by dipping the part in a solution containing organic dyestuffs or dyeing. The anodic coating can also be sealed with PTFE or oil, etc. to further enhance its tribological behaviour. The 'Ano-lube' process is designed to increase their wear resistance, reduce friction and repel moisture by flooding a freshly formed unsealed oxide coating with an aqueous dispersion of PTFE, thus allowing it to drain and then drying it at room temperature.

Anodising has been successfully applied to aluminium bicycle wheel sets manufactured by Mavic. Their construction involves the use of 6106 aluminium alloy for the rim, which has to withstand cyclic loading as well as the sudden impact forces often encountered in such a sport. The rim of the wheel is also subjected to impacts from the side as stones and gravel knock against it. The 6106 aluminium alloy has a comparatively low hardness of around 136 HV, and hence Mavic have chosen to anodise the wheel to increase its surface hardness to ~600 HV and hence its durability, and then to dye the surface black to impart a desirable aesthetic appearance.

Notwithstanding the fact that anodising is typically associated with aluminium and its alloys, some anodising processes have also been developed for titanium and magnesium alloys. Two methods of anodising titanium are used by industry: acid anodising and alkaline anodising. Both processes can produce a porous TiO_2 oxide film on the surface of titanium, but acid anodising can form a very thin oxide layer ($<0.1 \mu m$), while a relatively thick oxide film ($<4 \mu m$) can be produced by alkaline anodising. The purpose of anodising titanium is to enhance its wear resistance (alkaline anodising) and/or to impart desirable aesthetic appearance (via acid anodising). Vivid colours from magenta to cobalt blue can be obtained by anodising titanium in slightly acidified solution at different terminal voltages or thickness, due to the formation of strong interference colours. Anodising of titanium alloys can reduce friction and wear and can, to some extent, prevent galling and seizing by conjunction with dry film lubricants (e.g. Canadizing®).

Based on anodising technology, a new emerging technique (bearing many different names: anodic spark deposition,[19] micro-arc oxidation[20] or plasma electrolytic oxidation;[21] KERONITE™[22]) is emerging. A plasma arc discharge is produced on the surface of the component in an electrolyte using a high voltage and a high density of current. A much thicker oxide layer on aluminium, titanium or magnesium surface can be produced to enhance its resistance to abrasive wear and/or corrosion.

7.3.3 Thermal spraying

Thermal spraying is a general term used to cover a wide variety of processes in which a material is heated rapidly and simultaneously projected at high speed onto a surface, thus building up a thick overlay coating. A major advantage of thermal spray processes is that a wide variety of materials (metallic, ceramic and polymeric) can be sprayed without unduly heating the substrate ($<50°C$). The basic steps involved in most thermal spray processes are as follows:

- The spray material is heated to near, or somewhat above, its melting point.
- The molten or nearly molten droplets are accelerated by a stream of compressed air or other gases and projected against a substrate surface.

- On impact, the droplets flow into thin lamellar particles which overlap and interlock as they solidified.
- The total coating thickness is built up by multiple passes of the spray device.

Thermal sprayed coatings have a lamellar structure parallel to the interface, comprising bonded 'splats', which result from high rate impact and rapid solidification of high flux of melted particles with size of 10–100 microns. Adhesion is fundamentally a mechanical process, although diffusion takes place in some cases. Adhesion strength <70 MPa, which is as good as, or better than, any other non-diffusion coating processes. Porosity is often present as a result of outgassing, shrinkage or shadowing, the quantities of which vary from <0.5 to 10% depending on the process used. Therefore, it is necessary in many applications (where corrosion resistance is a major concern) to seal the pores or reduce the degree of porosity by a finishing treatment through using such sealant materials as waxes, epoxies and inorganics, or via re-melting the surface using electron beams and laser beams. Another limitation associated with thermal spraying is its line-of-sight nature, i.e. it can only coat what the torch or gun can 'see'.

Thermal spraying has found widespread applications and is used in virtually every industrial sector: to enhance the wear resistance of general engineering components, to increase corrosion resistance of large structures, to improve the biocompatibility of biomaterials, to impart oxidation resistance to aerospace engines, and to enhance performance of sports equipment. Although there are numerous thermal spraying processes available commercially, they can be classified into four groups according to the heating source or methods: flame spraying, electric arc spray, plasma spraying and detonation-gun (D-gun). The characteristics of these thermal spray processes are compared in Table 7.2.

The most widely used thermal spray process in the sports equipment industry is plasma spraying. This is because the use of plasma spraying, especially vacuum plasma spraying, has provided many advantages over other

Table 7.2 Comparison of principal thermal spray processes

	Flame	Electric arc	Plasma	D-gun
Velocity (m/s)	150	200	400	1500
Temperature (K)	3000	5000	12000	4000
Porosity (%)	10–15	10–15	1–10	1–2
Bond (MPa)	5–10	10–20	30–70	80–100
Deposit rate	High	Very high	Low	Very low
Cost	Low	Very low	High	Very high

spraying in terms of higher plasma temperature, independent control of the atmosphere, improved adhesion strength of the coating to the substrate (through the effects of preheating and the sputter cleaning of the surface to be coated), as well as through the elimination of contamination of the molten particles.

To maximise friction with the braking pads and to minimise the wear of a mountain bike wheel rim made of the 6160 aluminium alloy, a ceramic coating consisting of aluminium oxide and titanium oxide ($Al_2O_3 + TiO_2$) has been sprayed onto the sidewalls of the rim using the plasma spraying method. This ceramic coating serves as a thermal insulator as well as significantly improving the hardness of the aluminium rim by 30 times. As a result, more efficient braking can be achieved, as evidenced by significantly reduced braking distance (20% in dry weather and 50% when raining); furthermore, this extremely hard coating provides superior resistance to abrasive wear.[23]

Plasma-sprayed ceramic coating has also been adapted as a thermal barrier coating in the design of lightweight motorcycle brake discs consisting of Al–Si cast alloy matrix reinforced by silicon carbide particulate. Preliminary studies have shown that discs without the thermal barrier coating cannot be used on high performance racing motorcycles since excessive surface temperature would culminate in local melting of the aluminium matrix and brake failure. The ceramic thermal barrier coating can limit heat transfer effectively to the disc, and hence the disc temperature can be reduced from 340 °C to about 310 °C. The ceramic coating also allows the friction to be controlled by the surface finishing of the coating. During the 1996 season, these ceramic-coated discs were evaluated on selected 125 cc and 250 cc Grand Prix motorcycles in the UK.[24] Thermal barrier coatings are also plasma sprayed onto the bottom surface of exhaust valves used in racing car engines to protect the hardened steel valves from oxidation and softening caused by the high temperature (700~800 °C) highly corrosive exhaust gases.[25]

7.3.4 Physical vapour deposition (PVD)

PVD is one of the most widely used surface engineering technologies to deposit thin coatings from the vapour phase derived by physical means. In any PVD process coatings are formed via the following three steps: (a) creation of material vapour; (b) transport of the vapour and (c) condensation of the vapour and growth of the coating. According to the methods used to generate the vapour, all PVD processes fall into three general categories: evaporation, sputtering and ion plating. PVD processes are carried out at a relatively low temperature between 100 and 500 °C, hereby avoiding or eliminating any adverse effect on the underlying substrate material. Unlike thermally sprayed coatings, the surface finish of PVD coating may be as good as that of the initial substrate.

Many coating materials are available, ranging from pure metals to ceramic compounds, from super hard diamond to soft solid lubricant. Although PVD treatments of general engineering components and cutting tools have met with great success and have become routine processes in production, e.g. TiN coating on a high speed steel substrate, the PVD treatment of sports equipment has found limited applications to date for sports goods and is still undergoing development. In this respect, TiN and carbon-based coatings are the most widely researched and applied coatings, largely because of their excellent tribological properties in conjunction with a pleasant aesthetic appearance.

As with most refractory nitrides and carbides, TiN exhibits a high hardness (2300 HV), high tribological compatibility in terms of low friction (0.2–0.4) and high wear resistance (especially resistance to abrasion), thus reflecting their ceramic nature. Meanwhile, the pleasant gold colour has made TiN a very competitive coating material for sport equipment. For example, to increase the resistance to the abrasion and to confer a nice golden colour, Teer Coatings Ltd have coated golf clubs and the guides for the fishing lines of high-quality fishing rods with a TiN coating.[26] TiN coating as also been adopted in the manufacture of Ohlins MX/Enduro racing motorbike front forks, which are designed as high-end components, albeit for racing motorbikes. The suspension forks undergo a large degree of abrasive wear during their service lifetime, so as low a coefficient of friction and as high a surface hardness as possible would be beneficial. In this instance, Ohlins have steel upper tubes with a 'super-hard polished titanium nitride surface' to maximise the life of the component by reducing the amount of wear on the surface of each tube; TiN also has a relatively low coefficient of friction that will facilitate smoother movement of the two surfaces over each other.[27]

PVD arc-evaporated TiN coating has been used to improve the wear resistance of titanium steering racks for Formula One racing cars. However, the arc-evaporated TiN coating only achieved very limited success partially because of the lack of support from the relatively soft substrate, and partially related to the poor adhesion of the coating due to inadequate surface depassivation. Consequently, a modified PVD process, Nitron®, was developed by Tecvac Ltd in conjunction with Surface Engineering Group at Birmingham University to address the above problem. Prior to the evaporation of titanium, ions sputter clean the titanium surface to be coated at a higher-than-normal temperature (700 °C vs. 450 °C) for a longer time (2 h vs 0.5 h) in a nitrogen/argon mixture rather than argon. This removes the oxide film and forms a thin nitrogen diffusion layer as well as preheating the substrate that, in turn, enhances diffusion. Thicker TiN coatings produced by the process have demonstrated significant improvement in load bearing capacity, and the Nitron® treated titanium steering racks have routinely been used by some Formula One racing teams.

In recent years, carbon-based coatings (diamond and diamond-like carbon) have drawn more and more attention owing to their unique tribological, chemical and physical properties. The hardness of carbon-based coatings depends on the arrangement or bonding of carbon in the coatings, ranging from 1000–3000 HV for DLC to 6000–8000 HV for diamond coatings. The friction coefficients of the carbon-based coatings are extremely low (0.05–0.2) when operating against most engineering materials, and their resistance to wear is very high. In addition, most carbon-based coatings acquire a nice shining black colour, which is attractive for some sports equipment. One of the physical properties associated with the carbon-based coatings is their hydrophobic nature, which may be used to reduce the drag force between sports equipment surfaces and snow or ice. High-quality carbon-based coatings are produced by some surface coating specialists, notably DLC and GRAPHIT-iC™ coatings by Teer Coatings and BALINIT®C by Balzers.

The unique tribological properties characterised by extremely low friction and high wear resistance have been the main driving force for the industrial acceptance of carbon-based coatings for Formula One and other competitive motor sports, especially for components in transmission systems and engine valve trains. For instance, Teer Coatings Ltd routinely coat many engine and other parts with their low friction, wear-resistant carbon-based coatings, including cam followers, piston rings and injector nozzles for many motor racing organisations.[26,28] Balzers[29] have successfully applied their carbon-based coating, BALINIT®C, to many types of competitive racing car, motorcycle and engine components, including cam shafts, wrist pins, plungers for fuel injection pumps, racks, bevel gears and gear selector fork guides, etc. For example, the low friction properties of the coating can increase the performance of a Suzuki 600 supersport motorcycle by 1.5 HP, and increase the output torque by c. 1.7%. It has also been reported that advanced PVD carbon-based coatings can effectively enhance racing performance of 'Team 8' of the Camaro Cup, which is a racing class in Sweden. During the races in 1997, Team 8 had to change the gearbox three times and differential once due to the undue wear. Consequently, Team 8 decided to coat both the gearbox and the differential with BALINIT®C to address the problem. During the 1998 season, with coated components, they did not have to change these components, and they achieved good racing performance with podium results in each of the eight races they completed, and in the Camaro Cup Championship their performance gained second place.

In addition to the successful applications in motor sports, carbon-based coatings have also been used to coat ice skates and fishing equipment[26] and golf club heads (see Case Study I in Section 7.5).

7.4   Surface modification

7.4.1   Mechanical surface modification

Mechanical surface modification is a general term embracing the treatments aimed at modifying the surface properties and/or surface texture of a material or article using continuous or dynamic mechanical action. The main types of mechanical surface modification processes used in the manufacture of sports equipment are shot peening, knocking-in and blasting.

Shot peening

Shot peening is a cold working process in which the surface of a metal object is bombarded with a stream of small spherical particles (metal, glass or ceramic) called shot under controlled conditions. Shot striking the material acts like tiny peening hammers, thereby increasing the hardness of the surface layer, inducing residual compressive stresses as well as creating a uniform dimpled texture on the surface. These compressive stresses are beneficial in increasing resistance to fatigue and fretting fatigue.[30]

Shot peening has been adopted in the manufacture of sports equipment to enhance performance and to produce attractive finishes. A good example is shot peening of titanium mountain bike frames by a titanium frame builder, Independent Fabrication.[17] Independent Fabrication manufacture custom mountain bike frames from Ti-3–2.5 alloy with the aim of reducing the weight of mountain bikes without sacrificing their strength. Over the life of the bicycle, the frame will undergo numerous cyclic stresses of varying magnitude, due to the nature of mountain biking, and so the best possible fatigue life is essential for such an application.[31] Processing of the titanium alloy tubing involves butting (to attain correct tube geometry) and welding – both of these processes may build in residual tensile stresses and cause micro-cracks that would severely compromise the strength of the frame. For instance, heat generated by the welding process often will produce tensile stresses approaching the yield strength of the material. These harmful induced stresses in the heat affected zone contribute to poor fatigue life of welds and thus premature failure of the titanium mountain bike frames. Therefore, Independent Fabrication use shot peening to build up compressive stresses in the frame and at welds to counteract the residual tensile stresses generated during the process of cutting, grinding, butting and welding, thus preventing surface micro-cracks from propagating and effectively doubling the fatigue life of the titanium frame. The process of shot peening will also work harden the surface of the tube while giving it a finely textured surface, thus creating an attractive surface finish that is highly resistant to scratches. The textured surface glitters in the sun in a manner similar to that of a pearl paint finish, and so imparts an aesthetic effect of pearl lustre to the surface of the alloy.

Knocking-in

The principles of mechanical surface hardening have also been applied to wooden cricket bats through a process called 'knocking-in' by Sayers *et al.*,[32] in an attempt to prevent premature surface damage, i.e. attrition or wear during service. The wood from the willow tree is the best material for the manufacture of cricket bats, owing to its very resilient and shock-resistant characteristics. However, the soft and fibrous nature of the wood make it vulnerable to surface damage when hit by an energetic new cricket ball in service. To address the above problem, most good-quality new bats need to be knocked-in prior to use by hitting the bat with a wooden bat mallet or cricket ball for a period of time to compress the wood fibres and thus harden the surface of the cricket bats. Hardness measurements showed that the hardness of the bat face could be doubled by knocking-in for a period of 4 h, and detailed examination with a scanning electron microscope revealed that the surface wood fibres became increasingly enmeshed and compressed with knock-in duration time.

Blasting

Blasting is a process for cleaning or finishing metal surfaces with an air blast or centrifugal wheel, which throws abrasive particles against the surface of the object. Small, irregular particles are used as the abrasives in *grit blasting*; sand, in *sand blasting*; and steel, in *shot blasting*. Blasting differs in its primary purpose compared to shot peening, although both involve the impingement of small particles onto the surface being treated.[33] Whilst the main purpose of shot peening is to introduce compressive residual stresses and thus increase fatigue properties of the material peened, blasting is performed mainly to remove such undesirable surface layers as oxides and contaminates, to roughen surfaces as a pre-treatment for painting or coating, to increase friction and to create a uniform attractive surface texture for cosmetic purpose. For example, prior to painting steel bicycle frames, Independent Fabrication use sand blasting to remove contaminants and to create microscopic anchoring points for the paint to enhance its bond with the steel substrate.[34] Blasting processes have also been applied to the face or hitting surface of some golf club heads to increase the friction between the hitting surface and the golf ball in order to facilitate more ball spin;

some tennis rackets made of light alloys (such as titanium alloys) are blasted before being painted or coated with a transparent organic coating to enhance the bonding between the coating and the substrate, and also to impart pearl lustre aesthetic effect to the surface.

7.4.2 Thermochemical processes

Thermochemical treatments are relatively traditional processes for hardening metallic materials, which comprise the diffusion of interstitial elements such as nitrogen, carbon, boron or oxygen into the surface from a gaseous, solid, molten salt bath or plasma medium. Of these elements, carbon and nitrogen are the most widely used interstitial elements for case hardening of steels while nitrogen and oxygen are the most promising for titanium alloys, as they have the highest solubilities and diffusion rates.

Carburising

Carburising is a surface-hardening process whereby carbon is introduced into the surface of low carbon steel by holding the metal at a temperature in the austenite region (normally 850 °C–950 °C) in contact with a carbon-bearing medium. Hardening is normally achieved by quenching after carburising to form a hard, high carbon martensite structure and associated beneficial compressive residual tresses in the case with high fatigue and wear properties, supported by a tough, low-carbon martensite core structure. Several carburising processes have been developed, including pack carburising, salt bath carburising, gas carburising, vacuum carburising and plasma carburising. Among them, gas carburising is by far the most widely used process, particularly in high volume production. Vacuum and plasma carburising are relatively new processes, which have the advantages of causing less distortion to components and are more environmentally friendly.

Carburising is the only thermochemical process for ferrous materials that can produce very deep hardened cases (up to 4 mm) with high surface hardness (700–900 HV), good wear and fatigue properties and very high load-bearing capacity. Consequently, carburising is utilised in many types of sports equipment, particularly for motor racing, where combined properties of wear, fatigue and load-bearing capacity are required. For example, carburising is used extensively on components for the motor sports industry, including gearbox parts, engine components (crankshafts, gears, camshaft and followers) and car steering components and bearings. The most commonly used conventional carburised materials are carburising steels (such as SAE8620, 832M13, S156, etc.), normally containing 0.1–0.25 wt % carbon and some alloying elements (Mn, Cr, Ni, Mo).

Nitriding

Nitriding is also a widely used thermochemical process in which nitrogen is diffused into the surface layer of a solid alloy by holding it at a suitable temperature in contact with a nitrogenous material.

Nitriding of steels dates back to the early 1920s, and is used in many industrial sectors, including sports equipment. The nitriding temperature is normally below the iron–nitrogen eutectoid temperature (590 °C) without a phase transformation taking place during the process. Nitriding of steel involves several reactions, including:

- the reactions between nitriding media and the steel surface to generate reactive nitrogen species
- the diffusion of active nitrogen atoms into the material, forming a nitrogen diffusion zone
- the reaction of nitrogen in the diffusion zone with alloying elements to form very fine nitride precipitates, producing precipitation hardening
- a chemical reaction at the surface to form an iron nitride compound layer.

As with carburising, nitriding can be carried out in any of the four states of matter (solid, liquid, gas and plasma). Pack nitriding is not a reliable and reproducible process, whereas gas and liquid nitriding are not environmentally friendly, and tend to produce thick and brittle compound layers. Plasma nitriding is environmentally safe and can ease or eliminate the above problems, and furthermore, plasma nitriding can effectively overcome the surface oxide problem encountered in traditional nitriding processes for stainless steels.

Because of the absence of a quenching hardening mechanism and the comparatively low temperatures (400–570 °C) employed in this process, nitriding produces less distortion and deformation than carburising, which is very important for some precision wear-resistant components. Most nitriding steels contain 0.2–0.4% carbon and sufficient amount of nitride-forming alloying elements (Cr, Mo) to facilitate the desirable precipitation hardening. A typical nitriding steel is 720M24 (En40B), which contains about 3% Cr. High strength maraging steel is also used in order to achieve the maximum properties possible for components used in Formula One cars and racing engines.[28] Applications for carburising and nitriding may overlap to some extent in motor sports (for such applications as crankshafts, gears). However, the load-bearing capacity is much higher for the carburised components than for the nitrided one since the maximum thickness of the nitrided case is normally <0.6 mm compared with ~4 mm for carburising; on the other hand, much better scuffing resistance can be achieved by nitriding than by carburising.

Low-temperature plasma surface alloying of stainless steels

All stainless steels can be hardened to some degree by gas nitriding, because of their nitride-forming elements, such as chromium, vanadium, molybdenum or aluminium. However, it should be pointed out that gas nitriding of stainless steels requires certain surface preparation that is not required for nitriding low-alloy steels. This is because of the stabilised passive chromium oxide film on the surfaces of stainless steels, which, on the one hand protects stainless steels from corrosion and oxidation in service, but, on the other hand, impedes surface reactions and mass transfer of nitrogen in the nitriding process. Consequently, it is necessary to remove the film of chromium oxide prior to gas nitriding, which usually involves the use of harmful acids. Furthermore, although conventional plasma nitriding increases the surface hardness and wear resistance of stainless steels, it decreases general corrosion resistance by combining near surface chromium with nitrogen to form chromium nitrides.

Novel low-temperature plasma surface alloying (with nitrogen or carbon) processes have recently been investigated by Bell *et al.* at Birmingham Surface Engineering Group,[35,36] which can achieve combined improvements in tribological, corrosion and fatigue properties. This is mainly because, during plasma nitriding (or carburising) at a relatively low temperature (<420 °C), nitrogen (or carbon) can diffuse into the surface of stainless steel, whilst chromium is hardly able to diffuse any appreciable distance to form chromium nitrides (or chromium carbides). Accordingly, significant hardening can be achieved through solid solution hardening associated with the formation of a nitrogen (or carbon) supersaturated expanded austenite phase called 'S-phase'. This can be achieved without evoking the formation of chromium nitrides and the associated chromium depletion, thus maintaining or even enhancing the corrosion resistance of the nitrided case. As shown in Fig. 7.3, after being etched by a very corrosive solution, the substrate is corroded but the surface modified case remains unattacked, revealing the improved corrosion resistance; the indentations produced in the white layer are much smaller than those in the substrate, indicating strong hardening effect. These novel low-temperature plasma surface alloying processes show significant potential for improving both the wear and corrosion resistance of such stainless steel sports goods as golf club heads and carving runners for skis.

Plasma nitriding of titanium alloys

Nitriding is used successfully to impart good tribological properties and a aesthetic yellow golden colour to the nitrided titanium alloys. Most of the gas nitriding of titanium is carried out in the temperature range 800 to

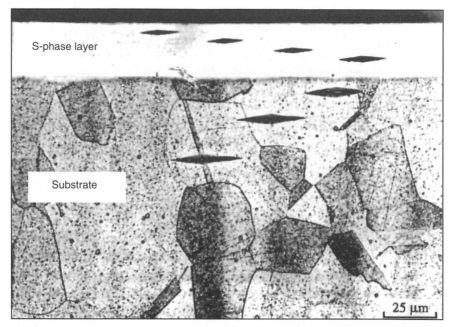

S-phase layer

Substrate

25 µm

7.3 Optical microstructure with Knoop hardness impressions showing improved corrosion resistance and hardening effect.

1000 °C, which significantly decreases the fatigue life by 20 to 35% and results in undesirable shape distortion. One solution is to use plasma as an alternative medium for nitriding of titanium alloys, and hence has received considerable attention recently. The sputtering of nitrogen ions can remove any oxide film, which prevents ingress of nitrogen, thereby increasing the processing rate and requiring a lower treatment temperature.

The microstructures resulting from plasma nitriding typically consist of a thin superficial layer of golden TiN (2200 HV, up to $2\,\mu$m), followed by a relative thicker layer of silvery Ti_2N (1500 HV, up to $10\,\mu$m) and beneath that a nitrogen-rich solid solution (from around 1000 HV down to the substrate hardness). A considerable improvement in tribological properties, especially the scuffing resistance, results from plasma nitriding. However, once again the high temperatures (~850 °C) give rise to a reduction in fatigue strength and undesirable distortion for components with complex geometries. Intricate components with both thin and thick sections cannot be treated above 700 °C without shape distortion. For example, attempts were made to plasma nitride the titanium wheel hulls for a Formula One car to enhance their tribological properties; however, so far unacceptable distortion was observed. It is reported that plasma nitrided titanium steering racks have been used successfully in motor racing cars.[28,37]

7.4.3 Thermal oxidation of titanium alloys

Recently, a wholly innovative surface engineering process based on thermal oxidation, designated the TO process, has been successfully developed by Dong *et al.*[38] Although the TO technique is essentially a thermochemical process and should be included in the above subsection, this novel process is worth discussing separately owing to the fact that this process is very effective in improving the tribological properties of titanium alloys, and is cost-effective and fully environmentally friendly.

The TO process is carried out in a oxygen-containing atmosphere at temperatures ranging from 500 °C to 700 °C and in essence converts the titanium surface into the ceramic rutile, TiO_2 which has excellent tribological properties. A typical cross-sectional structure of Ti6Al4V specimen treated using the proprietary TO treatment comprises a thin rutile oxide layer (~2 μm) supported by the oxygen diffusion zone (~20 μm) (Fig. 7.4).[39]

The friction-reducing capability of the TO technique is demonstrated in Fig. 7.5. The friction trace of the untreated Ti6Al4V material can be seen to fluctuate widely throughout the whole testing period, indicative of the 'stick-slip' adhesive behaviour of titanium and its alloys when sliding

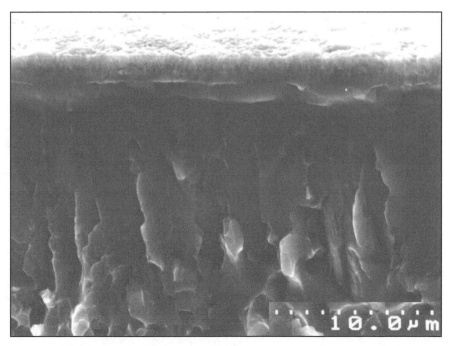

7.4 SEM micrograph showing fracture section of TO-treated Ti6Al4V.

against most engineering materials. By contrast, it is clear from Fig. 7.5 that TO treatment has not only significantly reduced the friction value but also made it stable, thus providing an attractive friction-reducing effect of TO treated Ti6Al4V alloy. Likewise, the wear resistance of Ti6Al4V alloy is enhanced significantly following the TO treatment. The wear rate of the TO-treated specimen is dramatically reduced by more than two orders of magnitude over the untreated material and it was even lower than that of the hardened steel counterpart by a factor of more than 10 (Fig. 7.6). Scuffing is a form of severe sliding wear characterised by unacceptably high friction and a high degree of surface damage. The anti-scuffing capacity of the TO-treated Ti6Al4V samples has been evaluated by obtaining the critical load-to-failure during the stepwise loading process of oil-lubricated sliding wear tests. The critical load of Ti6Al4V alloy can be improved from <20 N for the as-received material to 1850 N for the TO-treated material,[40] which makes it an ideal surface-engineered sports material for the design of sports equipment.

In summary, TO treatment is an effective surface engineering technique for enhancing the tribological properties of titanium and its alloys in terms of a significantly reduced friction coefficient, enhanced wear resistance, and greatly increased anti-scuffing capacity (Table 7.3). It can be seen therefore that the TO process has considerable potential for the surface engineering

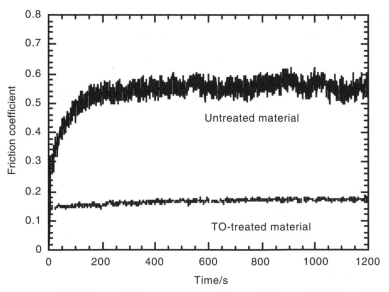

7.5 Effect of TO treatment on the coefficient of friction of Ti6Al4V.

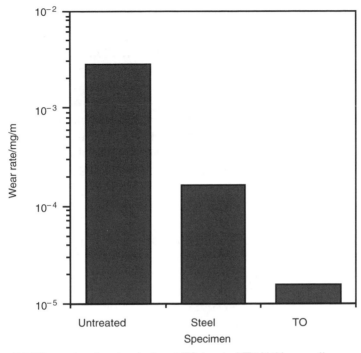

7.6 Wear rate of untreated and TO-treated Ti6Al4V as well as hardened steel for comparison.

Table 7.3 Improvement in tribological properties of Ti6A14V titanium alloy

Property	Untreated	TO-treated
Friction coefficient	0.4–0.5	0.1–0.2
Wear rate ($mg\,m^{-1}$)	2.76×10^{-3}	1.57×10^{-5}
Anti-scuffing critical load (N)	<20	1850

of titanium and its alloys. It provides for the possibilities of developing a low friction, high wear resistant titanium surface without compromising the corrosion resistance. Furthermore, the TO process not only provides titanium alloys with technological improvements but also it is a cost-effective and environmentally friendly process. Clearly, the TO process has opened up many opportunities for the design and manufacture of titanium components for sports equipment (see Case study II in Section 7.5.2).

7.4.4   Laser beam surface modification

Lasers have been widely used for many years in materials processing, however, it was not until the late 1980s that attention was paid to its potential in the surface engineering of motor sports equipment. Laser surface modification involves the use of the laser, as an intense energy source, to produce rapid heating in the solid state; on removal of the energy beam, rapid cooling via conduction of heat to the substrate can achieve deep-case (up 1 mm) hardening of ferrous materials, which are suitable for load-bearing and wear-resistant applications and is a valuable alternative to classical induction hardening technology.

Laser surface melting

The principle of laser beam surface melting is such that, when a laser beam irradiates a surface, a certain depth (normally between 0.1 and 1 mm) of the surface is rapidly heated to a temperature above the melting point of the material. When the beam is removed, the melted region rapidly solidifies as it is quenched by the substrate material beneath it (self-quenching), thus leading to microstructural refinement, extended solid solubility and formation of metastable phases. Laser surface melting has been used successfully to harden the surface of cast iron camshafts selectively used in some high performance sports cars. Laser surface re-melting can not only convert the coarse graphite into very fine Fe_3C but can also form a carbon-saturated matrix. Consequently, following laser surface melting, the hardness of the surface of grey cast iron cam can be increased from 200 HV to 750 HV, imparting significantly improved wear resistance to the cam surface.[41]

Laser surface alloying

Laser surface alloying is similar to that of surface melting except that an alloying addition is introduced into the melt pool during melting of the surface. Hence, surface alloying not only inherits the rapid solidification effects conferred by surface melting but also creates new effects resulting from the alloying elements. The most widely documented process using an energy beam to surface engineer titanium alloys is laser gas alloying with nitrogen, commonly referred to as 'laser nitriding'. It has yet to find widespread industrialisation in the auto sport and other sports sectors.

This process usually involves the use of a CO_2 laser to melt the surface in a nitrogen-containing atmosphere. A number of microstructures, hence different hardnesses, can be obtained by altering the amount of nitrogen take-up by the melt pool. The dominant component of the microstructure

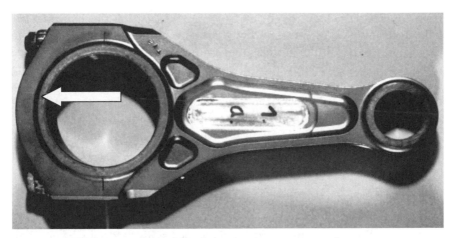

7.7 A laser surface-alloyed titanium connecting rod.

is, in order of decreasing nitrogen content, TiN dendrites, N-rich alpha-Ti dendrite or N-rich basket-weave. It has been found that the higher nitrogen concentration in the melt, the harder is the resultant microstructure. Developments are in hand to improve the wear resistance of titanium connecting rods (Fig. 7.7).[42]

7.5   Surface engineering case studies

7.5.1 Case study I: friction control of golf clubs

Over the past three decades, great progress has been made to improve the performance of golf equipment through innovation in the structural design, e.g. hollow heads, and by the introduction of advanced materials, e.g. titanium and composites.[43] Meanwhile, some efforts have also been devoted to enhance the performance of golf equipment further by adopting advanced surface engineering technologies. In this case study, the role of surface engineering in tailoring the friction of golf clubs will be demonstrated by way of example.

Role of coefficient of sliding friction

A golf club normally consists of a shaft jointed to a head that provides a ball-striking surface. According to their main function, the golf club head can be divided into five categories as follows: (a) drivers (as-far-as-possible shots from the tee) (b) fairway woods (controlled long-distance shots); (c) irons (medium distance and accuracy); (d) wedges (short-distance shots from sand and/or deep rough) and (e) putters (getting the ball into the hole from the green). An important part of a golf club is the face of the club head or the hitting surface, and one of the most important surface properties that significantly affect the performance of golf equipment is the coefficient of friction between the hitting surface and the golf ball. This is mainly because it is known from golf theory and practice that the sliding friction between the hitting surface and the ball, to a large extent, determines the spin rate and velocity of the ball after impact,[44] which ultimately affects the quality of shot in terms of flying distance and directional accuracy.

The effect on golf performance of friction between the hitting surface and the ball is closely related to the skills used in the game. Essentially, the game of golf can be classed into two major skills, the long game and the short game. Whilst the long game involves such skills as 'driving' a golf ball as far as possible from the tee (drivers) and 'hitting' a golf ball straight long distances from fairway, the short game involves skills such as 'pitching' a golf ball from a location near the green and 'putting' the golf ball into the hole on the green.

Generally, high friction between the hitting surface and the ball for long shots (especially for driving) is undesirable. This is mainly because tolerance for mis-hits on the sweet spot decreases with increasing friction between the hitting surface and the ball owing to the increased likelihood of 'hooking' or 'slicing' a given shot. Thus, the distance that the golf ball can fly would be reduced because energy is wasted on the fast side-spin of the golf ball and heat generated from the friction. Meanwhile, such undesirable shots also affect directional accuracy. Clearly, low friction between the hitting surface and the golf ball is desirable for long-distance shots.

On the other hand, however, high friction between the hitting surface and the golf ball is needed for club heads used to produce short-distance, accurate shots. For example, an important feature of a wedge is to form repeatable and high-rate back-spin on the ball as it comes out of the sand trap or from grass, which makes the ball land softly and stop quickly. Meanwhile, high rates of back-spin would contribute to straight shots because of the gyroscopic effect on the ball.[45]

Low-friction golf club face

The simplest way to reduce the friction coefficient between a golf club face and a golf ball is to apply a lubricant coating or film to the face or hitting surface. The trajectory of a golf ball can be changed by applying a friction-reducing substance to the club face prior to striking the ball, thereby reducing heat and spinning that a golf ball normally experiences after impact. The preferred lubricant coating is composed of silicon dioxide although other friction-reducing lubricants can also be used including silicone, silicate, PTFE, waxes and water-based lubricants.[46] The friction-reducing material is preferably in the form of a liquid so that it can be conveniently dispensed from an applicator and spreads to a generally uniform and even coating. Notwithstanding the fact that this is a low-cost, very flexible surface technique to be used to reduce the friction of club heads, there are several disadvantages to the technique. For example, some lubricant will surely pass onto the golf ball, thus causing the ball to fly through the air unevenly. It is also inconvenient to apply a liquid lubricant prior to a shot and to remove it afterwards.

A surface engineering technique to apply a permanent low-friction titanium nitride coating to a golf club head has been patented by Buettner.[47] Cast golf club heads made of 316 austenitic stainless steel are coated with a thin (5–10 μm) hard attractive coating of titanium nitride by conventional physical vapour deposition. The titanium nitride coating provides a very hard (1500 ~ 2000 HV), low-friction and wear-resistant surface to the club head. In addition, this titanium nitride coating doped with about 10 wt% niobium possesses a very attractive lustrous gold appearance. Accordingly, the low-friction characteristic allows the ball to fly further and straighter owing to the reduced spin rate and thus loss of energy and the low likelihood of hooking or slicing. Meanwhile, the low-friction surface can prevent dirt, grass and mud effectively from sticking to the club surface, and allow the club to swing through grass without losing speed unduly. Furthermore, the ceramic coating is highly wear-resistant, thus enabling the corners of the grooves in the hitting surface to remain sharp and maintaining driving power imparted to a golf ball during extended use. Coating golf club heads presents certain difficulties, particularly the distortion of the bore in the club head at a relatively high temperature (~500 °C) during the coating process. This problem can be effectively addressed by inserting and supporting the club head hozzle bore with a carbon pin during the PVD process.

In 1999, Diversified Technologies in USA introduced a new surface engineering technique called plasma immersion ion processing (PIIP) to coat golf club heads with a diamond-like coating (DLC).[48] A patent (US Pat. No. 2002/0004426) has recently been granted to Lin et al.,[49] which disclosed a PVD technique used to apply DLC on golf club heads. The main advantages of DLC over TiN as a coating for golf club heads are as follows. First, the friction coefficient of DLC is measured between 0.05 and 0.1, which is much lower than that for TiN. The extremely low friction or slickness of the DLC-coated face of the club may help further reduce the side spin imparted to a golf ball as it is hit. Second, DLC coatings are very hard (~3000–4000 HV), thus making the treated club heads extremely scratch resistant and maintaining a shining attractive appearance indefinitely. Third, DLC coatings can be deposited at a relatively low temperature (as low as

200 °C), which makes almost no changes to the shape or dimensions of the club heads. For example, it was reported that 50 titanium drivers were coated with non-hydrogenated DLC coating about 3 µm in thickness using the cathodic arc method. After using the DLC-coated clubs, most golfers felt that their golf performance, in particular, their driving distance (due in part to lower degree of hooking or slicing) had noticeably improved.[49]

High-friction golf club face

It should be indicated that controlling or tailoring the friction between a golf club face and a golf ball includes both reducing the friction for long-distance shots and increasing the friction for short-distance shots. As has been discussed above, high friction between the club hitting surface and the ball will impart large amounts of back-spin, which can promote quick stop or soft landing of the golf ball and contribute to a straighter shot. Some efforts have been made to achieve the desirable high friction between the club head and the golf ball.

Sand or bead blasting has long been used to increase surface roughness and thus surface friction, and to provide cosmetic uniform surface finishing of golf club heads, especially for wedges. However, the durability of the effect of such treatment is poor for the extended use, owing to the wear caused by the impact and sliding of the golf ball, although it is much softer than the hitting surface.

To form a durable high-friction hitting surface and thus to produce more consistent spin control, Carbite Golf has developed titanium/diamond inserts on the face of irons and wedges (e.g. Check-Mate wedge) using a powder metallurgy (PM) approach. Very fine titanium powders and fine diamond particles (of a few microns in size) are blended and then pressed-and-sintered.[50] The hard particles protrude slightly above the hitting surface, thus providing a desirable abrasive club surface and plenty of friction. The inserts have an initial surface roughness of 4 µm, well within the limits set by the USGA. During the lifespan of the club, the insert maintains a high level of surface roughness, since a fresh playing surface is continually uncovered. This insert technology offers a durable hitting surface with the maximum allowable surface roughness. Indeed, the success of the surface insert technology can be seen in its popularity among professional players.

In addition to irons and wedges, Carbite Golf launched a series of high-friction surface woods, beginning with Gyroseven, to achieve good spin control resulting in straighter and softer landing shots. Gyroseven combines a surface titanium nitride treatment with a keel-shape sole and offset head. Carbite Golf has also developed a surface coating process to apply high-friction composite coatings to irons with performance benefits similar to an insert, but less expensive to apply. This surface engineering technology has been applied successfully to an over-size wedge called the Viperbite, which prolongs the high-friction life compared to sand-blasted surfaces commonly used on conventional wedges to impart large amounts of back-spin.[51]

In summary, advanced surface engineering technologies offer great design freedom to tailor (either increase or decrease) the friction between a golf club face and a golf ball, which is required by different skills to enhance performance. Indeed, friction has also played an important role, either positive or negative, in virtually all sports products (such as racing cars, bicycles, skis, roller skates, boats, surfboards, canoes, tennis and squash rackets, sports shoes, swimming clothing, and athletics equipment, to name but a few). Clearly, advanced surface engineering technologies have great potential for tailoring friction on the contact surfaces of sports equipment, and are expected to find more and more wider application in the design and manufacturing of sports equipment.

7.5.2 Case study II: surface-engineered titanium components for motor sport

The driving force

The ever-increasing engineering performance required by motor sports has been the main driver for substituting titanium components for steel components in racing cars and engines. This is because titanium and its alloys possess the highest strength-to-weight ratio of all metallic materials and in nearly all racing environments, lighter weight will translate to higher performance and thus to faster lap times. Weight reduction can be divided into two distinct categories: (a) vehicle weight reduction and (b) reducing reciprocating masses. Vehicle weight reduction allows the designer to redistribute the weight advantageously, thus leading to handling improvements. Reduction in the reciprocating mass of an internal component would allow greater acceleration of the lightened component or assembly. The value of making these moving components lighter is more than the inherent weight reduction. For example, in the valve train system, small weight reductions can bring a significant increase in the toss speed or the upper limit of the engine speed, hereby leading to high engine performance.[52]

The technical barriers – poor tribological behaviour

Technically, titanium alloys are characterised by poor tribological behaviour, in terms of high and unstable friction, severe adhesive wear and strong tendency to scuffing. Therefore, when rubbing against most engineering materials, especially under pure sliding conditions, titanium alloy surfaces would be immediately damaged and transferred to the counterface owing to its inherent high chemical activity. This is determined by its electron structure, its rapid growth of the real contact area due to its crystal structure, and the complete ineffectiveness of conventional lubricants.[53] Clearly, the bad reputation of titanium alloys for poor tribological behaviour has been the main barrier to their successful application in the motor sports industry.

The solutions

To realise the great potential of titanium alloys in motor sports, the tribological limitations of titanium alloys, which are closely related to their inherent surface nature, need to be addressed. Surface engineering has proved to be a most promising way to enhance the surface-related performance of titanium alloys. Developments in surface engineering of titanium alloys have targeted tribological property improvement, which provides titanium designer surfaces with technically enhanced performance. Some novel surface engineering technologies have been developed recently, which provide the necessary basis for realising the full potential of titanium alloys in achieving high racing performance. The present case study demonstrates, by way of example, a number of major steps towards titanium designer surfaces.

Thermal oxidation (TO) treatment of racing engine components

As discussed in Section 7.4.3, the TO treatment can effectively enhance the tribological properties of titanium alloys as evidenced by significantly reduced friction and effectively enhanced wear resistance under light to moderate loads. As a result, this TO treatment has been applied successfully to many types of titanium engine components, especially valve train components. All these components have successfully passed trials on a cam rig, which comprises of a dummy crank driven by an electric motor, complete heads and valve train, gear chest and full oil system, i.e. the top half of the engine. The cam rig is used for endurance running of new designs and batch testing of critical valve train components. The rig is programmed to simulate race conditions taken from a rigorous race lap. Consequently, one British racing engine specialist routinely uses the surface-engineered titanium engine components, and up to now more than 25 000 components have been used, which have significantly enhanced the competitiveness of its racing products in the global racing engine market.

Duplex treatment of titanium gears

Notwithstanding the fact that thin PVD ceramic coatings have been proven to be successful to protect steel from wear, they are not so effective when applied to titanium alloys. Catastrophically premature failure will occur, resembling the so-called 'thin-ice effect' (Fig. 7.8),[54] when the soft substrate plastically deforms under the applied load. Although the novel TO processes are quite effective in enhancing the tribological properties of titanium and its alloys, it should be pointed out that the load-bearing capacity of TO-treated titanium alloys is not high enough to withstand the high stresses encountered in such transmission components as gears in racing cars in view of its relatively shallow hardened case. Therefore, deep-case hardening is necessary to provide high load-bearing capacity.

However, surface engineering practice with titanium has also indicated that most single surface engineering processes, which are widely used commercial processes for ferrous materials (such as induction hardening), are not so successful when applied to titanium and its alloys. This arises because no significant hardening effect can be achieved, as with steel, by quenching; rather softening of titanium alloys takes place when quenched martensitic structures are formed. Laser surface alloying by SiC demonstrates high Hertzian strength but poor friction behaviour.

A new oxygen diffusion (OD) process is being developed for deep case hardening of titanium and its alloys,[55] i.e. equivalent of carbon deep case hardening of ferrous materials. This family of processes involves the controlled oxidation of a titanium alloy to produce a dense, adherent oxide layer on the surface, followed by a high temperature vacuum diffusion treatment of the oxygen from the layer into the substrate. An effective hardened case of about $300\,\mu m$ can be achieved (20 times that of TO-treated

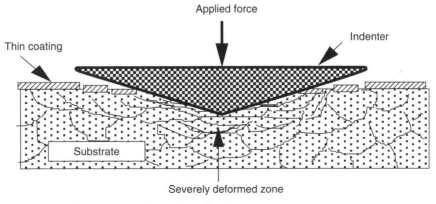

7.8 Schematic of 'thin-ice effect'.

material), which confers the high load-bearing capacity needed for heavily loaded titanium components. To reduce the surface friction novel duplex systems combining a low-friction, high wear-resistant outermost surface layer (such as diamond-like coating or TiO_2 rutile) with OD deep case hardening have been designed. The load-bearing capacity trials on the duplex-treated, as well as DLC-alone coated, lubricating trials on Ti6Al4V have demonstrated more than five-fold improvement in load-bearing capacity following novel duplex treatment, combining OD treatment with DLC coating. Another duplex treatment, OD treatment followed by TO treatment, also showed excellent load-bearing capacity.

Stimulated by the very promising laboratory results, the novel duplex systems have been applied to cam gears used in racing engines. The cam gear specifications are given in Table 7.4. Three gears were given an OD treatment and two of them were then TO-treated. The gears were tested on a cam rig comprising a dummy crank driven by an electric motor, complete heads and valve train, gear chest and full oil system. The rig is programmed to simulate race conditions taken from a rigorous race lap.

The single OD-treated gear failed (Fig. 7.9), leading to severe damage of both meshing surfaces. This was to be expected since while the OD treatment produces a significantly hardened depth of surface case, there is no change in coefficient of friction and no change in the surface energy of the titanium alloy, and accordingly it retains its poor tribological behaviour with conventional lubricants. By contrast, a duplex-treated (OD + TO) cam gear was tested at speeds up to 17000 rpm, at full load for the equivalent of 100 laps. The gear survived (Fig. 7.10) and its performance was considered as satisfactory. The testing conditions utilised while running this cam gear within the cam rig represent a major leap forward within engineering. Inspired by such an encouraging achievement, a much higher loaded (more than 1000 Nm) titanium main drive gear for Formula One racing car has been designed and survived gear rig tests, although further development

Table 7.4 Technical specification of the cam gear

Parameter	Specification
Modulus	2.1167 mm
Load	30–40 Nm for both the driver and the driven
Face width	9.2 for both the driver and the driven
Pressure angle	$\alpha = 20°$
Speed	the driver 8250 rpm and 8708 rpm for the driven
Lubrication	oil (splash)

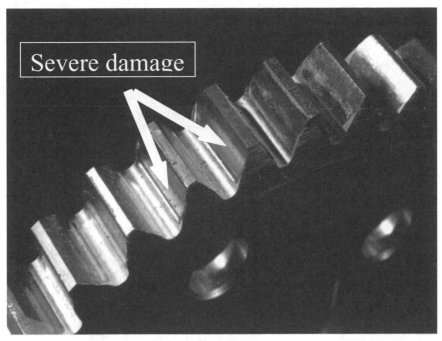

7.9 Severely damaged single OD-treated Ti6Al4V cam gear.

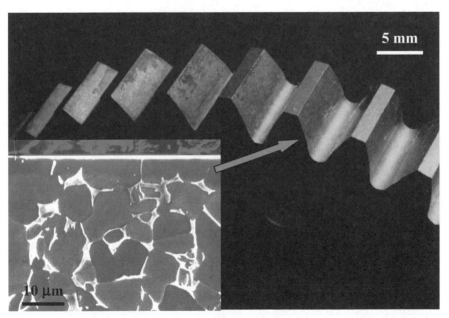

7.10 Test-survived duplex-treated (OP + TO) Ti6Al4V cam gear (the insert showing no plastic deformation of the near surface region along the pitch line).

will be needed before it can be put in service under the demanding conditions of a Formula One racing circuit.

7.6   Summary

Surface engineering is a wide-ranging, enabling technology, which embraces numerous surface coating and surface modification techniques and processes, and is capable of producing diverse combined surface properties to meet the challenging design requirements of advanced sports products. In addition, surface engineering has the ability to supply added value and thus add profit to advanced sports products. To date, surface engineering technologies have found several applications in the sports equipment industry and have produced a strong influence on sport equipment in terms of design changes and economic benefits.

Looking to the future, limitations to the further advance of the sport equipment industry are more likely to be surface related, and surface engineering can undoubtedly be instrumental in creating sustained competitive advantages to the sports equipment industry by generating superior sports products. The techniques and applications summarised in Table 7.5, which are by no means exhaustive, demonstrate the many areas where surface engineering will have an underpinning role in the sports equipment industry in the future.

7.7   Acknowledgements

The author would like to thank his colleague Professor T. Bell for his help in reviewing and preparing the manuscript. The author is also grateful to Blackwell Science Ltd for their kind permission to reproduce some photographs/figures from a paper by H. Dong, T. Bell and A. Mynott (1999) entitled: Surface engineering of titanium alloys for the motorsports industry published in *Sports Engineering*, **2**, 213–219.

Table 7.5 Some future applications of surface engineering in sport

Function	Surface engineering technology	Application
Friction control	Polymer coatings	Boats, canoes
	Self-lubricating coatings	Roller skates, golf club heads
	Ion beam modification of polymers	Skis, grip tapes, surfboats, snow board
Wear protection	Nano composite coatings	Roller skaters, ice skates
	Thermal oxidation of Ti surfaces	Racing car and engine components
	Oxygen diffusion of Ti	Ti transmission components

7.8   References

1 Easterling K E, *Advanced Materials for Sports Equipment*, London, Chapman & Hall, 1993.
2 Dong H, Bell T and Mynott A, *Sports Engineering*, 1999, **2**, 213–219.
3 Blau P J, 'Glossary of terms', in ASM International, *ASM Handbook*, Vol. 18, *Friction, Lubrication and Wear Technology*, Ohio, The Society, 1–21.
4 Haweky R, *Sport Science*, London, Hodder & Stoughton, 1981, 53.
5 Cox R W, *The Science of Canoeing*, Cheshire, Coxburn Press, 1992, 19.
6 Easterling K E, *Advanced Materials for Sports Equipment*, London, Chapman & Hall, 1993, 97.
7 Tabor D, 'Wear – a critical synoptic view', *Proceedings of International Conference on Wear of Materials*, New York, ASME, 1979, 1.
8 Kreighbaum E F and Smith M A, *Sports and Fitness Equipment Design*, Leeds, Human Kinetics, 1996, 65.
9 Tyrkiel E and Dearnley P, *A Guide to Surface Engineering Terminology*, The Institute of Materials, London, 1995.
10 Bell T, 'Surface engineering: a rapidly developing discipline', *European Journal of Engineering Education*, 1987, **12**(1), 27–32.
11 Bell T, 'Surface engineering: past, present, and future', *Surface Engineering*, 1990, **6**(1), 31–40.
12 Robert W H, 'Surface engineering and tribology in general engineering', in Morton P H, *Surface Engineering and Heat Treatment – Past, Present and Future*, London, The Institute of Metals, 1991, 12–42.
13 Gossner J P, 'Painting', in Cotell C M, Sprague J A and Smidt F A, *ASM Handbook* Vol. 5, *Surface Engineering*, ASM International, 1994, 421–447.
14 Oil and Colour Chemists' Association, *Surface Coatings*, Vol. 2, *Paints and Their Applications*, Randwick Australia, Tafe Education Books, 1987, 735–739.
15 Levinson S B, *Application of Paints and Coatings*, Federations of Society for Coatings Technology, Philadelphia, USA, 1988, 41–45.
16 Publicity Services J C, *Product Knowledge in Sport: A Guide from the FSGD*, Federation of Sports Goods Distribution Ltd, 22–27.
17 Independent Fabrication: http://www.ifbikes.com/how2/shotpeen.html.
18 Henley V F, *Anodic Oxidation of Aluminium and its Alloys*, Oxford, Pergamon Press, 1982.
19 Shnezhro I A, Tikhaya I S, Udovenkp Y E and Chernenko VI, 'Anodic spark deposition of silicates with AC', *Protection of Metals*, 1991, **27**, 346–350.
20 Wang Y K, Sheng L, Xiong R Z and Li B S, 'Study of ceramic coatings formed by microarc oxidation on Al matrix composite surface', *Surface Engineering*, 1999, **15**, 112–114.
21 Yerokhin A L, Nie X, Leyland A and Matthews A, 'Characterisation of oxide films produced by plasma electrolytic oxidation of a Ti-6Al-4V alloy', *Surface and Coatings Technology*, 2002, **130**, 195–206.
22 Isle Coat, company promotion brochure.
23 Mavic: http://www.mavic.com/servlet/srt/mavic/vtt prod_fiche?produitsid=59&technosid=17&lg=uk.
24 White J, Racing into production, *Materials World*, October 1997, 578–579.
25 Private communication with Dr M Cope of Ilmor Engineering Ltd, 2002.
26 Private communication with Dr D Teer of Teer Coatings Ltd, 2002.

27 Ohlins: http://www.ohlins.com.
28 Chester G W, 'Advanced coatings for high performance applications', in Austin K, *Surface Engineering Handbook*, London, Kogan Page, 1998, 29–31.
29 Private communication with Dr A. Bloyce of Balzers Limited, 2002 and their promotion materials.
30 Metal Improvement Company, Inc., *Shot Peening Applications*.
31 Morgan J E, 'The recurrent failure of modern cycle components', in Haake S J, *The Engineering of Sport*, Blackwell Science Ltd, Oxford, 1998, 153–162.
32 Sayers A T, Koumbarakis M and Sobey S, 'Surface hardness of cricket bats following 'knocking-in''', in Subic A J and Hake S T, *The Engineering of Sport: Research, Development and Innovation*, Blackwell Science Ltd, Oxford, 2000, 87–94.
33 Glossary of terms, *ASM Handbook*, Vol. 5, *Surface Engineering*, 944–973.
34 Independent Fabrication: http://www.ifbikes.com/how2/painting.html.
35 Zhang Z L and Bell T, 'Structure and corrosion resistance of plasma nitrided stainless steel', *Surface Engineering*, 1985, **1**(2), 131–136.
36 Bell T and Sun Y, Process for the treatment of austenitic stainless steel articles, US patent (US6238490), 2001.
37 Dong H and Bell T, unpublished work.
38 Dong H, Bloyce A and Bell T, 'Surface oxidation of a titanium or titanium alloy article', US Patent 6210807 (April 2001).
39 Dong H, Bloyce A, Morton P M and Bell T, 'Surface engineering to improve tribological performance of Ti-6Al-4V', *Surface Engineering*, 1997, **13**, 402–406.
40 Dong H and Bell T, 'Designer surfaces for titanium components', *Industrial Lubrication and Tribology*, 1998, **50**, 282–289.
41 Olaineck C and Luhrs D, 'Economic and technical features of laser camshaft remelting', *Heat Treatm of Metals*, 1996, 17–19.
42 Birmingham Surface Engineering Group, Unpublished work.
43 Strangwood M, 'Materials in golf', Chapter 6 above.
44 Ekstrom E A, 'Experimental determination of golf ball coefficient of sliding friction', in Cochran A J and Farrally M, *Science and Golf II*, Hertfordshire, UK, Aston Publishing Group, 510–518.
45 Shira C and Froes F H, 'Titanium golf clubs', in Froes F H, Allen P G and Niinomi M, *Non-Aerospace Applications of Titanium*, The Minerals, Metals & Materials Society, 1998, 331–343.
46 Sharpe G D, 'Systems for altering the coefficient of friction between a golf club face and a golf ball', US Patent No 5885171, 1999.
47 Buettner D, 'Coated golf club and apparatus and method for the manufacture thereof', US Patent No 5531444, 1996.
48 http://www.divtecs.com/PRESS/GolfClubsResistWear41399.htm.
49 Lin F S, Pai Y L and Sung C-N, 'Diamond-like carbon coated golf club head', US Patent No 2002/0004426, 2002.
50 Shira C and Froes F H, 'Advanced materials in golf clubs: the titanium phenomenon', *JOM*, May 1997, 35–37.
51 Shira C and Froes F H, 'Advanced materials in golf clubs', in Subic A J and Haake S J, *The Engineering of Sport*, Oxford, Blackwell Science Ltd, 2000, 51–59.
52 Allison J E, Sherman A M and Bapna M R, 'Titanium in engine valve systems'. *Journal of Metals*, 1987, **39**(2), 15–18.

53 Dong H and Bell T, 'Towards designer surfaces for titanium components', *Industrial Lubrication and Tribology*, 1998, **50**, 282–289.

54 Tyrkiel E and Dearnley P, *A Guide to Surface Engineering Terminology*, The Institute of Materials, London, 1995.

55 Dong H, Bloyce A, Morton P H and Bell T, 'Methods of case hardening', European patent (EP1000180), 2001.

8
Materials and tennis strings

R. CROSS
University of Sydney, Australia

8.1   Introduction

The literature on the physics and engineering of tennis is concerned mainly with the behaviour of the racquet frame, the dynamics of the interaction between the racquet and the ball, aerodynamics of the ball, and the interaction between the ball and the court. The strings tend to be somewhat incidental and are commonly treated simply as a stiff elastic membrane that dissipates little or no energy during an impact with the ball. Tennis players themselves also tend to be more interested in the properties of racquets, balls and courts than in the properties of their strings. Most players show an interest in their strings only when they break a string, in which case their primary concern is usually the cost and the inconvenience of a restring. Given that most recreational players break or replace strings only once every few years, tennis strings are not a major concern for this group of players.

Elite tennis players take a much greater interest in their strings. Top professional players can spend around $500 per day during a tournament just on tennis strings. A professional player may bring five or six racquets to each match, and each racquet is freshly restrung just hours before each match so that they are all at the same tension and so that they feel like new. It helps that sponsors pay for the racquets and strings when the player has a world ranking in the top 200. Some professionals will change racquets during a match at each ball change to ensure that the strings do not break. Other players with a world ranking in the top 1000 are not as extravagant, but they too will typically have three or four match racquets and a few practice racquets which are restrung every few days during a tournament. They tend to break strings at least twice a week and they do not like losing a point by breaking a string in the middle of a rally. The cost and durability of a string is therefore a major concern to these players. Clay court players break strings more frequently since sand gets into the strings and since players use more topspin, which causes the strings to rub against each other.

8.2   String types

Natural gut is still the preferred string of most of the top professionals but it is generally too expensive for players with a world ranking below the top 200. Natural gut strings are made from the intestines of cows. The intestines of three cows are needed to string one racquet. The intestine of a cow is long enough to string three racquets but the part of interest, the serosa, is very thin. Pig or sheep serosa is used to make sausage skins. Cats have never been used to make tennis strings or any other strings despite the fact that the term cat-gut is still used to describe the best quality violin strings. In this respect, the word cat has an obscure origin probably derived from the name of an ancient musical instrument or the town in Germany where gut was made. Making a string from a cow is a labour-intensive process, with the result that it costs about $80 to string a racquet with natural gut, compared with about $30 for a cheap nylon string.

These days, the most popular type of string by far are strings made from nylon. Elite players don't particularly like nylon strings, but the rest of the tennis players in the world buy more nylon strings than all other varieties put together. There are hundreds of different nylon varieties, depending on construction techniques, coatings, texture, number of filaments and string gauge. Manufacturers tend to make extravagant claims about the performance, durability, playability, softness and power of each different string, but laboratory tests show that all nylon strings have similar physical properties, especially when compared with other string materials. In order of increasing stiffness, the most common string materials are natural gut, zyex, nylon, polyester and kevlar. Nylon is commonly called 'synthetic gut'. Zyex strings are relatively rare. There is a factor of about seven difference in the stiffness of kevlar and natural gut strings of the same diameter.

8.3   The function of strings in a racquet

The primary role of the strings in a racquet is to reduce the impact force on the hand and arm, in the same way that jumping on a trampoline feels nicer than jumping on concrete. A thin elastic membrane would also achieve this result but it would be harder to swing a racquet with a solid membrane due to the increased drag force through the air. The transverse stiffness of the string plane is typically about 20 kN/m or about the same as the stiffness of a tennis ball. The stiffness of the string plane is proportional to the string tension, being proportional to the total number of strings in the racquet and it is inversely proportional to the length of the strings. String plane stiffness increases dynamically during each ball impact, by an amount that depends on the stiffness of the strings in the longitudinal direction. For this reason steel strings are not suitable for use in a racquet. They would also cut the ball to pieces.

Tennis strings vary in diameter from about 1.2 to 1.4 mm and are strung at a tension typically between 25 kg (55 lb or 245 N) and 30 kg (66 lb or 294 N). These are nominal tensions in the sense that the tension is measured or set in the stringing machine by pulling on each string as it is installed, but the actual tension in a strung racquet is always less than the nominal value, sometimes by as much as 10 kg (Cross and Bower, 2001). The primary reason is that all strings are viscoelastic, with the result that the tension decreases with time immediately after it is tied to the frame. The tension decreases rapidly at first, typically by about 4 kg in the first 20 minutes, which is about the time it takes to string a raquet. Most players and racquet stringers are unaware of this effect since it is an effect that can be observed properly only in the laboratory. Nevertheless, elite players usually specify their preferred string tension to within 1 kg.

Professional players generally prefer to string their racquets at high tension, while recreational players generally prefer a lower tension. The reason is partly due to the fact that professionals consistently hit the ball in the middle of the strings while recreational players are not as consistent and often hit the ball towards the edge of the frame. The middle of the strings is the sweet spot region containing the vibration node and the centre of percussion, each a few cm apart (Brody, 1987; Cross, 1998). An impact at the vibration node results in no vibration of the frame. An impact at the centre of percussion generates minimum shock in the hand and arm. Consequently, professionals are not overly concerned by the amount of shock and vibration in a racquet, while recreational players notice it more. Shock and vibration are both reduced by reducing the stiffness of the string plane and by impacting the ball in the sweet spot region.

Racquet manufacturers advise players to string at high tension for improved ball control or at low tension for increased power. Both of these effects are relatively small and are probably not detectable by the average player. Decreasing the tension from 60 lb to 50 lb adds about 0.7% to the serve speed and about 2% to the speed of a groundstroke (Cross, 1999). Increasing the tension acts to decrease the impact duration and hence the racquet frame will rotate through a smaller angle during the impact with the ball. The frame rotates in several different directions during the impact. For example, an impact near the top of the frame in a forehand stroke produces a torque about the long axis, resulting in a shot that passes higher over the net than intended. This effect can be reduced slightly by increasing the string tension. A more significant reduction is achieved by using a racquet with a large moment of inertia about the long axis. Such a racquet typically has a large head but the price to pay is that the stiffness of the string plane is reduced (at any given string tension) as the head size is increased. A consequence is that recreational players tend to prefer light racquets with a large head but professionals prefer heavier, more powerful racquets with a relatively small head.

8.4   Frame stiffness

Shock and vibration in a racquet can be further decreased by increasing the frame stiffness (Cross, 2000b; Brody *et al.*, 2002). It is partly for this reason that all modern racquets are made from graphite or include a high proportion of graphite. Wood and aluminium racquets are now obsolete. Graphite composite materials allow the manufacturer to produce a wide variety of racquet frames, all of which are stronger, lighter and stiffer than old wood racquets and all have a larger size head. It is often claimed that modern racquets are also more powerful, but this is not correct. Racquet power increases with racquet weight at any given racquet speed. The advantage of modern racquets is that they allow the player to swing the racquet faster without losing control of the ball. The result is that the modern game of tennis is played at a significantly faster pace than in the wood racquet era.

Professionals tend to avoid very stiff, wide-body frames since the effective string area is reduced for oblique impacts and since vibration of the frame provides useful and instant feedback on whether they hit the ball cleanly. If a player attempts to return a ball with topspin, then he or she needs to ensure that the ball is incident obliquely on the strings rather than at right angles. This is most commonly achieved not by tilting the racquet head but by swinging the racquet in an upwards arc to impact the ball. A wide-body frame is typically about 30 mm deep in the direction perpendicular to the string plane, resulting in a frame that is about three times stiffer than a 20 mm wide frame. A ball incident obliquely may clip a wide-body frame before striking the strings or as the ball rebounds off the strings.

Frame stiffness can be tested qualitatively by tapping the frame with one finger and listening to the handle vibrations with the handle close to one ear. It helps to hold the handle lightly at the vibration node about 16 cm from the end to avoid damping the vibrations. Old wood racquets vibrate at about 85 Hz. Wide-body graphite racquets vibrate at about 180 Hz. The racquets used by professionals vibrate at about 140 Hz. A frame that vibrates at a frequency above 200 Hz does not vibrate to any significant degree at all when used to hit a tennis ball since the impact duration is too long. The impact duration is typically about 5 ms. Such an impact will excite vibrations at frequencies up to about 200 Hz but the frequency spectrum of the impact does not extend much beyond 200 Hz. Higher frequency modes are effectively damped. For that reason, the only significant vibration mode in a tennis racquet is the fundamental mode.

8.5   Laboratory testing of tennis strings

There are no rules in tennis concerning the physical properties of tennis strings and there are no standard methods of testing them. This contrasts with the rigid specifications for tennis balls. The only rules of significance concerning the strings are that (a) they must be woven or interlaced in a single plane and (b) no object can be placed in the hitting area. The first rule prevents players from imparting excessive spin to the ball. Strings that are not interlaced can stretch within the string plane and impart very high spin to the ball when they spring back into their normal position. The second rule has a similar effect but it allows players to insert a string damp-ener in the strings near the frame. This changes the ping sound to a dull thud but it does nothing else. It has no effect on frame vibrations or on racquet power or on any other aspect of racquet performance. Neverthe-less, some players refuse to play without them and some refuse to play with them, presumably because it is what they are used to. The sound of the impact is an important psychological factor but string dampeners have no other known physical significance apart from the fact that some players can feel a slight tingling sensation through the handle as a result of string vibra-tions persisting well after the initial impact is over.

The author has conducted many different tests on many different strings, partly in an attempt to discover what physical properties might be impor-tant to a player. The tests were devised originally simply to determine the main properties of strings so that more realistic models of the ball–racquet interaction problem could be developed. In the process it was discovered that tennis strings have more properties than was originally thought possi-ble. The most surprising property is than no string can ever be tested twice with the same result. Once a string is stretched, it never regains its original properties. The best one can do is to retest another sample of the same string. This will usually generate the same result but it depends on the uni-formity of the string along its length. Fortunately, manufacturing techniques are usually sufficiently good that this not a major concern.

Experienced stringers can easily pick the difference between a 1.25 mm diameter string and a 1.30 mm diameter string just by eye. String diameter is usually controlled to within about 0.02 mm but there is no rule in tennis that says it has to be. It is simply a rule of marketing that the best strings on the market and those that are marketed best will sell the best. The people with the greatest influence on marketing are not the manufacturers but the racquet stringers since they are the ones who deal directly with their customers.

8.6   Quasi-static stretch tests

If a string is stretched by applying a fixed load, the amount of stretch increases while one is attempting to measure it. This is a property of all viscoelastic materials and is described as creep. Under a constant load a string will stretch forever but the rate decreases with time. There are several ways to minimise this problem. One is to stretch the string at a slow, constant speed in a materials testing machine. Another and more relevant approach is to stretch it very rapidly before it has a chance to creep very far. A third method is to stretch it slowly by a given amount and then to measure the amount it 'unstretches' when the load is reduced. The advantage of the third method is that it generates a result that is almost the same as a rapid stretch (Cross, 2001d). Even though the rate of creep is rapid during the initial stretch, it is much reduced during an unstretch cycle since the high initial load acts to break most of the low energy bonds in the material. Creep in polymer materials is primarily due to slippage and untangling of long chain molecules followed at an exponentially decreasing slower rate by breaking of higher energy bonds.

Typical results of slow stretch and unstretch cycles for the four main string materials are shown in Fig. 8.1. Unlike a metal spring, the loading and unloading curves for a tennis string are always non-linear and different, and

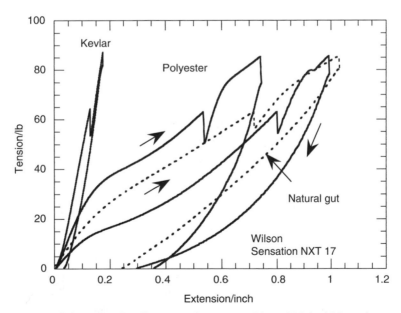

8.1 Loading and unloading stretch tests on 20 cm (7.9 inch) length samples of four different tennis strings. The Wilson string is a popular multifilament nylon string.

the string remains permanently stretched after unloading. Kevlar is clearly much stiffer than nylon or natural gut. At tensions above about 40 lb all strings become stiffer the further they are stretched. Natural gut stiffens the least. At tensions up to about 60 lb, nylon strings stretch further than natural gut but this is of no relevance to the performance of a tennis string. The relevant factor is the additional stretch when a ball impacts the strings. In this situation, natural gut is a softer string since it stretches further for any given increase in tension above that used to string the racquet. In practice, the impact of a ball of at any given speed stretches all strings by roughly the same amount. It is the increase in tension during the impact that varies the most. For a very fast serve, the tension in natural gut increases by about 20 lb before dropping back to its pre-impact value. For a kevlar string the increase in tension may be as high as 100 lb during an impact. A steel string would increase in tension by over 300 lb during a fast serve.

The test results in Fig. 8.1 were obtained using a materials testing machine programmed with the following three steps.

(a)   Stretch the string at a rate of 50 mm/min until the tension reaches 63.0 lb (280 N), then hold the string at this length for 100 seconds. During the 100 seconds the tension dropped by several lbs depending on which string was tested. This step simulates what happens when a string is installed in a racquet. It takes about 20 minutes to string a racquet but the biggest drop in tension occurs during the first 100 seconds.

(b)   Stretch a bit further at 50 mm/min until the tension reaches 85.5 lb (380 N), then hold at the new length for 10 seconds. During the 10 s wait, the tension dropped a few more lbs. This step simulates hitting a ball a few thousand times. Each time a ball is hit, the strings stretch and the tension increases to about 80 lb or more, depending on the type of string and how hard the ball is hit. When a ball is hit, the tension might start at 60 lb (the pull tension), rise to 80 lb and fall to 59.98 lb, due to the fact that a few bonds will break during the impact. One hit lasts only 0.005 s. But, if a ball is hit 2000 times, then the tension will drop by a total of a few lbs.

(c)   Decrease the tension back to zero at 100 mm/min. The last step is significant since one can measure the decrease in length without any significant creep effects occurring. A convenient and standard way to do this is to measure the decrease in length when the tension drops from say 70 lb to 50 lb. If one measured the change in length from, say, 40 lb to 20 lb, then the answer would be different and it would not be very relevant since strings are used in the range from about 50 lb to 70 lb or a bit higher. For example, suppose that the extension drops from 1.002 inch at 70 lb to 0.915 inch at 50 lb. The dynamic string

Table 8.1 Typical dynamic stiffness values (20 cm length string)

String	Type	Diameter (mm)	k (lb/in)
Babolat VS Power 17	Gut	1.24	115
Babolat VS Power 16	Gut	1.29	125
Tecnifiber NRG2 17	Nylon	1.23	220
Tecnifiber NRG2 16	Nylon	1.34	215
Prince DNA Helix 17	Nylon	1.25	230
Wilson Xtreme Control 17	Nylon	1.23	239
Wilson Sensation NXT17	Nylon	1.24	244
Wilson Syn Xtreme Gut 17	Nylon	1.25	260
Wilson Polylast 16	Polyester	1.28	397
Prince Control Freak 16	Kevlar	1.27	629
Gamma Infinity 15L	Kevlar	1.38	797

stiffness, k, is the change in tension divided by the change in length. In this case, $k = (70 - 50)/(1.002 - 0.915) = 230$ lb/in, which is the result obtained for the Prince DNA Helix string listed in Table 8.1.

A representative set of results for dynamic stiffness is given in Table 8.1. Longer strings are proportionally less stiff since doubling the length of a string is equivalent to connecting two lengths in series and hence the extension is doubled at any given tension. An unexpected result is that large diameter strings are not necessarily stiffer than small diameter versions of the same string. For example, the 16-gauge version of the Tecnifiber string is thicker than the 17-gauge version but it is a softer string in terms of its dynamic stiffness. The 17-gauge version stretches further when starting from zero tension but it gets stiffer more rapidly than the 16-gauge version at tensions above 50 lb due to the higher stress. Similarly, if two identical strings are connected in parallel and tested as a single string, then one strand has a dynamic stiffness that is typically about the same as two strands in parallel. Another surprising result is that a single strand usually creeps at a lower rate than two strands in parallel. This is recognised by players who say that thin strings tend to hold tension better than thick strings.

8.7   Energy loss in a string

The graphs in Fig. 8.1 indicate that strings lose a lot of energy when they are stretched. Any graph that shows force or tension vs. distance can be used to calculate the work done in moving through that distance. The same is true of a string. Work has to be done to stretch a string, and that work is stored as elastic energy in the string. The string gives back some of that energy when it returns back to its original length, but it doesn't give back all the energy since some of it is used up in breaking bonds in the string. The amount of energy lost is represented by the area enclosed by each of the curves. The fattest curves are the polyester curves. They have the biggest area so they lose more energy than other strings. As a result, the drop in tension during the 100 second pause at 63 lb, and also the drop in tension during the 10 second pause is larger than that for any other string. The polyester curves are so fat that about 70% of the stored energy is lost when the tension drops back to zero.

The area of each curve is, in fact, of very little relevance to the situation where a string is used to hit a ball. In that case, the tension rises from about 50 lb to about 80 lb and then back to 50 lb (or maybe 49.98 lb) in only 0.005 seconds. This happens so fast that only a few bonds are broken. As a result, very little energy is lost. The curves in Fig. 8.1 were all obtained at a stretch rate of 50 mm/min. A tennis ball stretches it at 140000 mm/min during a fast serve. Only about 4 or 5% of the stored elastic energy is lost when a string is stretched rapidly by a sudden impact. This is demonstrated easily by dropping a heavy steel ball on the strings of a racquet when the head is clamped. The ball bounces to almost the same height as the drop height, regardless of string tension, string type or age of the strings (Brody, 1995; Cross, 2000a).

8.8   Perception of string properties

A common complaint among top players is that strings lose resilience over time. If one defines resilience as the ability of a string to return to its original length after stretching, then the results in Fig. 8.1 support the players' complaints. However, tennis strings are not used in that way. When a string is installed in a racquet, it stretches and is then tied at that stretched length. It stretches a bit further every time a ball is hit but it always returns to the original strung length after every impact. In that respect, tennis strings are 100% resilient. Furthermore, they return about 95% of the stored elastic energy regardless of the number of impacts. It seems likely therefore that players are actually complaining about the gradual loss in tension over time. Since this results in a softer impact, the racquet will feel different after a few matches and the strings may be perceived as being less lively or not as 'crisp' as a new string. The change in sound of the ball or the strings may also alter a player's perception of string performance.

Elite players also complain that strings feel dead after a few matches, referring to a perceived loss in power. This complaint is most often directed at nylon strings. The tests with a steel ball show that old strings do not lose power. In this case also it appears that players may be referring to the effect on racquet feel of a drop in string tension. A top player once told me that he uses a string dampener because it makes the strings feel stiffer. That is physically impossible but that was his strong impression.

Comments and complaints like these are hard to interpret and make it difficult to correlate physical properties of strings with those characteristics that are preferred by players. Furthermore, blind tests of strings such as those regularly reported in *Racquet Tech* magazine indicate that there is no universal agreement among players as to which strings are best. Most strings tested this way draw comments ranging from 'this is the worst string I have ever used' to 'this is the best string I have ever used'. Such comments indicate that there is no best string for all players, and that the best string for any given player is likely to depend on the playing style and the type of racquet used by that player. Similarly, no two tennis racquets feel the same and no single racquet stands out as being the best for all players.

8.9   Measurements of tension loss and dynamic stiffness

An extensive set of measurements of tension loss and dynamic stiffness has been undertaken using the apparatus shown in Fig. 8.2 (Cross *et al.*, 2000). This apparatus was designed to monitor changes in string tension with time and with repeated impacts, while leaving the clamped length of the string fixed. The string was clamped in metal jaws separated by a distance of 320 mm prior to stretching, and then stretched to a tension of 28 kg (274 N or 61.7 lb). The tension was adjusted manually, by rotating the tensioning nut with a spanner. The tension was measured using a commercial load cell and electronic indicator calibrated to read up to 100 kg with a resolution of 0.01 kg. As well as a digital readout, the indicator had a 0–10 V analogue output, which was connected to a storage oscilloscope and to a data acquisition system to monitor rapid changes in tension during each impact.

The tension was held at 28 kg for 10 seconds by manual adjustment of the tensioning nut. This simulates the procedure commonly used when stringing a racquet, where the string is tensioned for about 10 s before the stringer applies a clamp to hold the string in place. For strings that lost tension slowly, this required a slow increase in elongation in order to maintain constant tension. Some strings lost tension rapidly, in which case it was necessary to stretch the string at a much faster rate to maintain the tension at 28 kg. After this 10 s period, no further adjustment of the tensioning nut was made, and the tension was allowed to decrease for a period of 1000 s (16.7 min). During this time, the tension was recorded at 1 s intervals with the data acquisition system. A plot of tension vs. log(time) is linear after the first 100 s, even over a period of several days.

The clamps used in this apparatus were constructed from a fine file, cut into 38 mm lengths and clamped firmly onto the string with a screw in each corner. While this detail may seem trivial, many other methods of clamping the string were attempted and all failed for various reasons. Small

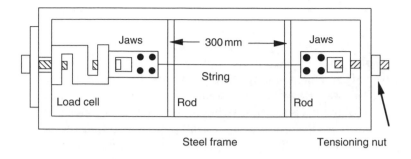

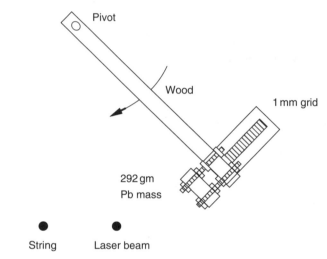

8.2 Apparatus used to measure string tension and the transverse displacement of a string when subject to a hammer impact.

slippages or creep within the clamps are difficult to distinguish from a genuine drop in string tension, making the clamping method crucial.

After allowing the string to relax for 1000 s, it was then subjected to ten impacts at the centre of the string using a hammer incident at 2.63 m/s and at right angles to the string. Two cylindrical rods were located near each of the jaws, as shown in Fig. 8.2, to restrict the transverse motion of the string to a 300 mm region between the rods. This ensured that tension was applied to the load cell along its axis, and it minimised possible damage to the string at the entry point to the jaws. The hammer consisted of a lead block of mass 292 g mounted at the bottom end of a light wood beam. The beam was pivoted at the top end by means of a ball race to minimise friction at the pivot point. The hammer was allowed to swing into the string, through a fixed angle, as a pendulum.

Under normal conditions, the strings of a racquet experience a peak transverse force of up to about 1500 N. Such a force, acting on a ball of mass 57 g over a period of about 5 ms, is required to change its velocity from +30 m/s to −30 m/s. The force is distributed over all the strings, but if one assumes that the brunt of the force is shared mainly by five mains and five cross strings, then the peak force on each string is about 150 N. In the apparatus shown in Fig. 8.2, the total effective mass of the 292 kg lead block, the wood beam and the attached optical grid was 0.45 kg. The hammer changed its velocity from +2.63 m/s to about −2.5 m/s over a period of about 30 ms, giving a peak force between 120 and 200 N on the string, depending on its stiffness.

Since a single string has a much lower transverse stiffness than the strings of a fully strung racquet, it is difficult to simulate, with a single string, the impact conditions encountered during normal use. To maintain the same peak force and impact duration, it would be necessary to impact the string with a projectile of mass much smaller than that of a tennis ball, travelling at a speed much higher than the normal speed of a tennis ball. Alternatively, one can maintain the same peak force using a low-speed hammer of mass larger than that of a tennis ball. In this case, the impact duration is longer than normal, but this has the advantage of simulating the cumulative effect of a number of impacts each of duration 5 ms.

Measurements of the velocity of the hammer and the transverse displacement of the string were made by passing a laser beam through an optical grid attached to the hammer, as shown in Fig. 8.2. The grid consisted of 50 1 mm thick parallel lines separated by 1 mm, photocopied onto an overhead transparency and mounted in a light aluminium frame. Since the beam diameter was slightly larger than 1 mm, the detected laser signal consisted of a series of sinusoidal fringes as described below.

8.10 Tension loss results

Figure 8.3 shows string tension as a function of time for a natural gut and a polyester string initially tensioned to 28 kg. Natural gut has one of the lowest rates of tension loss (some kevlar strings are lower) while all polyester strings lose tension rapidly. The rate at which a string loses tension can be decreased by holding the initial tension at 28 kg for a period of about 30 s, or by pre-stretching the string one or more times. This is generally not a practical proposition considering that a racquet stringer would have to charge an additional hourly rate to string the racquet. Alternatively, the string can be tensioned at a higher initial tension to compensate for tension loss after clamping. The effect of stretching a string to 28 kg for various time intervals, before clamping the string, is shown in Fig. 8.4. The effect is quite pronounced, and it also affects tension loss during subsequent impacts.

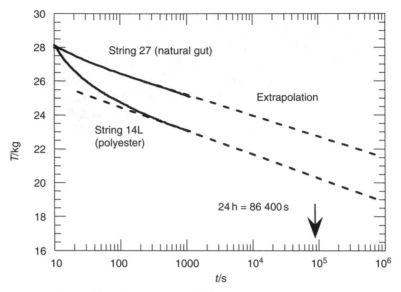

8.3 Tension vs. log (time) for two different strings initially tensioned to 28 kg for 10 s prior to clamping.

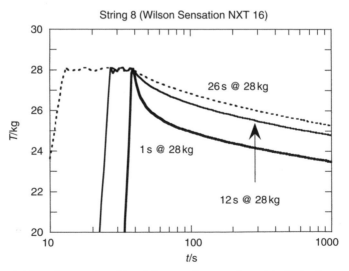

8.4 Tension vs. log (time) for a string tensioned to 28 kg for several different time intervals prior to clamping.

The effect of a series of ten hammer impacts on string tension is shown in Fig. 8.5. The tension rises during each impact, by an amount that depends on the dynamic stiffness of the string. The rise in tension during each impact is not properly recorded in Fig. 8.5 since the data was captured at only one point every 1 s. The tension rises by about 9 kg for natural gut, about 18 kg for nylon strings, about 22 kg for polyester strings and about 45 kg for kevlar strings. The effect is the same as that for a static increase in tension, in that the tension starts to decrease rapidly as soon as the tension is increased. As a result, the tension immediately after an impact is less than that before the impact, particularly after the first impact. Successive impacts result in successively smaller drops in tension. The decrease in tension after each impact therefore depends on the previous history of the string, and it also depends on the magnitude of the impact. If the experiment is repeated at a lower hammer speed or with a lighter hammer, then the net loss in tension after a series of ten impacts is reduced. Conversely, the loss in tension is greater with a heavier hammer.

An interesting effect occurs between impacts, in that there is a slow recovery of tension in most strings. This effect can be seen in Fig. 8.5 for the polyester string. Given that a decrease in tension can be explained by the breaking of weak molecular bonds, then an increase in tension is presumably due to the formation of new bonds. Alternatively, one can describe the

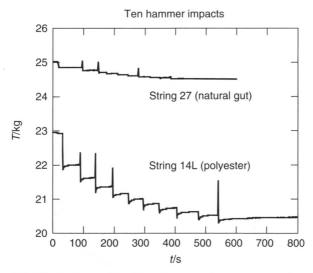

8.5 Effect of a series of ten hammer impacts on string tension for the strings shown in Fig. 8.3. Figure 8.5 is a continuation of Fig. 8.3, the first impact commencing at $t = 1020$ s (redefined as $t = 0$ in Fig. 8.5)

string as having a memory. If the tension is held at, say, 28 kg for a few minutes, and is then decreased quickly to, say, 15 kg and clamped at this value, then the tension will immediately start to rise back towards 28 kg. The tension will settle at a steady value, about 18 kg, when the rate of formation of new bonds is balanced by the rate at which old bonds break.

The rise in tension during an impact, and the corresponding laser signal used to monitor the displacement of the string, is shown in Fig. 8.6. Each new fringe maximum corresponds to an additional y displacement of 2 mm. The y displacement could be measured to within 0.1 mm by counting fractional fringes. The tension does not rise significantly until the string has deflected at least 3 or 4 mm. The tension reaches its peak value a few ms before the string is displaced its maximum distance, since the rate at which the tension drops due to breaking of bonds is a significant fraction of the rate at which the tension increases due to the additional stretch. This is particularly noticeable during the first few impacts and with strings such as polyester where bonds are easily broken. The laser signal in Fig. 8.6 reveals that the rebound speed of the hammer is about 92% of the incident speed. This result was typical of the first few impacts for all strings. After the first few impacts, the rebound speed increased to between 95 and 98% of the incident speed for all strings, indicating that the energy loss in all strings is very small. One nylon string was tested with 200 impacts. Despite a significant loss in tension after 200 impacts, the energy loss in the string during each impact remained neglible.

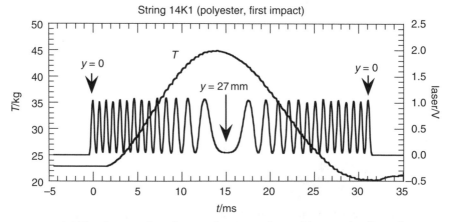

8.6 Tension vs. time for a polyester string subject to the first of ten impacts. Also shown is the corresponding laser signal used to measure the hammer speed and string displacement.

8.11   Impact dynamics

The deflection of a string at its midpoint is shown in Fig. 8.7. If the string has an initial length L_o at tension T_o, and if the length increases to L and the tension increases to T when the string is displaced a distance y at its midpoint, then:

$$L = \left(L_o^2 + 4y^2\right)^{1/2} \tag{8.1}$$

and

$$T = T_o + k(L - L_o) \tag{8.2}$$

where k is the dynamic stiffness of the string. The restoring force, F, acting at the midpoint is:

$$F = 2T\sin(\theta) = 4Ty/L \tag{8.3}$$

and hence the stiffness of the string in a direction perpendicular to the string is given by:

$$k_{\text{perp}} = F/y = 4T/L \tag{8.4}$$

In all cases of practical interest, y is much smaller than L_o, in which case $L \sim L_o + 2y^2/L_o$, $T \sim T_o + 2ky^2/L_o$ and

$$k_{\text{perp}} \sim (4/L_o)(T_o + \{2ky^2\}/\{L_o\}) \tag{8.5}$$

For small values of y, where $2ky^2/L_o \ll T_o$, k_{perp} depends only on T_o and L_o and is independent of k. The perpendicular stiffness would then be same for all strings of the same length and initial tension, even if they were made from steel. Differences in k_{perp} only become apparent for relatively large values of y. Even then, the differences are not as large as one might expect. For example, if a mass impacts on a string with a large value of k, then k_{perp} will be relatively large but the y deflection will be relatively small.

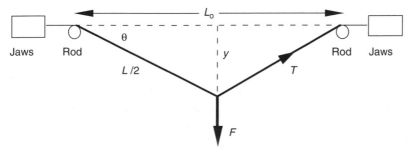

8.7 Geometry of a string deflected by a distance y at the centre.

Inspection of equation [8.5] indicates that k_{perp} is not simply proportional to k and that there is a possibility that k_{perp} could be almost independent of k if ky^2 remains approximately constant. Similarly, k_{perp} could also be independent of T_o if the increase in y at low T_o is large enough. The impact dynamics can be determined by solving the equation of motion for a mass M impacting on the string in the y direction at initial speed v_1. The equation of motion is then given by:

$$d^2y/dt^2 = -F/M = -4Ty/(ML) \qquad [8.6]$$

assuming that there is no energy loss in the string. Numerical solutions of equation [8.6] are shown in Fig. 8.8 for a case where $M = 0.45\,kg$, $L_o = 0.3$ m and $v_1 = 2.63\,m/s$, corresponding to the experimental conditions described above. Results are shown for three different values of T_o and for a range of k values from 10 to 150 kN/m. Of all the strings tested, natural gut had the lowest k, about 20 kN/m, and kevlar had the highest k, about 140 kN/m. For these parameters, all strings deflect between 22 and 33 mm and the impact duration, t, varies from 24 to 36 ms. The increase in tension, DT, is smallest for natural gut and largest for kevlar. As a result, the peak force on the string (or on the hammer) is smallest for gut and largest for kevlar. For any given k, the peak force is almost independent of the initial string tension. The total impulse on the hammer, latex $\int Fdt$, is the same for all impacts since the rebound speed of the hammer is the same as the incident speed in this model. The force waveform is roughly a half sine wave at low k, but is more nonlinear at large k, having a narrower and more peaked or bell-shaped waveform.

Individual test results for about 300 different strings have been published in the *USRSA Racquet Tech* magazine, in February 2002 and June 2003. A summary of the results for the initial set of 90 strings, all 16 gauge (1.30 mm diameter), is given in Figs 8.9 and 8.10. Natural gut strings have a dynamic stiffness $k \sim 20$ kN/m. Polyester strings have k values from about 40 to 60 kN/m, while kevlar strings have k values from about 90 to 140 kN/m. Gut and nylon strings elongated by about 7–15% when tensioned to 28 kg. Polyester strings elongated by about 4% and kevlar strings elongated by only 1 or 2%.

Figure 8.10 shows the net loss in tension resulting from the ten impacts as a function of the tension 1000 s after clamping the string. It is reasonable to assume that a string with strong molecular bonds will lose tension slowly with both time and repeated impacts. The majority of strings fit this description, but kevlar strings are exceptional. Most of the kevlar strings lost tension slowly with time but the tension loss during each impact was larger than that for other strings. The low loss in tension with time is observed at tensions up to 28 kg. The high loss in tension due to each impact is related to the fact that the increase in tension, DT, during each impact is much

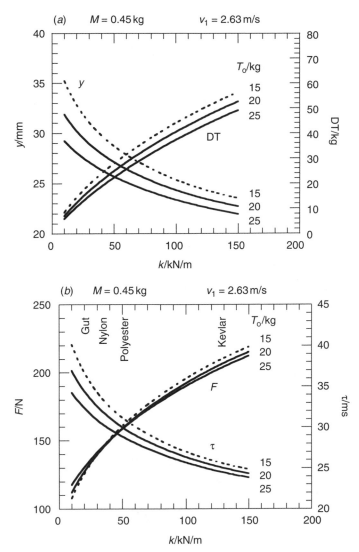

8.8 Numerical solutions of equation [8.6] for strings at an initial tension T_o = 15, 20 or 25 kg, showing (a) the maximum string deflection, y and the increase in string tension, DT (b) the peak force, F and the impact duration, τ.

larger than the rise in other strings. Tests at lower hammer speeds indicated that for any given hammer speed, the impact tension loss in a kevlar string is always larger than the loss in other strings and that the tension loss is roughly proportional to the hammer speed.

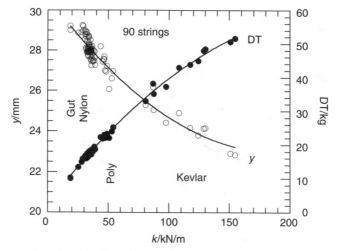

8.9 Results showing the maximum *y* deflection and the increase in tension, DT, vs. dynamic stiffness, *k*, for 90 different strings subject to ten impacts with a hammer. *y*, DT and *k* are all averaged over the ten impacts.

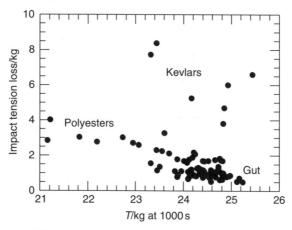

8.10 The total decrease in tension resulting from ten impacts, as a function of the tension recorded 1000 s after clamping, for the same impacts as those in Fig. 8.9. All strings were tensioned to 28 kg for 10 s prior to clamping.

8.12 Coefficient of friction

When a ball impacts obliquely on the strings of a racquet, the rebound angle depends on the coefficient of friction between the ball and the strings, for the same reason that the rebound angle off the court surface depends on whether the court is fast or slow (Cross, 2000b). The coefficient of friction between the ball and the strings has not yet been measured very precisely, but it is likely to depend on a number of factors such as the number and diameter of strings, as well as the surface roughness and composition of individual strings. With topspin or backspin, the ball generally slides along the cross strings and across the main strings. The coefficient of friction for a ball that is dragged along a set of parallel strings glued to a flat surface was measured and found to be significantly lower than the coefficient of friction when the ball is dragged across the same set of strings. It is clear that both coefficients increase as the surface roughness of the strings increase, and both contribute to the total friction force on the ball. Consequently, it was decided to include some simple measurements of string friction when comparing the 90 different strings. The procedure, as described below, measures the coefficient of friction along rather than across the strings.

Measurements of the coefficient of sliding friction between a string and the cloth material of a tennis ball were made by gluing tennis ball cloth to a 60 mm diameter, 150 mm long PVC pipe so that the cloth completely covered the exterior surface of the pipe. The pipe was mounted horizontally so that it could not move or rotate. Two turns of a 1 metre length of string were wrapped around the cloth, a mass $M = 0.15$ kg was clamped onto the lower end of the string, and the upper end was attached to a spring balance. The force required to lift the mass vertically at a constant low speed was measured by the spring balance. The required force varied from 0.6 kg to about 10 kg (98 N), depending on the coefficient of friction between the string and the cloth. The force, F, required to lift the mass is given by:

$$F = (Mg)e^{2\pi\mu N} \qquad [8.7]$$

where $g = 9.8$ m s^{-2} is the acceleration due to gravity, N is the number of turns and μ is the coefficient of sliding friction. Most strings had values of μ in the range 0.15 to 0.18. Some polyester strings had μ values as low as 0.11. A small number of strings had a larger surface roughness or a different texture, with μ values in the range 0.2 to 0.36.

String friction affects not only the rebound angle but it is also likely to affect string tension. When a stringer strings a racquet, one part of the string is held in a clamp and the free end is pulled through grommet holes. Friction between the string and the grommet holes will result in different tensions between the clamped end and the pulled end, as described by equation [8.7] with $N = 0.5$. From this point of view, strings with low μ are likely to be at a more even and higher overall tension, while strings with large μ will provide a better grip on the ball.

8.13   Discussion

The best string to use in a tennis racquet is the one that the player likes best, but many of the top professional players prefer natural gut. Gut is a highly elastic string and it maintains tension better than most other strings. A significant number of top players actually prefer strings such as polyester which have the opposite properties. Polyester is a relatively stiff string, and it loses tension relatively quickly. It appears that the two undesirable properties of polyester might tend to combine to produce a good string. The tension drops rapidly, immediately after a racquet is strung with a polyester string and it drops significantly with repeated impacts. The tension increases substantially during each impact with a ball, due to the high stiffness of polyester. The average tension during an impact with a polyester string may then be similar to that with a gut string. However, even if the average tension is the same, the force and the impact duration are not necessarily the same. As shown in Fig. 8.8(b), the force on a gut string will always be smaller than the force on a polyester string (for the same racquet, the same number and gauge of strings and the same impact speed) regardless of the initial tension. Similarly, the impact duration is usually longer on gut than polyester, although the duration may be almost the same if gut is at quite high initial tension and polyester is at quite a low initial tension.

The magnitude, as well as the duration of the force acting on the strings of a racquet will affect the feel of the strings, and it will also affect ball control. The 'feel' is not a quantity that is measurable with instruments available to scientists, but the physical effects of the force waveform are clear. For example, a large force on the strings will generate a large force acting on the hand, at least for impact points well removed from the sweet spot area of the strings. For an impact at the sweet spot, the force on the hands and the arm is much reduced (Cross, 1999), in which case all strings are likely to feel much the same. If the total impulse (i.e. the time integral of the force) is held constant, then the effect of a large force acting for a short time might be expected to have the same effect as a small force acting for a longer time. However, there are two significant differences. Firstly, a short duration impact will excite high frequency vibrations of the racquet frame more efficiently than a long duration impact (Cross, 2000b). The second difference is that, for a long duration impact, the racquet will rotate through a larger angle while the ball is in contact with the strings. For example, if the average angular velocity of the racquet during the impact is say 20 rad/s, then the racquet will rotate through an angle of 5.7° during a 5 ms impact, or by 6.9° for a 6 ms impact. A change in impact duration will therefore result in a different rebound angle of the ball. Similarly, if the ball impacts the strings at a point towards the edge of the racquet frame, the racquet will rotate about the long axis through the handle, and the ball will be deflected away from the desired trajectory. For a given impulse, the rotation angle is proportional to the impact duration (Cross, 2000b). Consequently, some players may prefer polyester strings since the impact duration is shorter and hence ball control is improved, while others may prefer gut since it has a softer feel. In either case, the energy loss in the string is usually negligible, and there should be almost no difference in the speed of a ball rebounding off gut or polyester strings, despite the slightly enhanced losses in the ball and the slightly enhanced frame vibration losses when a ball impacts on stiff strings (Cross, 2000b). For many players, nylon strings seem to offer the best compromise in terms of feel, control, price and durability. Gut and polyester strings are not noted for their durability.

8.14   Oblique impacts on tennis strings

The performance of a string in a racquet can be measured by firing a ball at the strings and measuring the rebound speed and angle of the ball (Knudson, 1997; Bower and Sinclair, 1999). Manufacturers do this routinely but in most cases the ball is fired perpendicular to the string plane in order to measure the performance of the racquet. Some manufacturers fire balls at an oblique angle in order to measure the durability of the strings when subject to repeated impacts. It is only very recently that the spin has been measured using a high speed video camera. This was done in 2001 by Simon Goodwill at the University of Sheffield as part of his PhD thesis, by firing balls at a clamped racquet at a speed of 29 m/s and an angle of incidence of 40 degrees to the normal. Some of Goodwill's data are shown in Fig. 8.11 and the main conclusions are as follows.

(a)   The rebound spin of the ball does not depend significantly on string tension or string type or string gauge or spacing between the strings.

(b)   The rebound speed of the ball increases slightly as the tension is lowered.

(c)   The rebound angle depends on string tension, string type and the spacing between strings. The ball rebounds at an angle closer to the normal as the string tension is reduced.

All three parameters depend on the spin of the incident ball as shown in Fig. 8.11. A ball bounces off the court with topspin but is incident on a racquet with backspin, typically at around 500 rad/s. In order to determine the behaviour of a ball bouncing off a hand-held racquet, the data shown in Fig. 8.11 needs to be incorporated into a model that takes into account the speed and direction of the racquet, as well as the speed and direction of the ball.

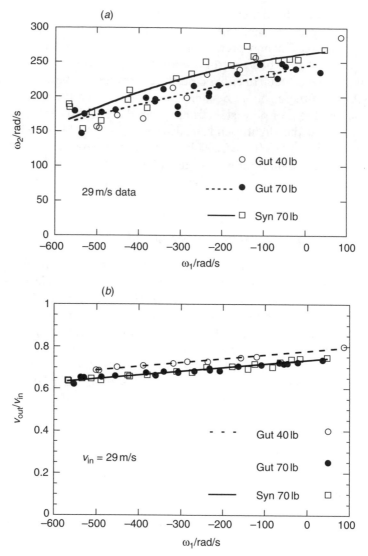

8.11 Rebound spin, speed and angle for a ball incident as 29 m/s and at 40 degrees to the normal vs. incident ball spin. The racquet head was clamped to prevent it rotating. Results are shown for natural gut and nylon (Syn) strung at 40 lb or 70 lb. Data points are omitted in (c) for clarity, as are best fit curves in (a) and (b), θ_2 is the rebound angle to the normal.

8.15   Conclusions

Tennis strings can be categorised by dynamic stiffness, by tension loss with time and with repeated impacts and by the coefficient of friction between

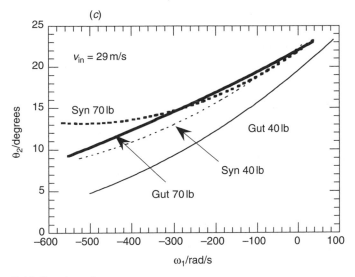

8.11 Continued

the string and the ball. The energy loss in a string is too small to make any significant difference in energy loss between different strings, even after many impacts. There are four broad types of string making up the bulk of strings available on the market. In order of increasing dynamic stiffness, these types are natural gut, nylon, polyester and kevlar. The most common type of string on the market is nylon, which is available in many forms depending on whether it is constructed from a solid core, or whether it contains many small diameter filaments and whether it contains other composite materials or resins or coatings. The different construction methods for nylon account for a small spread in properties, but there is no significant overlap with either natural gut or polyester, which are distinctly different. Kevlar strings are also very distinctive in that they are very stiff and they suffer a larger impact tension loss than other strings. From this point of view, it is difficult to compare one nylon string with another. All nylon strings have similar physical properties. The differences between any two nylon strings are relatively small compared with the relatively large differences between nylon and other types of string.

Tennis players themselves have developed a language of their own to rate strings, using terms such as touch, feel, playability, control, power, comfort, crispness and liveliness. Players respond not to the physical properties of the strings but to the effect of those properties on motion of the racquet. The resulting force, on the hand, represents a combination of rotation, translation and vibration of the racquet handle, and is difficult to describe in common English words. The perception that strings lose resilience or power over time is not supported by laboratory experiments. Further studies will be needed to determine how player perception of string properties relates to the physical properties of a string and to what extent string plane stiffness and impact sound affect the perception of string properties.

8.16   References

Bower R and Sinclair P (1999). Tennis racquet stiffness and string tension effects on rebound velocity and angle for an oblique impact. *Journal of Human Movement Studies*, **37**, 271–286.

Brody H (1987). *Tennis Science for Tennis Players*. University of Pennsylvania Press, Philadelphia.

Brody H (1995). How would a physicist design a tennis racquet? *Physics Today*.

Brody H, Cross R, and Lindsey C (2002). *The Physics and Technology of Tennis*. Racquet Tech Publishing, USRSA San Diego.

Casolo F and Lorenzi V (2000). On tennis equipment, impact simulation and stroke precision. *Tennis Science and Technology* (Ed. Haake, S.J. and Coe, A.O), Blackwell Science, pp. 83–90.

Cross R (1998). The sweet spots of a tennis racquet. *Sports Engineering*, **1**, 63–78.

Cross R (1999). Impact of a ball with a bat or racquet. *American Journal of Physics*, **67**, 692–702.

Cross R (2000a). Physical properties of tennis strings. *3rd ISEA International Conference on the Engineering of Sport*, Sydney 2000 (Ed. by A.J. Subic and S.J. Haake). Blackwell Science, pp. 213–220.

Cross R (2000b). Flexible beam analysis of the effects of string tension and frame stiffness on racquet performance. *Sports Engineering*, **3**(2), 111–122.

Cross R (2000c). Do strings affect power? *Racquet Tech*, September.

Cross R (2001a). String gauges: are thin strings softer or stiffer? *Racquet Tech*, Feb.

Cross R (2001b). Measuring string tension: Part 1. *Racquet Tech*, June.

Cross R (2001c). Measuring string tension: Part 2. *Racquet Tech*, July.

Cross R (2001d). Stretch tests on strings. *Racquet Tech*, September.

Cross R and Bower R (2001). Measurements of string tension in a tennis racquet. *Sports Engineering*, **4**, 165–175.

Cross R, Lindsey C and Andruczyk D (2000). Laboratory testing of tennis strings. *Sports Engineering*, **3**, 219–230.

Knudson D (1997). The effect of string tension on rebound accuracy in tennis impacts. *International Sports Journal*, **1**, 108–112.

Lindsey C (2002). String selector map. *Racquet Tech*, February.

Lindsey C and Cross R (2000a). String testing. *Racquet Tech*, June.

Lindsey C and Cross R (2000b). String test results: what do they mean? *Racquet Tech*, July.

8.17   Further reading and other resources

www.racquettech.com is the home page of the United States Racquet Stringers Association, which publishes a monthly magazine devoted to racquets, strings and stringing techniques. It has about 7000 members.

Calder CA, Holmes JG and Mastry LL (1987). Static and dynamic characteristics of tennis string performance, ITF technical report ITF–L–015.

Cross R (1997). The dead spot of a tennis racquet. *American Journal of Physics*, **65**, 754–764.

Cross R (1998). The sweet spot of a baseball bat. *American Journal of Physics*, **66**, 772–779.

Cross R (1999a). The bounce of a ball. *American Journal of Physics*, **67**, 222–227.

Cross R (1999b). Dynamic properties of tennis balls. *Sports Engineering*, **2**, 23–33.

Cross R (2000a). Effects of friction between the ball and strings in tennis. *Sports Engineering*, **3**(2), 85–97.

Cross R (2000b). Dynamic properties of tennis strings. *Tennis Science and Technology* (Eds S.J. Haake and A. Coe), Blackwell Science, published as the Proceedings of the 1st International Congress on Tennis Science and Technology, London, August 2000, pp. 119–126.

Cross R (2000c). The coefficient of restitution for collisions of happy balls, unhappy balls and tennis balls. *American Journal of Physics*, **68**, 1025–1031.

Cross R (2000d). Tension loss along a string. *American Journal of Physics*, **68**, 1152–1153.

Cross R (2001a). Customising a tennis racquet by adding weights. *Sports Engineering*, **4**, 1–14.

Cross R (2001b). Why bows get stiffer and racquets get softer when the strings are added. *American Journal of Physics*, **69**, 907–910.

Materials and tennis rackets

H. LAMMER AND J. KOTZE

Head Sport AG, Austria

9.1   Introduction

One does not have to go too far back in tennis history to notice the radical differences in racket design and the impact it has made on the way the game is played. Most of us probably still have memories of relentless duels fought between the likes of McEnroe and Borg, with wooden rackets that now look more representative of current squash rackets. Deeper investigation into the archives reveals even more radical design changes, ranging from the more pragmatic to the absolute bizarre. It is interesting, however, that most of the lasting technologies were either related directly to changes in the materials, or to design and manufacturing changes made possible by new materials. Changes in materials affect racket sizes, shapes and weight, which in turn improve stability, comfort and power.

This chapter attempts to show how different materials have assisted in changing the way we perceive the game of tennis at present. A brief time-line, focussing on remarkable models with a lasting impact on designs, is presented, followed by a detailed section on different material properties and the application thereof to benefit designs. The last section describes the current racket manufacturing process as used by most modern tennis factories, highlighting the influences of different materials on manufacturing processes.

9.2   Influence of materials on racket technology

9.2.1   Basic material history

A few people might be surprised to know that the first rackets were human body parts. According to the first records, dating back as far as the twelfth century, tennis was first played with the palm of the bare hand, hence the early name 'le jes de paume'. Subsequently, various gloves, bats and paddles were used before the first wooden construction, with strings, was introduced in the sixteenth century.

In the years following, different types and combinations of wood dominated, until as recently as the mid-1900s. Initially, frames were made of a single solid piece of ash wood, soaked in cold water, boiled to make it pliable and bent into the desired shape while still hot. The wood was not cut, but split and shaved along the grain, producing continuous fibre propagation along most of the racket frame, requiring careful selection of suitable wood specimens (Clerici, 1976; Robertson, 1974; Kuebler, 1995).

At first, these rackets were very weak in the throat area but failures were reduced by wrapping combinations of canvas, vellum and bindings around the critical areas. Another problem was warping of the frame when it was exposed to wet conditions. This was improved using hickory and strips of metal reinforcement in the throat. The next advance was in the 1930s, with the development of laminated frames consisting of an arrangement of the layers at different angles, hence achieving directional stiffness. Synthetic cements and formaldehyde were used to bond the odd layered frames, which consisted of up to 11 layers. Additionally, more types of wood, like beech, were introduced as alternate layers with ash wood, resulting in a combination of the strengths of both materials.

The introduction of a single leather laminate allowed more geometric freedom and increased strength, but was soon replaced in the 1960s with Black Walnut, Vulcan fibre (a resin impregnated in paper) or plastics, like Bakelite. Later, glass and graphite fibre laminations were also introduced, increasing frame strength even more (Bodig and Jayne, 1982; Kuebler, 1995).

In the meantime, metals were also making headway in different variations. In the early 1920s solid extruded aluminium frames started making their appearance but were substituted by cast magnesium alloys about five years later. It was only in the mid-1960s, when hollow extruded profiles made it the market for both aluminium and magnesium alloys, and lasted until the late 1980s. The high strength-to-weight ratios of these profiles opened up new opportunities for designers, i.e. increased head sizes, which led to the revolutionary oversized rackets in the early 1970s. Subsequently, aluminium was also used as a cold drawn tube up to the late 1980s and is currently used for its low price and in some junior rackets (Kuebler, 1995; Polich, 1995).

During the 1970s composites of glass fibre in epoxy were entering the market, which paved the way for what was probably the greatest revolution in tennis rackets to date. Initially, glass fibres were mixed with carbon fibres but later evolved in rackets with carbon fibres as the main component. Rackets were hollow, or filled with foam, and the carbon fibres made it possible to obtain stiffer, lighter and longer lasting rackets. From 1980 till the mid-1990s, polyamide was also used in frames, either as a thermoplastic injection with carbon fibre reinforcements, or as braided filaments combined with graphite fibres (Haines *et al.*, 1983).

Currently, composite rackets consist predominantly of carbon fibres as the main component, complemented with anything from glass, boron, ceramics and Kevlar to titanium and copper fibres, applied in strategic areas to provide the optimal combination of properties. The stiffness-to-weight ratio of these modern materials, as well as the versatile manufacturing process, enables designers to incorporate more effective racket designs with better control, power and vibration characteristics (Easterling, 1993; Brody, 1995a; Polich, 1996).

9.2.2 Classic and trend-setting racket models

During the history of tennis rackets there has been a number of rackets of significant importance, which can be attributed to material developments. In the following section we attempt to provide a selection of these rackets, while highlighting their significance (Fig. 9.1).

'Scaino' racket

The first racket reported to have a frame and strings, similar to how we know it today was in the mid-1500s. It is often referred to as the 'Scaino' racket since it was described in detail by Scaino, an Italian priest and doctor. Prior to this, people were mainly playing with gloves and bats, making it a radical breakthrough for its time. The racket consisted of an almost tear-shaped wooden frame, with a diagonal stringing pattern. The racket's head was relatively large, and with no throat it was remarkably similar to rackets being used in racquetball today (Kuebler, 1995).

Dunlop Maxply

The Dunlop Maxply racket is often referred to as the most famous racket ever. The Maxply was the first multi-ply wooden racket to be made, providing superior strength properties to its predecessors, which had consisted only of single bent strips. The racket was constructed from Vermont ash, cherry and hickory layers, glued together with water-repellent glue. The handle plates were made of bass-wood and the entire racket was finished off by experienced craftsmen to produce a very elegant product. It was introduced in 1931 and used for almost 50 years, before being beaten off the shelf by more modern constructions made from lighter materials (Kuebler, 1995).

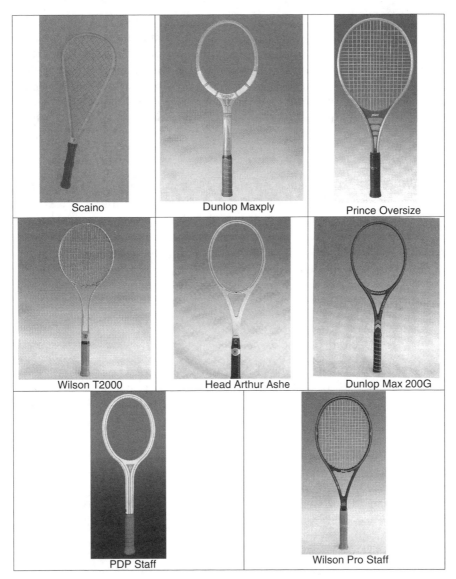

Scaino	Dunlop Maxply	Prince Oversize
Wilson T2000	Head Arthur Ashe	Dunlop Max 200G
PDP Staff		Wilson Pro Staff

9.1 Remarkable rackets up to the 1980s. (Images from Kuebler, 1995.)

Prince Oversize

In 1974, the development of extruded aluminium profiles allowed Howard Head, former owner of Head Ski Company, to change the design of rackets forever. He invented and patented the first oversize racket, which had a 50% larger string area and was claimed to have a four times larger sweet spot. Another advantage, and probably a more important one, was the 50% increase in resistance to twisting in the hand when hitting off-centre shots, resulting in a more stable and less strenuous racket to play with. The racket also had a typical polyamide throat piece, which was much stronger than the construction for wooden rackets. It instantly became a world bestseller and was used until 1988, with its design remaining almost unchanged throughout the entire period (Fisher, 1977; Arthur, 1992; Brody, 1995b).

Wilson T2000

The Wilson T2000, introduced in about 1979, was by far the most popular steel racket ever on the market. The racket had a tubular frame with a cross-sectional shape like a number 8. It was plated with chromium and a unique patented steel wire system for threading and attaching the strings. The latter consisted of a bent steel wire creating hoops all along the inner circumference of the head, through which strings are strung. It is in turn attached to the frame by spiral-wrapping a thinner wire around it and the frame (Kuebler, 1995).

Head Arthur Ashe Competition

The Head Arthur Ashe is a classic example of a pressed racket, utilising the same technique as for making skis. It was developed in 1979 and consisted of a plastic core, made with a BMC process, and bonded on both sides by aluminium sheets (Kuebler, 1995).

Dunlop Max 200G

In 1980 Dunlop developed a unique injection moulding process, used for their popular Max 200G. The hollow racket frame consisted of Polyamide 66, with carbon fibre reinforcements. To manufacture the hollow frame, a bismuth–tin alloy was used as its core. The metal has a melting point lower than the polyamide and is melted and removed after moulding. The process was less labour intensive than contemporary processes and produced a racket with very good dampening qualities, which was very popular right through the 1980s (Haines et al., 1983; Haines, 1985; Kuebler, 1995).

PDP Staff

The world's first completely fibre glass racket was introduced by the PDP Sports Company in about 1975. Apart from the grommets and foam-covered handle, the racket frame was made as a single piece, without any parts and composed entirely of glass fibre (Kuebler, 1995).

Wilson Pro Staff

Towards the 1990s most manufacturers had standardised to carbon fibre composite rackets, manufactured with a bladder-mould process. A lay-up, consisting predominantly of carbon layers orientated in optimal fibre directions, is rolled round a long foil tube, which is pressurised inside the mould to produce a thin-walled racket with optimised strength properties. A more detailed description of the process is provided later in this chapter. One of the first classic rackets to be manufactured in the way was the Wilson Pro Staff, which was introduced in the late 1980s and is still being used today (Kuebler, 1995).

Prince Vortex

Another significant racket, based on thermoplastic materials, was the Prince Vortex, introduced in 1991. Its hollow frame consisted of Polyamide 6, braided with graphite fibres, and the manufacturing process utilised a tubular mould with a pressurised silicone bladder. Besides the higher temperatures, the process was similar to a thermoset matrix racket (Beercheck, 1991; Prince brochure, 1993; Kuebler, 1995).

Wilson Profile Hammer

Wilson's next series, the Profile Hammer system, introduced in 1989, took specialised composite designs to the next level. The racket, weighing about 280 g, was based on two major patents; the Profile and Hammer systems, both made possible by a combination of specialised fibres such as Kevlar and boron fibres. The Profile system specified the racket to have a maximum width in the middle of the frame, tapering down to the handle and tip of the racket. This resulted in the highest stiffness at the point of maximum bending. The idea of the Hammer system is to move the mass to the head of the racket by reducing the weight in the handle, while maintaining overall strength and stiffness. This racket propelled Wilson Company to top of the world market, by being a best seller for three years (Beercheck, 1991; Wilson brochure, 1992; Wilson, 2002; Kuebler, 1995) (Fig. 9.2).

Head Twin Tube

In 1996 Head developed a revolutionary twin tube system, consisting of a polyamide sleeve wrapped around the frame of the racket's head just before moulding, producing a head with a polyamide outer layer. The tube reduces vibrations in the head and can be laser printed with complicated

9.2 Remarkable rackets from the 1980s to 2002.

graphics before moulding to apply the graphics. Traditionally, detailed graphics comprise a very high percentage of racket costs, since it is very labour intensive and time consuming but with this system virtually any graphics can be applied to the racket head at no extra expense. Since the graphics are printed underneath the polyamide, it is protected against abrasion by the outer layer and is therefore longer lasting than traditional cosmetics (Head brochure, 1996; Head, 2002).

Head Titanium

Probably Head's biggest success story to date was the introduction of titanium technology in 1998. The racket's throat included an outer weave of carbon fibre and titanium wire, which stiffened the throat and allowed for a large reduction in weight. The Ti.S6 racket, weighing 225 g unstrung, was a world best seller for three consecutive years (Head brochure, 2001; Head, 2002).

Head Intelligence

A couple of years later, in 2000, Head was the first company to use piezo ceramic fibres in their Intelligence series. The piezo ceramic fibres are moulded into the outer layer of the throat area, on both sides of racket. The polarised ceramic fibres, sandwiched between printed electrodes, convert and dissipate the impact energy capture as electrical energy. Each of the fibre units is connected to a self-powered circuit board located in the handle, which stores the impact energy and returns the inverted signal back to the throat, fast enough to stiffen the racket frame, dampen up to 50% of the vibrations after impact and increase the power (Head brochure, 2001; Head, 2002; Crawford, 2000b).

Wilson Triad

The Wilson Triad series, introduced in 2001, dampens vibrations by dividing the racket into three parts, hence isolating the handle from the head. The components include: the hoop, comprising the head with a thin triangular throat piece, the handle terminating in a V-shaped throat piece and a V-shaped elastomer separating the two pieces. The hoop and handle are manufactured separately and bonded to either side of the elastomer with a very strong adhesive. The throat pieces for the hoop and the handles have two common grommet wholes on either side, through which four main strings pass, improving the locking between the three components. The system isolates the impact shock and dampens up to 60% of the vibrations (Crawford, 2001a; Wilson brochure, 2002; Wilson, 2002).

9.3   Frame materials

During the history of the game, various materials have been experimented with, in combination with countless innovations. Nonetheless, through a natural selection process, which tends to maintain the unique balance between simplicity and functionality, only a few material concepts have had a lasting influence on designs.

As with most other products, the cycle began with wood, which has made the most lasting impact so far on designs. It was only a few hundred years later that manufacturers started experimenting with metals, but these never managed to dominate the market before being replaced by composite materials. Today, most rackets consist of a carbon fibre-based composite frame, combined with various other materials to enhance specific design intentions.

9.3.1 Wood

Wooden rackets are now virtually obsolete, with only a few still available on the market. However, from the beginning of the game, wood was a dominant material in racket frames for almost 400 years. Wood is a natural composite material, consisting of elongated cells distributed in its own natural resin. It is therefore anisotropic (i.e. much stiffer along the grain) and the composition of the structure is dependent on the type of tree and its growing conditions. These diverse properties were utilised in the development of laminated frames, which could combine the strengths of the different grain directions and wood types. The harder woods, like birch, maple, mahogany, hickory and beech were used to stiffen rackets while ash or maroti, softer and more resilient woods, produced more flexible rackets and walnut, sycamore, maple, birch, cedar, mahogany and holly were used as the outer layer for their cosmetic appearance (Table 9.1).

Wood was gradually phased out by other materials owing to a number of disadvantages:

Table 9.1 Mechanical properties of various wood types

Wood type	Specific gravity	Modulus of rupture GPa	Flexural modulus GPa	Shear strength (parallel to grain) MPa
Ash black	0.49	0.087	11.0	10.8
Ash white	0.60	0.103	12.0	13.2
Hickory pecan	0.66	0.094	11.9	14.3
Maple red	0.54	0.092	11.3	12.8
Oak, Red willow	0.69	0.100	13.1	11.4
Sycamore	0.49	0.069	9.8	10.1
Cedar incense	0.37	0.055	7.2	6.1
Beech	0.64	0.103	11.9	13.9
Birch yellow	0.62	0.114	13.9	13.0
Walnut black	0.55	0.101	11.6	9.4

Note: All properties are typical for woods grown in America with 12% moisture content. Taken from Green *et al.* (1999), Bodig and Jayne (1982).

- Instability and warping when moist limited the racket's outdoor life.
- Natural defects weakened the material or required expensive quality control.
- Low ultimate and fatigue strength caused collapsing under high impact and string tension.
- Difficulty in manufacturing lightweight and hollow thin-walled frames.
- Relatively weak mechanical bonding limited design variations.

To improve on the mechanical properties, wooden laminates were combined with leather, metals, polyamides and finally composites like glass fibre and graphite until wood was eventually replaced completely by composite frames, thus ending the legendary wooden era (Kuebler, 1995; Brody, 1995a; Polich, 1996; Green *et al.*, 1999).

9.3.2 Metals

During the early nineteenth century metal frames started to appear on the scene but it was not until the end of the 1960s that they really made a noticeable impression. The major advantage of metals over wood was the higher shear and fatigue strength and the ability to produce complex profiled frames. Initially, frames consisted of solid and later hollow extruded profiles, which resulted in even stronger and lighter frames. For the first time an oversized frame could be constructed that would withstand the high impact forces and string tension. Another major benefit with metals was fixing the throat piece. This was always a weak area in the wooden rackets but could now be made of virtually any suitable material and riveted to the frame, at a very low cost, without considerable weakening (Brody, 1995b; Kuebler, 1995; Polich, 1995).

The only metal alloy used frequently in racket frames is Aluminium 6061, which is subjected to specific heat treatments to produce desired characteristics (Table 9.2). Aluminium is still used today in low-cost racket filling and holds an important position in the market.

Table 9.2 Properties of Aluminium 6061 with different heat treatments

Treatment	Yield strength (MPa)	Tensile strength (MPa)
Annealing (0)	55	125
Matured hardening (T4)	145	240
Returned hardening (T6)	275	310

Note: Taken from eFunda (2002).

9.3.3 Composites

There are numerous types of composite rackets available today, and most of them employ the latest space-age composite materials. Composites either consist of fibres or filaments from very strong materials (graphite, glass, boron, ceramic) or consist of a lamination of various materials with unique properties such as wood, glass, aluminium. The current processes allow improvement of the properties of the material in terms of flexibility, structural strength, weight, etc. Moreover, because of the fibrous nature of the material, the fibres can be orientated to give strength and stiffness in one direction and allow some flexibility in another.

The so-called graphite or carbon rackets consist mainly of carbon fibre-reinforced composites with an epoxy matrix, while glass and aramid fibres are used only to a small extent in some rackets. Additionally, thermoplastics are used mainly as a matrix in the form of polyamide.

More information on calculating composite materials properties is beyond the scope of this chapter but can be found in the literature, such as Kelly and Mileiko (1983).

Fibres

Carbon fibres

The raw material for carbon fibres is a fibre made out of polyacrylnitril (PAN). The precursor, consisting of 12000 filaments, is commonly used in rackets after undergoing heat and stress treatment. This process, known as pyrolysis, predominantly determines the properties of the final fibre. For instance, the temperature and the stretching during this process define the Young's modulus and the strength. PAN fibres are a good precursor for carbon fibres, due to its all-carbon backbone, which forms a ladder polymer when heated to between 200 °C and 300 °C. The subsequent heat treatment at between 1000 °C and 2400 °C causes oxidation and dehydrogenation, producing fibres of high strength (HST), high modulus (HM), and intermediate modulus (IM). The alternative basis for carbon fibres is pitch, resulting in high modulus fibres, which is more cost effective. Since the deformation of a racket is very large as a result of stringing and impact, the most common fibres used are HST fibres.

Glass fibres

Glass fibres were the first fibres used as reinforcement in rackets. Fibres consist mainly of silicium oxide and are spun from molten glass. The most important type of glass is a so-called E-glass, with the 'E' representing electric, due to its original used in electrical applications. Compared to carbon, the advantages and disadvantages are easy to distinguish; it exhibits high tensile stress but low stiffness and high compression stress, combined with a higher density.

Aramid fibres

Based on an aromatic polyamide, spun from a solvent, aramid fibres have very good impact behaviour. Their disadvantage is a lack of compression strength, which precludes the broader use of this fibre. In addition to the impact behaviour, their low density is another benefit.

Boron fibres

The very stiff and brittle boron fibres have been used to some extent in the past, but are not used currently in racket designs (Table 9.3).

Matrix

In order to fabricate fibre-reinforced composite articles, it is necessary to impregnate fibres with a matrix. The fibres usually also have a sizing, for better adhesion between the fibre and the matrix.

Thermosets

Epoxide resins are the most favoured for use with carbon fibres and in high-performance applications, because of their good mechanical properties, low shrinkage and the ability to bond to other materials. All epoxies are characterised by the presence of the epoxide group: two carbon atoms and one oxygen atom arranged in a three-membered ring (Fig. 9.3).

The reactivity depends on the position of the group in the molecule and on steric factors. The opening of the epoxide ring by a curing agent leads to cross-linking and ultimately to the production of a hard, insoluble solid. When fully cured, all the epoxide groups should have reacted, but this probably does not occur in practice where the epoxy often gets brittle owing to the high reaction ratios (Kelly and Mileiko, 1983).

To cure the epoxide, it is necessary to use a hardener and possibly an accelerator and often to heat the constituents, in the correct proportion, for an hour or more to between 100 °C and 120 °C. For 100 parts resin, between 10 and 80 parts of hardener are required. Curing cycles to maximise a given property are usually determined empirically by the resin manufacturers, and will vary not only for different resin types but also for differing desired properties (Kelly and Mileiko, 1983).

Table 9.3 Filament properties for common composite materials

Material	Density (×10³ kg m⁻³)	Long tensile mod. (GPa)	Long tensile strength (MPa)	Trans. tensile mod. (GPa)	Shear modulus (GPa)	Compression strength (MPa)	Strain at fail. (%)	Major Poisson's ratio	Diameter (μm)	Long. coeff. of thermal exp. (×10⁻⁶ °C⁻¹)
E glass	2.54	70	3100	70	28.7	1750	2.5–3.0	0.22	10.0	5
Carbon										
*VHM	2.0	517	1860				0.38	0.25	8.4	−1
HM	1.9	350	2000	12.1	13.7		0.5	0.28	11	−0.5
HT	1.78	230	2900	20.4	24.0		1.3	0.26	8.0	0.5
A	1.76	215	2400				1.27	0.26	8.5	1.0
Aramid	1.45		2800	5.38	2.0*	250*	2.0–3.0	0.34*	12.0	−2.0
Boron	2.63	420	3400	420	180	2300	0.7	0.13		2.8

Note: *VHM = very high modulus; HM = high modulus; HT = high tensile (strength); *indicates results for a 60% composite. All figures are approximate and derived from manufacturers' data, taken from Kelly and Mileiko (1983).

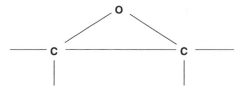

9.3 The epoxide group.

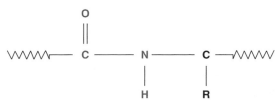

9.4 The amide group.

Table 9.4 Material properties of common polyamides

Property	PA 6	PA 6-GF30	PA-CF30	PA 66	PA 11
Density (g/cm³)	1.13	1.36	1.28	1.14	1.04
Yield till breakage (N/mm²)	40	100	240	65	50
Elongation till breakage (%)	200	4–5	1.5	150	500
Young's modulus (MPa)	1400	5000	16000	2000	1000
Bending strength (MPa)	50	130	330	50	70
Maximum temperature of use (°C)					
briefly	140–180	180–220	180–220	170–200	140–150
continuously	80–100	100–130	110	80–120	70–80
Melt temperature (°C)	220	220	220	255	185

Note: Taken from Kelly and Mileiko (1983) and eFunda (2002).

Thermoplastics

As mentioned earlier, the only thermoplastic materials used in racket manufacture up to now have been polyamides (Fig. 9.4).

Polyamides have good strength and toughness, with excellent fatigue resistance. However, they are prone to absorbing moisture, ranging from 8–10% for PA 6 and PA 66 to 2–3% for PA 11 and PA 12. The PA 11 and PA 12 have similar characteristics and a number of advantages including:

Table 9.5 Comparing the properties for a range of materials used in racket frames

Material	Density (g/cm³)	Tensile strength (GPa)	Young's modulus (GPa)	Specific tensile strength	Specific Young's modulus
HST-Epoxy	1.5	1.9	130	1.27	87
HM-Epoxy	1.6	1.2	210	0.94	119
E-Glass Epoxy	2.0	1.0	42	0.5	21
Aramid Epoxy	1.4	1.8	77	1.3	56
Nylon (PA 6)	1.13	0.04	1.4	–	–
Steel	7.8	1.0	210	0.13	27
Titanium	4.5	1.0	110	0.21	25
Aluminium	2.8	0.5	75	0.17	26

Note: Taken from Kelly and Mileiko (1983) and eFunda (2002).

Table 9.6 Material properties of elastomers

Property	NR Natural rubber (cis-polyisoprene)	SBR Butastyrene (GR-S)	IR Synthetic (polyisoprene)	CR Cloroprene (neoprene)
Specific gravity (ASTM D 782)	0.93	0.91	0.93	1.25
Tensile strength (MPa)				
Pure gum (ASTM D 412)	17–24	1–2	17–24	21–28
Black (ASTM D 412)	24–31	17–24	24–31	21–28
Elongation (%)				
Pure gum (ASTM D 412)	750–850	400–600	–	800–900
Black (ASTM D 412)	550–650	500–600	300–700	500–600
Recommended temperature Range (°C)	–51–82	–51–82	–51–82	–40–116
Hardness (durometer)	A30–90	A40–90	A40–80	A20–95

Note: Taken from Perry and Green (1997).

good temperature stability, creep and pressure strength, and good chemical resistance.

In addition, the generally known injection moulding process with PA 66 and PA 6 was also used in the form of fibres, allowing it to be comingled with carbon fibres, forming a kind of a thermoplastic prepreg (pre-impregnated fibres) (Kelly and Mileiko, 1983) (Table 9.4).

Table 9.5 provides a comparison between the properties of the most important materials used in racket frames.

9.3.4 Elastomers

In the racket itself, or as an accessory, all types of elastomer are used by the different brands, with the main benefit being dampening. There are two types of rubber: natural rubber, which comes from the latex and contents of some trees and other plants; and synthetic rubber, which is an oil by-product. The main characteristic of this material is that it can be stretched to many times its original length and can bounce back into its original shape without warping. Table 9.6 contains relevant properties for commonly used elastomers.

9.4   Materials for accessories and special parts

With frame developments pushing the boundaries further every year, accessories are also becoming more sophisticated to help improve performance in any way possible. In these, development materials often have a very important role to play, with almost any material from metals to elastomers being used.

9.4.1 The handle

The handle itself is often made of polyurethane foam, injected or glued to the handle after the moulding process. The foam improves vibration dampening and provides a cost-effective solution to manufacturing rackets with different grip sizes. Many manufacturers also incorporate some form of elastomer as the core of the handle, isolating it from the rest of the frame to dampen impacts. The principle is not to have a solid connection between the hand and the frame but rather to have the moment transferred through a rather thick absorbing material. In the newer lightweight rackets many of these systems have been omitted though, in order to save weight. Butt end caps are usually made as injection-moulded parts out of various thermoplastics (Head brochure, 1995; Beercheck, 1991; Polich, 1995).

9.4.2 The grip

The grip is the interface between the racket and the player and therefore has a very important role as the last frontier to affect a player's perception of an impact. Hence, the purpose is to minimise the shock and vibrations transferred to the players, and provide a firm grip so that the racket does not slip in the hand, especially when wet with perspiration. The first wooden rackets had no grip and relied on the shape and surface texture of the handle for a firm grip, while the natural properties of the wood provided the dampening. Often, softer strips of woods were attached to the outside of the handle to improve matters. Soon after that, leather grips were introduced and, although not that popular, are still in used today (Kuebler, 1995).

Leather grips have been replaced mostly by sophisticated materials such as rubbers, polyurethanes and polyesters. Most current grips consist of combinations of these materials, in many cases with a thick textured polyurethane outer layer providing shock absorption and a firm grip. This layer is often perforated to channel the perspiration to a second polyester felt layer, where it is absorbed.

Many players use temporary over-grips made from thin polyurethane. These grips are applied over the normal standard grip, mostly to prevent slipping. The grips are cheaper and can be changed regularly, even between sets, to ensure a fresh grip during match situations and to protect the standard grip underneath (Brody, 1995b; Head brochure, 2001; Wilson brochure, 2002).

9.4.3 The grommets

Traditionally, grommets were only holes drilled through the racket frame for connecting the string to the frame, but with the development of metal frames it became necessary to have some form of protection for the strings from abrasion against the harder metal frame. Later, with the introduction of thin-walled composite frames, it became even more important to have grommets, which in this case prevented the strings from cutting through the thin carbon walls, under high tension. It would also assist in the stringing by providing a guide through the hollow frame. Further functionality was added by using the grommets on the tip of the racket to protect the frame against abrasions when in contact with the ground.

Not too long ago most grommets were still simple round tubes with a tight fit through both the inner and outer frame walls, not allowing too much movement of the string inside the frame during impact. In an attempt to enlarge the string surface, without increasing the head size, various manufacturers have been moving to grommets with larger holes on the inside of the frame, allowing the strings to extend their movement as far as possible to the outside wall. Manufacturers have also utilised grommets to give extra flexion or to dampen vibrations to the string bed by adding softer materials or designing the string bed so that it would act as a spring during impact (Crawford, 2000b, 2001b; Wilson brochure, 2002).

These multi-functional grommets require a very tough but flexible material, which can be manufactured to very accurate specifications for a perfect fit into the frame holes. Tolerances on grommets are very tight to ensure on the one had that they do not fall out and, on the other, that they are not too difficult to fix, since it is a manual process. The materials found most suitable for this purpose are polyamides such as PA 11 or PA 6.

9.5   Current manufacturing process

9.5.1 Composite rackets

As mentioned before, all high-priced rackets these days are manufactured predominantly from thermoset carbon fibre composites, with other materials placed strategically for optimum performance. This is due mainly to carbon fibres providing the best combination of material strength and manufacturability, with current technology. Except for the small deviations, the manufacturing processes are fundamentally the same for most major brands, even more so since most use the same manufacturers in the Far East. Filament winding is also used to produce rackets, but such rackets are only a small part of the market.

Production of composite rackets

Prepreg

Prepreg (pre-impregnated fibres) are either bought in ready prepared roles, or made in-house by drum winding. The latter process entails winding resin-impregnated fibres onto a large drum, producing 0 degree prepreg (fibres have zero degree direction to the long axis of the sheet or the racket), which can then be cut into the desired sheet sizes. Although prepreg are manufactured in endless combinations, the basic carbon prepreg used for most rackets is mainly produced with the following specifications, which can be varied for optimum designs:

- carbon fibre content (grams per square metre)
- resin content (a percentage of the total mass).

The sheets are cut by hand, or machines, at different angles (0°, 30°, 45°, 60°, 90°) and widths to produce layers with specified fibre angles. Most layers are placed on top of an identical one but one with the fibres orientated in the opposite direction, producing a layer with fibres aligned symmetric to the long axis.

Lay-up

The lay-up refers to the positioning of the different prepreg layers to form the basic frame. These layers are cut to the correct sizes and are then positioned on a flat-heated bench to make the prepreg tackier, so that they stick better to the adjacent layers. Although all companies have their own trade secrets and patents distinguishing their lay-up from those of the others, the basic lay-up for most rackets is very similar, with an example shown in Fig. 9.5.

The basic principles are to use zero degree prepreg for bending stiffness and ± 45 degree prepreg for torsional stiffness and anything in-between depending on the desired combination. Additionally, most manufacturers add extra material, often glass, at the racket tip for the high impact forces in this region.

Pre-preg layers	Parts	Material	Fibre angle
	1	PA Foil	direction
	2	Glass	±20°
	3	Carbon	±30°
	4	Carbon	0
	5	Glass	0
	6	Carbon	±30°
	7	Carbon	0
	8	Carbon	±30°
	9	Carbon	0
	10	Glass	±30°
	11	Carbon	±30°
The main tube			
	12	Carbon	±30°
	13	Carbon	±30°
	14	Carbon	±30°
	15	Glass	±20°
	16	Carbon	±30°
	17	Carbon	90°
	18	Glass	±30°
	19	Glass	±30°
The throat piece and reinforcements			

9.5 A typical lay-up for a composite racket.

The main tube constitutes most of the racket's frame and is prepared first on a flat table and then rolled or folded around a polyamide foil tube, which is pulled over a rod. The rod is then removed, producing a hollow prepreg tube rolled around the foil, which, once inside the mould, will be filled with air to provide the internal pressure.

The throat piece is prepared separately by wrapping prepreg around anything from rubber pieces or expandable foam. This is necessary since there is no easy way to get air pressure inside the throat piece during moulding. The two methods function in one of two ways to create internal pressure; the air or the foam inside the bags expands due to the heat, while the rubber pieces are made fractionally too large for the mould and are therefore compressed when the mould is closed.

Moulding

Before moulding, all the pieces are assembled on a template to obtain the basic racket shape. Simultaneously, the final prepreg pieces are also added to the strategic areas. The main tube is bent around a shape with an inner diameter similar to that of the racket face, and the ends are pressed together and wrapped with a prepreg layer to form the handle. The throat piece is then fitted and attached with small supporting prepreg pieces. The air hoses are then connected to both ends of the tube and the finished lay-up placed inside the mould, which is closed and, depending on the prepreg and the desired cycle time, the correct temperature cycle and internal pressure are applied to set the prepreg.

Finishing

Almost half of the racket's manufacturing costs are its finishing and cosmetics after moulding. The first step is to deflash the racket, removing the excess resin from the mould seam. It is then cut to the desired length and the foil removed to reduce the racket weight. The rackets are then sanded to roughen up the surface for better lacquer adhesion, followed by the drilling of the grommet holes. Next, small pit holes and minor defects are filled with body putty and sanded for a perfectly smooth surface. The frames are then painted with an electrostatic system, which applies a small electrical charge to the racket, thus attracting the lacquer with opposite charge for better adhesion. Subsequently, the rackets are heated in a ventilation room to harden the lacquer, after which the detailed aesthetics are applied. These are mostly in various forms of printing, i.e. silk and tampon printing and the more labour-intensive water decals (decorative transfers). The rackets are then ready for all the accessories such as the end cap, grip, grommets and strings to be fitted.

Most manufacturers have different quality control points at various stages in the process. Random selections are also made from the finished rackets, which are subject to more stringent testing to ensure high quality. The quality process is discussed in further detail later on in this section.

9.5.2 Aluminium rackets

Aluminium rackets first made their appearance in the 1960s and were later replaced by composite rackets for the high-end of the market. Composites allow for the manufacture of better performance rackets but with higher material and manufacturing costs. There is still a very significant market for cheaper, low performance, rackets though, especially with junior rackets, where performance is not that important. This market is catered for by aluminium rackets, with a manufacturing process radically different from that for composite rackets.

An aluminium beam is extruded with the desired profile, cut to size and annealed to soften the material for bending. It is then bent around a template, which forms it into the basic racket shape, the holes drilled and the frame hardened again with a heat treatment. The rackets are then lacquered by hand and all aesthetics applied, as for composite rackets. Next, the throat piece is simply screwed or riveted into place and the two ends of the tube forced together with a ring at the throat and riveted through the bottom for the handle. The racket handle is then placed into a mould and filled with polyurethane foam for the basic grip shape. The final steps are fitting the accessories, by a process similar to that used for composite rackets.

9.5.3 Racket testing

Although most manufacturers and racket companies have their own specialised tests, there are also several similar tests used by most manufacturers.

Non-destructive tests

In most manufacturing processes the mass, balance point and swing weight are monitored at different stages to pick up problems in the process as early as possible. These properties, as well as frame stiffness (measured using the standard RA test) are used as basic indicators for manufacturing specifications and are usually measured for randomly selected, or all finished, rackets.

A specialised three-point bending test is performed to profile the bending stiffness in the impact direction along the length of the racket face. The face is supported at two locations on each side of the racket; the first at 5 and 6 o'clock and the second at 10 and 11 o'clock (referring to the face as an analogue watch). A load contacting both sides of the frame is applied at discrete points along the face. The deflections for all the points are measured and compared for different rackets.

The torsion test is similar to the three-point bending test except for the support, which is not at both sides of the face but at alternate sides, i.e. at 5 and 11 o'clock. It creates torsion in the frame, and measuring the deflection provides an indication of the torsional stiffness of the racket.

The tip deformation test applies a two-point load, at 1 and 11 o'clock on the tip in the direction of the handle, indicating the radial stiffness of the face at the tip of the racket. This is a critical area on the face, experiencing very high loading from the mains during impact. Again, the deflection is measured for an applied load.

Destructive tests

A destructive version of all non-destructive tests is also performed. During these tests the racket is deformed until it breaks and the ultimate breaking force is measured.

The tip impact test again is a dynamic test for the tip region. For the test, the racket is dropped from incrementally increasing heights onto its head until it breaks. The sum of the drop heights up to the failure is used as the indicator of the tip's impact resistance.

The tendency of frustrated players to throw the racket on the ground and on the net often results in failures at 3 and 9 o'clock on the face, forcing manufacturers to design and test for the condition. The racket is fitted like a swinging pendulum, rotating around the handle and is dropped from the horizontal position to collide with the solid round edge in the vertical position. The racket can be loaded with weights and, as with the tip impact tests, the weights are increased with every impact until failure. In this case the total weight of all the masses used up to the failure is used as the indication of the resistance to side impacts.

Rackets also need to be temperature resistant and are therefore subjected to a temperature test. The racket is strung and placed inside an oven at about 80 °C for approximately 4 hours and then checked for any defects.

Another interesting test is designed to evaluate the grommet strength. A string is threaded through two grommet holes and both ends are pulled on the inside of the frame until the grommet or the frame wall fails owing to the high shear force. The force is measured and the maximum force causing failure is used as an indication of the strength.

Additionally, rackets are also subjected to fatigue tests in various embodiments, including anything from hitting a number of serves to a dynamic version of the tip deformation test.

9.6   Design criteria

The design criteria for rackets depend very much on the style and quality of the players they are designed for. As a result, there are various different kinds of rackets on the market aiming to meet the needs of most users. Criteria include properties such as power, control, comfort and mechanical failure. Ideally, one would like to have all these properties maximised in all rackets but for the optimum results it's often a trade-off between the more important ones, with power and control mostly being the two opposing properties to design for.

9.6.1 Power

Probably the first rule of racket design is 'for more power, use the strings'. Strings return about 90% of their deformed energy after impact, while the ball alone (when rebounding on a hard surface) only returns about 45% of the energy (Brody, 1995b). Additionally, the energy in the deformation of the racket frame is not returned to the ball in time to add energy, which means any racket characteristic maximising string deformation during the impact, rather the ball or frame deformation, would increase the power. Relating it to frame parameters would include achieving longer strings and a stiffer frame. After the introduction of the oversized rackets in the 1970s, rules have been introduced to limit the size of the string bed. The current rules specify maximum dimensions for the string height and width, which are measured to the inside of the frame of the face. Subsequently, attempts at maximising the string bed were mostly focused on squaring face shapes, rather than on a simple sphere shape. This results in the longest possible length for all the strings, but has structural limitations where frames become difficult to manufacture, are not very strong and not pleasing to the eye. Soon, more moderate shapes were adapted with more subtle features. In the last decade though, various manufacturers have been playing around with larger grommets to allow the effective size of the string bed to be extended to the outer side of the frame. Another factor affecting power is mass distribution, for which it is difficult to achieve an optimum but the principles are straightforward: a lighter racket can be swung faster, increasing the speed transferred to the ball, while the heavier racket is swung slower but the increase in mass also causes an increase in ball speed, especially if more of the mass is located at the impact point. Hence, the challenge is to design a racket light enough to be swung at a high speed and for all strokes, but still have enough mass behind the ball for optimum rebound characteristics.

9.6.2 Control

A smaller head and frame of higher torsional stability usually provides better control because there is less angular deflection of the strings and the frame. A larger angular deflection for an off-centre impact will result in the ball coming off the strings at an angle larger than the perfect rebound angle that a flat surface would provide. This small difference in angle can cause a large error in the intended ball placement on the court. A higher resistance to polar rotation causes less twist of the racket in the hand, hence resulting in a more accurate shot. In order to achieve a smaller head, which is torsionally stable, additional weight is placed on the perimeter of the racket (3 and 9 o'clock). It may be useful also to note at this point that spacing of the strings can also have an effect on the control; more dense strings result in a higher effective string tension, decreasing the angular deflection for better control but also tending to reduce spin, which might be important for some players.

9.6.3 Comfort

Comfort is a very abstract concept linked to the player's perception. This is still a very complex and vague area, since different players will have a different perception of the same racket, resulting from various factors. Additionally, players have their own way of describing how they perceive a racket, which makes it very difficult to convert the feedback into concrete design parameters. So far, most of the attention has been given to minimise the amplitude of the impact and the resulting vibrations. Stiffer rackets tend to have vibration-dampening characteristics lower than those of the older wooden rackets that players used to compare with, and various additional systems were experimented with to provide additional vibration dampening. The stiffer rackets also have a higher-pitched sound, which is usually not preferred by most players, hence the use of string dampeners. Methods to minimise the impact shock include moving the centre of percussion to the face centre, and with perimeter weighting. Also related to comfort would be the surface finish used and the grip material used on the handle, which is very much standard for most manufacturers.

9.6.4 Mechanical failure

Rackets are designed to last as long as possible within demanding restrictions imposed during high performance play. The introduction of carbon fibre materials resulted in a dramatic leap in racket performance and strength. Initial strength problems were related mostly to the junctions between the main racket frame and the throat piece and at the tip of the racket, where it is subject to the largest force from the main strings, but most manufacturers have resolved these critical areas with strategically placed reinforcements. The drive for lighter rackets in some markets has forced manufacturers to push their structural designs to the absolute limit, resulting in complex and very specific lay-ups, aimed at impact specific loading conditions. More ways of reducing weight without sacrificing too much in strength have been using prepreg with lower resin content, applying fewer lacquer layers and removing the foil and expandable material used to provide the pressure inside the mould. When subjected to abnormal loading during testing though, these rackets can often perform surprisingly weakly, since they are not designed to withstand abnormal loading conditions. Critical in this process is therefore proper testing methods to ensure rackets are designed for the correct loading conditions encountered during play.

9.6.5 Designing for consumer groups

The simple matter is that there is no perfect racket for all players. Players have different levels and styles of play, and different racket models are developed to suit as many individuals as possible.

Players mostly start off with a lighter racket with a larger head. Not having the strength and skills of the professional players, these players tend to hit more off-centre hits. The larger head compensates for this, allowing the player to make better contact with the ball and so assist in a steeper learning curve and hence in more enjoyment of the game. It also provides torsional stability during the off-centre hits, which saves the player a lot of energy, and larger heads mean more power, giving the player more of an advantage in these early stages.

The next major category is the club player, who plays regularly for recreation. Being in the development stages of their game, they tend to move to rackets with smaller head sizes for more control and a specific swing weight determined by the individual's style of play. Base-line players, who tend to have a longer swing, tend to go for rackets with a higher swing weight, while volley players will go for a lighter more manoeuvrable racket. Rackets designed for that range of players have a wide variety of swing weights to fit every player's needs.

The next step are the tour players, who are so well conditioned that they would rather sacrifice power for ultimate control, resulting in the exact placement of the ball to win the point. These rackets have a much smaller head and are substantially heavier than beginner rackets. Players at this level can also swing a heavy racket much faster for longer, hence creating more power themselves and not relying on the racket's lighter weight. Some top professionals have claimed to play with rackets almost twice as heavy as beginners' rackets! These rackets are usually custom weighted with lead tape for the individuals.

These different criteria for groups of players often lead to manufacturers marketing rackets under categories such as beginner, 'intermediate' and tour series, with each series having a selection of swing weights to choose from. Most manufacturers also have racket selection systems to assist players in their choice (Easterling, 1993).

9.7   Future trends

What will be the next technology in tennis, revolutionising the sport again like the examples mentioned in this chapter? This is the question every brand tries to answer on a regular basis when the new products are developed or presented at a trade show. The most important issue here was, still is and probably will be: the use of new materials or the use of a construction principle, which allows the use of advantageous materials. Criteria like power, control or comfort have not been designed to a maximum yet. Everyone is willing to get the latest racket, which will help him win the next game. Whether this is a more powerful or more controllable racket is up to the player.

An area for increased research is the relation between the racket and injuries like tennis elbow. Although everyone has been aware of the problem and has been designing rackets to it, the truth is that very little is known about what really causes tennis elbow. The International Tennis Federation launched a congress for the future of the game in 2001, which is an excellent basis for promoting information and discussing new tennis developments and would hopefully lead to finding all the right answers in order to develop the perfect racket.

9.8   References

Arthur C (1992). Anyone for slower tennis? *New Scientist*, 2 May: 24–28.

Beercheck R C (1991). Sporting goods win with high-tech materials. *Machine Design*, June: 62–66.

Bodig J and Jayne B A (1982). *Mechanics of Wood and Wood Composites*. Van Nostrand Reinhold Company Inc., New York.

Brody H (1995a). How would a physicist design a tennis racket? *Physics Today*, March.

Brody H (1995b). *Tennis Science for Tennis Players*. Pennsylvania: University of Pennsylvania Press.

Clerici G (1976). *Tennis*, 2nd edn. London: Octopus Books Ltd.

Crawford L (2000a). Hole-y Wars. *Racquet TECH*, 10, 19, 20.

Crawford L (2000b). Head's intelligence technology – the shocking details. *Racquet TECH*, November, 6,8,9,27.

Crawford L (2001a). Comfort Quest: Shock Absorption the Triad Way. *Racquet-TECH*, July, 4, 6,8,10,12.

Crawford L (2001b). Völkl's catapult reinvents the stringbed. *Racquet TECH*, August, 20–23.

Easterling K E (1993). *Advanced Materials for Sports Equipment*, School of Engineering, University of Exeter, Chapman & Hall.

eFunda (2002). eFunda Inc., PO Box 64400, Sunnyvale, CA 94088, USA, http://www.efunda.com, Accessed 2002.

Fisher A (1977). Super racket – is this the Shape of things to come in tennis? *Popular Science*, **44**, 46, 150.

Green D W, Winandy D E and Kretschmann D E (1999). Mechanical properties of wood. *Wood Handbook – Wood as an Engineering Material*. Gen. Tech. Rep. FPL-GTR-113. Madison, WI: U.S. Department of Agriculture, Forest Service, Forest Products Laboratory.

Haines R C (1985). Volume production with carbon fibre reinforced thermoplastics. *Plastics and Rubber Processing and Applications*, **5**, 79–83.

Haines R C, Curtis M E, Mullaney F M and Ramsden G (1983). The design, development and manufacture of a new and unique tennis racket. *Proc Instn Mech Engr's*, **197B**, 71–79.

Head brochure (1995). *The New Pyramid Power Technology – It Will Change Your Game*. Head Sport AG, Kennelbach, Austria.

Head brochure (1996). 'Tennis 1996/97', Head Sport AG, Kennelbach, Austria.

Head brochure (2001). 'Tennis 01–02', Head Sport AG, Kennelbach, Austria.

Head (2002). Head Tennis web page, http://www.head.com, Accessed 2002.

Kelly A and Mileiko S T (1983). *Handbook of Composites: Fabrication of Composites*, 2nd edn. Elsevier Science Publishers, Amsterdam, The Netherlands.

Kuebler S (1995). *Book of Tennis Rackets From the Beginning in the 16th Century Until About 1990*, Kuebler GmbH, Singen.

Perry R H and Green D W (1997). *Perry's Chemicals Engineer's Handbook*, 7th edn, McGraw-Hill Professional.

Polich C (1995). Tennis rackets. *Sport and Fitness Equipment Design* (Ed. Kreighbaum, E. F and Smith, M.A.). Champaign, IL, Human Kinetics, 85–95.

Prince brochure (1993). *Performance Tennis Collection 1993*, Prince Manufacturing, Inc., Princeton, NJ.

Robertson M (1974). *The Encyclopedia of Tennis*, ed. J. Kramer, New York, The Viking Press, Inc.

Wilson brochure (1992). *Tennis 1992*, Wilson Sporting Good Co Ltd, Southall, UK.

Wilson (2002). Wilson Tennis web page, http://www.wilsonsports.com, Accessed 2002.

Wilson brochure (2002). Wilson: Tennis 2002. Amer Sports Europe, Gräfelfing, Germany.

10
Materials in bicycles

J. M. MORGAN
University of Bristol, UK

10.1 Introduction

The request to write on the subject of materials in bicycles, tempts me to suggest that, before continuing to read any further, you, the reader, write down a list of all possible materials that might be used, or fitted, in one form or another in, or onto, a bike. Immediately such a question is considered, it becomes obvious that no single book, let alone a single chapter can possibly deal with all variations and possibilities of materials that might be used. No doubt, by now, you will have already mentally processed a materials list, and I imagine that this list will almost certainly include such materials as aluminium, steel, titanium, polymers, elastomers, ceramics and composites of all types. As a rhetorical question, however, how many of you included wood and papier-mâché in your list? And, of course, which materials did you include that I haven't yet mentioned? At this stage I have to confess that I have never heard of a papier-mâché bike, but nothing would surprise me. On the other hand, wooden bikes have a history . . . and apparently a future.

10.2 Wooden bikes!

I first became aware that wooden bikes had ever existed while watching the BBC television programme *Antiques Roadshow*, and, if I remember correctly, the bike I saw in this programme had been made in Italy in around 1940. One aspect of the bike I saw that immediately intrigued me was that even the wheels, including rims and spokes, were fabricated from wood! How on earth could such components be made from such an *impractical* material and possibly have survived the first ride down a cobbled road or mountain path? . . . surely such a material choice for construction would present a veritable challenge even for modern engineers and designers. However, for those interested, further research will quickly indicate that wooden bikes were not just items of antique curiosity but are alive and kicking today in the twenty-first century.[1]

The first two-wheeled bicycles, known as *swiftwalkers*, were built in the late eighteenth century and, not surprisingly, were made of that favoured engineering material of the time . . . wood, but steel bicycles had started to be manufactured by about 1870. Wooden bicycles, however, remained commonplace until the 1930s, and indeed, it can be argued that the development and manufacture of these wooden bikes around the turn of the century led directly to the success of the first powered airplane flight by the Wright brothers. It seems not unreasonable to suppose that the experience gained by the brothers, Wilbur and Orville in 1893, while operating the Wright Cycle Company, would have given them invaluable experience in the manufacturing, joining, repairing and stressing using wood and wire, which would certainly have been essential requirements for the successful construction of their wood and fabric flying machines in 1900.

10.2.1 Bamboo and hickory

Bamboo cycles, produced by the Bamboo Cycle Company of London, were popular from 1895 to 1901, and, during the same period in America, the Tonk Manufacturing Company of Chicago produced a variety of hickory bicycles for gentlemen and, to reflect the trend towards increased emancipation, also for ladies. However, by the early twentieth century the effective replacement of wood by a more modern high-tech material, (i.e. steel), only demonstrated the evolutionary trend that continues to this day of utilising to advantage each new generation of materials in bike production and manufacture.

Finally, before moving on from the subject of wooden bicycles, rest assured that, should you develop a sudden urge to own, or even ride, a wooden bike, a number of companies will provide you with such an item. Such manufacturers can be found on the world wide web, by searching for such names as Giuseppe Matera in Italy, Gota in Brazil, and The Amazing Wooden Bike Company in Indonesia, to name but three. However, if you have your heart set on buying a wooden bike, be prepared to pay from about $3000 upwards for the privilege.

10.2.2 Antelope skin?

Before getting around to discussing what may be regarded as more relevant bike materials that might be used in twenty-first century bike manufacture, let me briefly mention the early twentieth-century bike brought back from Africa by a British traveller, which was reported as having: a frame made of hewn wood, wheels made from lath and reeds, and the chain from knotted antelope skin . . . and knotted antelope skin was one material that definitely didn't feature in my *materials for the use of* list.

10.3   Material properties

Let us move on to discuss those *more relevant* materials that have already been mentioned briefly. The problem, however, is where and how to start. If you are a student of Engineering Science, or Materials, or similar, you will certainly be aware that a large number of textbooks are available, all of which list, in one form or another, the so-called *mechanical properties of materials*. However, those of you who have ever tried to use such material properties lists *in anger* may have come to the conclusion that obtaining confident design information about a material relevant to your particular problem is never as straightforward as is suggested. For example, given the apparently simple task of wanting to build a bike frame from *steel*, what useful information might we obtain from a standard list of mechanical properties? Discounting *theoretical* property values such as those for metal single crystals, etc., the first thing that we discover is that mechanical property values are given for a huge number of different steels, including plain carbon steels, low alloy steels, alloy steels (including stainless steels), special steels . . . and so on. Furthermore, all these steels can be manipulated by a variety of heat treatments, which, although your reference sources may naively imply a simple correlation to give predicted changes in tensile strength, yield strength, fracture toughness etc., the same sources may unfortunately omit to mention other associated physical changes that may well detrimentally affect the final mechanical strength of your frame. Finally, the influence of real physical factors such as grain size variation, etc. due to manufacturing processing can easily be forgotten simply by omission.

Lastly, to test our mental agility, strength values are likely to be quoted in a random mixture of units including, for example, psi, ksi, tons per sq. inch, Pa, MPa, N/mm², kg/mm² or whatever. Despite the fact that conversion factors can be found eventually somewhere (or can be worked out) to allow us to normalise strange units into whatever units we choose to work in, the task of getting the units right should never be considered to be trivial. As an example of the real danger, or otherwise, of trying to design by using the plethora of mixed units offered to engineers, one only needs to consider the £125 million failure of the Mars Climate Observer satellite, which was destroyed after the navigational system requested Newton–seconds of thrust to be applied to the altitude thruster system and unfortunately got pounds–seconds instead.[2]

10.3.1 A theoretical steel frame

Let us assume that we want to build a bike frame and that we are going to choose some sort of *steel*. Using the classic engineering technique of first identifying what has been tried, tested and proved by previous manufacturers, we decide that we want to make our frame from steel *tube*. After all, the idea of making the frame from solid bar might have some theoretical advantages, but would certainly not appeal to *informed* cyclists, the majority of whom seem to possess a universal fixation that bike parts have to be as lightweight as possible (. . . more on this theme later!). What type of steel do we choose for our frame? Since alloy additions to steel make negligible difference to a steel's density, and assuming that we can identify a suitable alloy to provide us with sufficient strength benefit for the extra cost incurred, the first design decision required is . . . what size of tube to buy? By considering previous successful designs, it appears that a typical steel bike frame is made from something like 30 mm diameter tube. Although intuitively we might imagine that a frame made of 150 mm diameter tube would be stronger than one made of 30 mm diameter tube, the 150 mm diameter tube would create certain inherent problems that might prove rather painful to the cyclist during use.

So, if we agree on using a traditional 30 mm diameter tube, we are left with the final design detail of specifying the actual wall thickness. Steel stockholders' data sheets identify that a 30 mm diameter steel tube with 1 mm wall thickness weighs 0.2 kg/ft (another mixed unit!), while the same tube with a 3 mm wall thickness weighs 0.6 kg/ft. Since a typical bike frame contains about 6½ ft of tubing, if we opt for the 3 mm thick tube, rather than the 1 mm tube, we incur an additional weight penalty for our bike frame of about 2½ kg (5½ lb).

It might now seem logical to further our deliberations by calculating what a theoretical minimum tube wall thickness might be in order for it to be strong enough to function in a bike frame without breaking. However, it is possibly more helpful at this point to delve briefly into the psychology of the cyclist, and to relate his or her perceptions and desires to the manufacturing, marketing, advertising and selling practices of bike manufacturers world-wide.

10.3.2 The lightweight paradox

Over the last decade the image of cycling has changed dramatically. Not only has the sport evolved to encompass mountain biking, stunt cycling, and other bicycle-linked *dangerous sports*, but the bike has increasingly become a fashion icon, reflecting a demand by cyclists for a product that projects an outward expression of owning and using a highly engineered, sophisticated, state-of-the-art machine. Part of the perception of those participating in this highly fashion-conscious sport is that a saving of a few grams in weight by utilising seductively designed components made from modern lightweight metal alloys will make all the difference between what can only be described as a good ride, or something rather less satisfying. On this contentious subject, bike *aficionados* assure me that often the two primary considerations that are taken into account when purchasing a bike, or bike components, are weight-saving and image, . . . and, usually in terms of priority, by a short head in that order.

The fact that saving a few grams in component weight may well be nullified as a result of some common cycling practices does not seem to diminish the demand for lightweight components. For example, stopping at a local hostelry and taking on fuel in the form of pasties, chips and a couple of pints will certainly increase the total weight of rider plus bike by far more than the few grams saved in buying the latest super lightweight component.

Paradoxically, in terms of cost to the cyclist, there also seems to be a general rule that the cost of a component is inversely proportional to its weight, since *lightweight* equates to fashionable, and, as we all know, fashion-related items can be surprisingly expensive; as can the cost of many modern bikes. It is now not unusual for a half-decent bike to cost as much as a half-decent second-hand car, a price equality that would have been considered out of all proportion ten or more years ago.

However, one should not be misled into thinking that a preoccupation with weight saving in cycling is a modern phenomenon, as a report on the 1888 Annual Stanley Exhibition of Cycles makes clear, which states

> . . . we thought that the limit of extreme lightness had been reached only last year when a racing bicycle was exhibited weighing only $17\frac{1}{2}$ lb only; but this year a racing cycle with a 32 inch driving wheel, exhibited by Messrs. Ashton brothers of Clapton, turned the scale at 11 lb . . . nothing is new![3]

10.3.3 How strong is strong enough?

To return to the problem of designing a bike frame in steel, in asking the question, what wall thickness should we chose for the tube?, we are really trying to answer the fundamental question of how strong must the frame be in order for it not to fail. To answer this question we need to know some precise details about the loading on the frame. Unfortunately, when we start to think about specific numbers, we actually have only a very rough idea of any loading values that might be of use to us. However, since we have accepted the challenge of designing a bike frame, perhaps as a short cut to obtaining a design solution, it might prove useful to address our design problem from another direction and ask a slightly different question, which is . . . how do we think the frame is likely to fail?

10.4   Failure by fatigue

On the basis that reducing a problem *ad absurdum* sometimes provides a clue to a possible solution, consider what would happen if we made our bike frame from thin aluminium foil, such as that used in cigarettes or chocolate bar wrapping. Being presented with such a frame, we wouldn't be surprised if the frame immediately collapsed either under its own weight or as soon as we got on it. On the assumption that our steel frame is going to be strong enough not to collapse immediately it is sat on, and even survive the first few miles of cycling, then, by a process of iteration, we are suggesting intuitively that failure will, in some way, be a time-dependent process, and, more specifically, that it will be related to the loading history that is applied to the frame during its lifetime. In other words, we would have good reason to suspect that, apart from any case of abnormal over-load, frame failure is likely to occur as part of what is described in most engineering textbooks as *fatigue failure*.

These same textbooks will also tell us correctly that fatigue failures are progressed over a relatively long period of time, where the failure is initiated by the generation of small cracks that are formed as a consequence of the repeated application of relatively small loads. Over a period of time, these small cracks grow until, finally, sudden catastrophic failure takes place through the slow reduction of sound material. Traditionally, fatigue cracks are nearly always initiated at surface defects, changes of section, or similar, i.e. at any part of a structure where a stress concentration or stress enhancement effect occurs.

10.4.1 Stress concentrations

Thus, in trying to ascertain how thick our steel tube has to be in order to make a bike frame from it, we need to have a good idea about the stress levels produced in all parts of the frame, while it is being used. In particular, we need to know about imposed stress levels at changes of sections, joints and fittings, bolt holes, welded or brazed areas, etc. or in any other region likely to be prone to fatigue crack growth. Furthermore, we need to know how these stress levels will vary for different types of riding terrain, such as on- or off-road, and also for different rider weights, since thin, fat or even extra large cyclists, will subject a frame to very differing loads.

The problem has suddenly become almost over-complex, and, unless you have unlimited confidence in techniques such as finite element simulation and fatigue life prediction, I would suggest that it is unlikely that we can confidently obtain any absolute values about the real loading conditions imposed on a bike frame subjected to multiple use conditions. And, at this stage, let us not forget that manufacturers and retailers of bike frames will only profitably sell a bike frame that appeals to the cycling fraternity; implicitly that means that it must be as lightweight as possible.

10.4.2 The metal of the gods – titanium

By now, I guess some of you are frothing at the mouth with impatience, and silently (or not so silently) voicing comments along the lines of . . . haven't you heard that we use titanium alloys nowadays? It is certainly true that titanium alloys (named, of course, after the first Greek gods, the Titans) are as strong as, or stronger than, alloy steels, and only weigh about half as much as steel for the same volume. Furthermore, in theory, the fatigue strength of titanium alloys is excellent, and is often quoted as being 10× better than steel. If one searches the web for information regarding the use of titanium in bikes, one finds that the bike manufacturers who use titanium alloys cannot find anything detrimental to say about titanium alloys. Claims such as 'titanium is the *only* material for bikes' will be made by most titanium alloy users, some of whom even offer a lifetime guarantee on their products. Unfortunately, when the small print of many of these guarantees is perused, it appears that such lifetime guarantees exclude failures due to wear and tear.

10.4.3 Scratches and similar

So, a word of warning. . . . Apart from being expensive, titanium alloys are difficult to machine, weld, heat-treat, join, etc. and any of these processes can seriously adversely affect textbook-quoted fatigue life values. For example, surface damage of the order of $500 \mu m$ in size (0.5 mm) typically may reduce the fatigue strength of common titanium alloys by half. In extreme cases, the fatigue strength of titanium alloys can even be reduced by up to two orders of magnitude if critical small scratches or defects are introduced into the surface of a structure.[4] Good advice is given by manufacturers of standard titanium alloys who recommend that:

> since fatigue properties are dependent on surface preparation (and often follow a log normal statistical distribution), in the absence of pertinent experience designers are well advised to develop their own fatigue data for the actual part configuration for use.[5]

Such advice should at least strike a note of caution in those cyclists preparing to invest in titanium alloy bikes.

However, in all fairness, I have to admit that the only failure of a titanium bike frame that I have ever seen occurred to a well-known make of bike when it was being ridden off-road for the first time by a large male student, who happened to be one of the university's leading oarsman. In his case the frame accommodated some of the random loading imposed on it by simply changing its shape. However, as the wealthier cyclists among you will know, top-quality titanium frames are manufactured all around the world, and these certainly don't change shape even when ridden by the heaviest rider. Unfortunately, the down-side is that, if you want to buy one of these frames, your bank balance may well be depleted by $2000 or more and, since this will only be enough to buy the frame, you may be left robbing your piggy bank to pay for the rest of the bike.

10.5   Bike failures – some case studies

By now, the more astute among you will have gathered that I am not convinced that simply considering a list of mechanical properties for different materials will help very much with untangling the puzzle of choosing suitable materials for use in bikes. If the antelope skin works for a bike chain, the moral of this story would seem to be, that within reason, any material might be OK. A more profitable approach might be taken, however, if, instead of trying to identify specific materials for a component, we consider the failure characteristics of a number of components regardless of material. Following this approach, it might be hoped that an examination of real failures will point to a commonality of cause, which can then be applied to any component, no matter what it is made of. Following this reasoning, a selection of case studies will therefore be detailed in order to identify any failure causation similarities.

10.5.1 A pedal-spindle failure

Figure 10.1(*a*) shows a composite photograph of a steel pedal spindle as fitted into the pedal housing. Figure 10.1(*b*) shows an almost identical pedal spindle that fractured during use (top), a similiar unbroken spindle (middle) and the fracture surface of the failed spindle (bottom). The characteristic tell-tale signs of fatigue failure of the spindle can be seen clearly in the bottom picture.

Manufacturing cost

The fractured spindle was part of a pedal set, originally fitted onto a mid-price range mountain bike. The bike had been ridden sporadically on

10.1(a) Pedal spindle as fitted.

normal roads for about a year before the pedal spindle broke. When the spindle broke, the cyclist fell off and subsequently had to undergo repeated surgery to realign a dislocated shoulder blade. Although originally fitted onto a reasonably expensive bike, somewhat surprisingly, identical replacement pedal sets were available from bike shops for as little as £5.95 per pair. Bearing in mind the profit margins added to this final sale price by the bike shop, the importer, other possible retailers, and the original manufacturer, etc., it is clear that the pedal set inherently had an intrinsic value of less than £1, of which the steel pedal spindle could be valued in pence.

A simple design calculation

The observant among you will have noticed that the failure of the spindle (by fatigue) occurred at a change of section, where the spindle diameter reduces from just over 9 mm to about 7 mm . . . and, armed with this information, you might care to do some 'back-of-an-envelope' calculations. If you consider the extreme to be a case where all the rider's weight acts on the outside edge of the pedal, you will be able to calculate the bending

10.1(b) Broken spindle, unbroken spindle, fracture surface.

moment along the spindle (which increases with distance from the pedal end), and with this information be able to carry on to find the maximum tensile outer fibre stress acting on the 7mm diameter spindle. (Note that this stress is obtained by dividing the product (My) by a function of the diameter of the spindle, raised to the power of 4.) If you compare your final calculated maximum stress with typical yield strength values for steel quoted in the textbooks, you may be surprised to find how close these two values are. And I should point out that no allowance has yet been made for any stress intensification effect created by the change of section itself.

A partial solution

This particular pedal spindle has now been redesigned so that the section reduction from 9mm down to 7mm now occurs much closer to the pedal end: a change in design that does something fairly drastic to the calculated stress value. However, apart from demonstrating that basic strength of materials theory can actually be quite useful in the design process, and that fatigue failures really do occur at surface defects, changes of section, etc., perhaps this failure also emphasises the fact that, although we might expect to get the quality that we pay for, . . . only a very small proportion of what we actually pay might in reality represent the design effort, or lack of the same, which has been put into the component . . . *caveat emptor*!

10.5.2 A handlebar failure

Figure 10.2 shows the middle section of a failed aluminium alloy handlebar, where the fracture surface of the broken tube indicated failure due to fatigue. The bike was being ridden downhill in rush hour traffic when the handlebar snapped in two. The cyclist (an engineering student) fell off, but considered himself fortunate in that the bus coming along behind him avoided running him over. The handlebar had been purchased from a reputable supplier for about £30. The student had chosen this particular handlebar because it was a few grams lighter than some of the alternatives on offer, and also because it had an attractive fashionable logo printed on it. When the student brought me the broken handlebar for my black museum, I asked him if, as an engineer, he would consider a 22 mm diameter, 1.5 mm wall thickness, aluminium alloy tube, costing about £1.50p per metre, as adequate to support an imposed 80 kg cyclic load applied 300 mm away from a rigid cantilever fixing. 'Probably not' was his reply. I think his answer was correct.

10.2 Failed aluminium alloy handlebar.

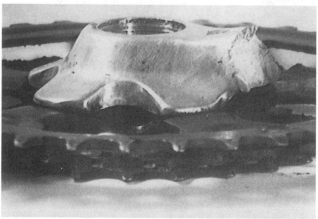

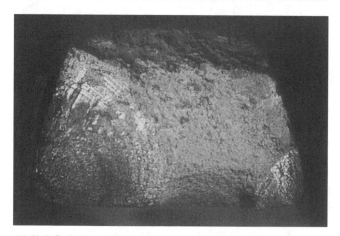

10.3(*a*) Failed crank and fracture surfaces.

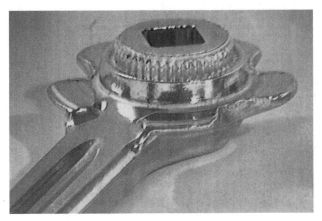

10.3(b) Unbroken crank.

10.5.3 The good-looking crank

Figure 10.3(*a*) shows a chain wheel crank that failed after relatively light use after being used only on normal roads. Again, the fracture surface (bottom) shows clear evidence that failure was due to fatigue. For comparison purposes, an identical undamaged crank purchased as a complete crank set for less than £20 is shown in Fig. 10.3(*b*).

The broken crank had been fitted as original equipment on a bike that had never been ridden in an off-road environment. As can be seen, the crank broke only a short distance along from the crank spindle end, and, upon examination, exhibited all of the classic signs of fatigue failure. Final catastrophic failure occurred after incremental crack growth had reduced the remaining cross-sectional area of the crank to a critical size. There is little doubt that the crack initiation site coincided with a designed-in undercut, which, in part, appeared to have been incorporated into the aluminium alloy casting solely to give the crank an appearance of possessing an expensive independent spider fixing onto the chain wheel. It is hardly surprising that this designed-in stress raiser (whose sole purpose seems to be aimed at giving an impression of mimicking a more expensive product) promoted the catastrophic fatigue failure shown in Fig. 10.3(*a*).

Unavoidable stress raisers

More typically, premature crank failures often show themselves as a fatigue failure across the pedal spindle end of the crank, as shown in Fig. 10.4. Here,

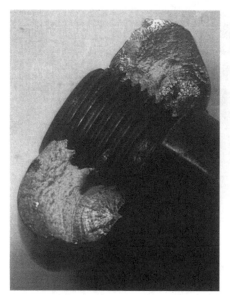

10.4 Small end crank failure.

a fatigue crack has propagated from the shoulder of the threaded pedal spindle hole through the small end of the crank.

In this case, the thread roots have provided a classic stress intensification site, a situation which can all too often be made substantially worse if the shoulder on the crank is damaged (scratched or scored), when the pedal spindle is screwed in and tightened into the crank. For those of you about to fit new pedals to your bikes, our in-house laboratory fatigue tests indicate that the frequency of this type of crank failure can be reduced greatly if some form of plastic cushioning washer is fitted on the pedal spindle shoulder before fitting the spindle into the crank.

Surprising crank failures

Nearly all crank failures tend to surprise the rider when they occur. However, some crank failures can also initially surprise any subsequent investigator, since failure by fatigue can occur at a position along a crank where no obvious stress raisers, or overload, are anticipated. Figure 10.5 shows an example of such a failure. Further discussion on these types, and other types of crank failure can be found in Reference 6.

10.5 Crank failure with no obvious stress raisers.

10.6   Pedal cycle injury statistics

Even before considering further examples of bike component failure, it is clear that cycling, like every other sport or pastime, has its hidden dangers, and that simple failures of bike components can lead easily to serious injury of the rider. Perversely, however, it appears that, to a large extent, cyclists accept these self-imposed injuries, and also routine component failure, as part of the cyclists' lot.

Department of Transport statistics for the UK show that, on average, about 25 000 people are injured each year as pedal cycle casualties, of which approximately 200 are fatalities.[7] Worryingly, however, the Department of Transport also suggests that pedal cycle injuries might be under-reported by as much as two-thirds, since it appears that cyclists often just don't report their accidents.

Unfortunately in their statistics, the Department of Transport do not suggest how many of the reported injuries may be due to pilot error and how many due to bike component failure. Furthermore, it seems to be par for the course that, even in the case of cyclist fatalities, the investigating authorities tend to assume human error rather than bike component failure as the principal cause of an accident. On the other hand, as remarked by police accident personnel, after a bike and cyclist have been in collision with, or run over by, a motor vehicle, little evidence of pre-accident component failure may remain.

However, with the increased use of bikes for local commuting and also for use in off-road environments, it is not surprising that, occasionally, the dangers of cycling are brought to our attention by the national press. A report in the *Daily Telegraph* titled 'Mountain bike trend sends injuries soaring' speaks for itself, where it was reported that 'the growing popularity of cross-country mountain biking has led to a rise in the number of ruptured organs, head injuries, and broken limbs'.[8]

British (and other) Standards

Logically, once the possibility of linking serious injury with bike component failure is accepted, the question that is raised immediately is: 'surely there must be some safety standards that bike components have to comply with before they can be sold?' . . . and, of course, the answer to this question is 'yes, there are'.

However, the subject of standards, even for bikes, is vast, and not surprisingly, complicated as well. Even so, to illustrate the apparent present state of confusion, it is probably worth briefly dipping our toes in this particular murky pond and considering how present standards relate to just one specific bike component . . . the crank.

British Standard BS 6102 part 1 1992, *Specification for safety requirements for bicycles* identifies, as might be expected, that the purpose of the British Standard is to ensure that 'bicycles manufactured in compliance with it will be as safe as possible'.[9]

With regard to the crank-pedal system, BS 6102 only requires a single dead load test of 1000N–1500N to be applied, where, with the crank fixed in a horizontal position, the dead load is applied vertically downwards for 15 seconds' duration.

The International Standard (ISO 4210) requires a similar dead load test, but also, in addition, a fatigue test where, with the cranks fixed at 45 degrees from the vertical, a cyclic load of 1400N is applied vertically to the pedals for 50000 cycles. For the maximum permitted loading rate of 25 cycles per second, this test can be completed in about half an hour.[10]

The Japanese Industrial Standard JIS D 9415–1993 specifies three tests: dead load, impact and fatigue. With regard to the crank-pedal system, JIS D 9415 identifies a single dead load test of 1500N where, with the crank fixed in a horizontal position, the dead load is applied vertically downwards for 1 minute. In addition, two impact tests are specified: one, with the crank horizontal, where a 100N load is dropped from 150mm onto a pedal, and the other, with the crank vertical, where the same load is dropped from 1000 mm onto a pedal. Finally, JIS D 9415 specifies a fatigue test, where, with the cranks fixed at 55 degrees from the vertical, a cyclic load of 1200N is applied vertically to the pedals for 100000 cycles. For the maximum permitted loading rate of 1 cycle per second, this JIS D 9415 fatigue test will last about 27 hours.[11]

What is clear from the above is that existing standards vary enormously in their requirements, and that some are far more demanding than others. From the tests described above, it would appear intuitively that, at the present time, the most arduous test for cranks is the Japanese Industrial Standard, followed by the ISO Standard, with the least arduous being the British Standard. However, one assumes that cranks sold in the UK have to pass only the British Standard test as a requirement. Further discussion on this topic can be found in Reference 6.

10.6.1 Crank fatigue tests

From fatigue tests carried out in the laboratory[12] it was observed that, when cranks were tested under JIS fatigue testing conditions, high-quality machined cranks would typically fail after about 30 hours, while high-quality forged cranks typically would fail after about 40 hours. For the cyclist, it is interesting to note that the forged cranks tested are now sold by the manufacturer on a free exchange basis, should they ever fail during use. As far as the author knows, none has yet been returned. However, life is never perfect, and some cranks, despite passing the JIS fatigue test, still fail by fatigue during normal cycling.

A test programme involving strain gauging the side face of a crank allowed the maximum stresses developed on the side face to be measured during each loading cycle of the JIS fatigue test. The results obtained during the JIS fatigue test (1200 N at 1 cycle per second) showed that, with the crank in a simulated bottom dead centre position, a maximum stress of about 90 N/mm^2 was produced on the crank's outer side face.[12]

10.6.2 Stress monitoring

An off-road mountain bike was fitted with identical strain gauged cranks as in the JIS test so that the maximum stresses in the crank could be monitored dynamically during *bouncing tests* in the laboratory, and also while cycling on real roads. A large male student, bouncing on the pedals, with the crank at bottom dead centre and the bike stationary, produced maximum side face stresses in the crank of about 120 N/mm^2: a stress value approximately 33% higher than the maximum stress induced during the JIS test. However, such extreme loading cannot really be described as representative for normal road cycling, even though similar *bouncing* activities by largeish riders may actually occur relatively frequently during mountain, cross-country and stunt cycling.

10.6.3 Road testing

To monitor crank stresses during normal road cycling, a circular 700 metre road route was identified around the University of Bristol, which included

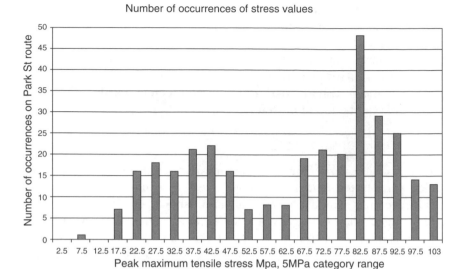

Number of occurrences of stress values

10.6 Maximum stress values per rev.

cycling from the bottom to the top of the 20% gradient (cardiac arresting) Park Street.

Figure 10.6 shows, for the crank side face, a histogram of the maximum stress values obtained per crank rotation. What can be seen from Fig. 10.6 is that, although the $120 N/mm^2$ *bouncing* stresses were not generated during normal road riding, maximum peak stresses of just over $90 N/mm^2$ were encountered regularly. Out of the 330 stress values obtained during the 700 m circular ride, approximately 50 of these values were above the 90 N/mm^2 threshold produced in the JIS fatigue test.

If a good-quality crank typically survives the JIS fatigue test for 30 hours, with induced side face stresses of $90 N/mm^2$ applied at a rate 1 cycle per second, the crank is said to have a fatigue life of about 100 000 cycles at this stress threshold. Over the 700 m normal cycling route, 50 crank rotations produced stresses of $90 N/mm^2$ or higher. Thus, at this sort of loading it can easily be shown that the fatigue life of even a good-quality crank may well be reached after cycling approximately only 1400 km. If you are a typical cycling commuter, averaging about 12 km a day, if you regularly cycle up 20% gradients, your cranks may fail by fatigue after about 6 months of use. On the other hand, they might last a year if you are lucky enough to be able to freewheel down the 20% gradient on your way home.

In a perfect world, an international standard would be defined that tested against crank fatigue failure under all cycling conditions. However, let us not forget the fixation that cyclists have with purchasing lightweight com-

ponents, nor the fact that, allegedly, the population is getting ever larger
... and heavier!

10.6.4 Who pays when it hurts?

Let us assume that you are cycling happily along and a bit of your bike
breaks. The consequences of the failure can be assumed to be one of three
alternatives. One, you survive without injury. Two, you are injured superfi-
cially, and limp back to repair the damage. Three, you are injured seriously
and can't limp anywhere! It could be that your particular component breakage is linked with a cycling *recklessness* on your part and your accident is
really of your own making. On the other hand, maybe the accident wasn't
your fault at all, and you think reasonably that the component shouldn't
have broken in the first place. What can you do to obtain compensation
from the bike retailer or from the component manufacturer?

Most of us will have vaguely heard of the Sale of Goods Act, and the
Consumer Protection Act, and within these contexts may also have read
about such concepts as product liability, not fit for purpose, sub-standard,
etc. The following two salutary tales illustrate examples of the Consumer
Protection Act in action, and relate to two recent cases of injury-related
bike component failure.

10.7   The exploding wheel rim (case 1)

Mr H. was cycling along a country path when the aluminium rim of his
front wheel exploded catastrophically and caused the front wheel of his
bike to jam in the front forks. The result was that Mr H. somersaulted
over the handlebars of his bike, after which he was taken unconscious to
hospital. Subsequently, as a result of his injuries, Mr H. was retired from his
work on grounds of ill health. In due course, Mr H. brought an action
against the manufacturers of the wheel rim, claiming substantial damages
for injuries received and loss of employment. Mr H.'s exploded rim is shown
in Fig. 10.7.

In an engineering sense it is clear that all wheel rims are generically the
same, i.e. circular and of standard dimensions, and perform the same fundamental engineering function, i.e. retain the tyre and spokes, etc. However,
wheel rims can be made from steel, aluminium alloy extrusions, or even
from carbon fibre. In addition, rims may also be supplied with a variety of
fashionable surface finishes, such as ceramic coatings, anodised layers, etc.
For the sake of this tale, however, we are concerned only with Mr H.'s rim,
which was made of aluminium alloy, and as such, bearing in mind today's
cyclist's preoccupation with all things lightweight, is representative of the
majority of most rims used on bikes today.

10.7 Mr H.'s exploded wheel trim.

10.7.1 Rim profiles

Diligent searching through the discarded scrap-heaps of various bike shops reveals that, although every make of rim differs in detail from every other rim, all rims appear, more or less, to belong to one of four generic types of extrusion. Figure 10.8 shows examples of these four rim types, which I have called (from top to bottom): (i) solid, (ii) arched-reinforced, (iii) side-tube reinforced and (iv) arched and side-tube reinforced. Types (ii), (iii) and (iv) all appear to be in fairly common usage, and, as can be seen, differ only in respect of the fundamental method of reinforcing and stiffening of the rim.

10.7.2 Loading of the rim

What is it that's going to allow a bike rim to explode catastrophically? Consider, in a simple way, the actions and reactions that a bicycle wheel might undergo during cycling. The tyre contains an inner tube, which is pumped up with air to keep it inflated, and therefore has a constant internal pressure applied to it. In addition, the tyre flexes sideways and up and down at every sequential point around its circumference as each point meets the

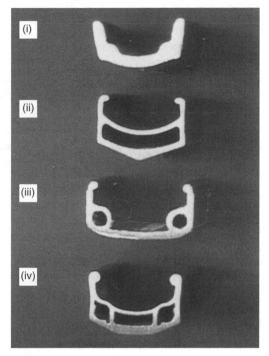

10.8 Four generic rim profiles.

road and supports the weight of the rider and bike. As well as reacting against this more or less steady cyclic loading, the tyre will also have to accommodate random loadings due to the tyre passing over bumps, pot-holes, stones, etc. The overall flexing of the tyre will, in some sort of way, transmit a sideways cyclic load into the side-walls of the wheel rim. An extra cyclic loading will also be transmitted into the side-walls of the rim when the brakes are applied. When pulled *on*, the brake blocks squeeze the rim side-walls together, while the rim passes between the brake blocks. Once a part of the rim has passed through the brake blocks, the rim side-walls recover elastically to their original dimensions. Naturally, it seems reasonable to hope that the design of wheel rims, inner tube and tyre arrangements are calculated to withstand all of the applied stresses and strains briefly described above.

10.7.3 Wearing of the rim

With use, the sideways bending strength of the rim side-walls decreases, primarily as a result of the brake blocks wearing the side-walls away each time the brakes are applied. How quickly this wearing away of the side-walls occurs in an aluminium rim is dependent on a number of factors such as how often the brakes are applied, how severe the braking is, what sort of terrain the bike is being ridden on, what sort of conditions exist between the brake blocks and the rim during braking, and what material the brake blocks are made of. However, wearing the rim away when the brakes are applied is as natural a consequence as night following day. In other words, the two actions are inseparable, and no one would ever suggest anything different. After all, what else could the result be when the brakes rub on the rim? The answer to that question depends on whether continued wearing away of the rim necessarily leads to catastrophic fatigue failure of the rim, of the sort experienced by Mr H.

And, . . . with regard to assessing liability, there are two more questions that need to be addressed at the same time. Firstly, can the amount of wear that has taken place be easily seen or measured; and secondly, does the industry issue any warnings that rim thinning can lead to catastrophic failure?

10.7.4 Weakness of the rim

From investigations of a number of explosive rim failures, it would appear that explosive failure takes place if the side-wall wear has occurred in such a way as to leave the side-wall thinned down or unsupported at the inter-section where the side-wall joins the reinforcing support. If, on the other hand, wear occurs, and the vertical side-wall remains well supported, then *safe* failure may well occur.

Figure 10.9 shows a number of worn rims where superimposed arrows have been drawn in to indicate the observed wear area.

In Fig. 10.9 the rim on the left-hand side is part of Mr H.'s rim, which failed in an explosive manner. As can be seen, wear of this rim has taken place to weaken the side-wall at the intersection of the arched support. For the other rims, however, none of which failed catastrophically, it can be seen that thinning of the side-walls has not left the rim side-walls unsupported, but rather has thinned the side-walls at some distance beneath the rein-forcing support.

Wear occurring beneath the reinforcing support point, leading to a safe failure, is amply illustrated in Fig. 10.10. Here it can be seen that the side-wall of the rim has, in parts, completely worn away (to the extent that a coin can be placed into the gap left in the extrusion). However, despite this extreme wear, the rim has still remained structurally intact.

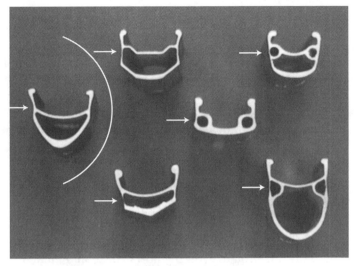

10.9 A selection of worn rims, maximum wear as indicated.

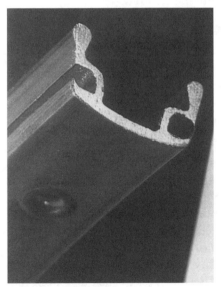

10.10 A 'safe' rim failure.

10.7.5 Measuring wear

It has to be admitted that measuring the amount of rim wear existing in a rim is more difficult to carry out than might be first thought. As the side-walls of the rim wear away, they become less stiff. However, the action of the internal pressure in the inner tube, which pushes the rims outwards, combined with the restraining action of the tyre bead, seem in practice, to keep the distance between the outside surfaces of the side-walls more or less constant. Furthermore, even removing the tyre from the rim does not allow easy measurement of the thickness of the side-wall, since the internal radius of the rim is often smaller than the radius of most commercially available ball-micrometer heads. This means that it is often impossible to actually make micrometer contact with both surfaces of a rim at the point where wear has taken place.

10.7.6 Failure to warn

As to industry warnings of possible catastrophic rim failure, all I can say is that, for all the rim failures I have examined, none of the bike instruction manuals make any mention that, during normal use, aluminium rims might, in a relatively short time, wear away to the extent that they can fail catastrophically. It follows, of course, that no industry instructions are given suggesting how the wear of such rims could be measured.

Lastly, before we see how the law judged Mr H.'s accident, it is relevant to point out that the phenomenon of exploding rims is not a new problem, and has been documented on and off in a number of public articles since 1993.[13]

10.8   The Consumer Protection Act

Six years after Mr H.'s accident, his case finally reached the Central London County Court.[14] Mr H.'s barrister argued that the basis of Mr H.'s claim centred on legislation encompassed in the 1987 Consumer Protection Act. To explain this legislation, Mr H.'s barrister referred specifically to the precedent of a recent, high-profile, test case in which a group of claimants, infected with hepatitis C virus after receiving blood transfusions, claimed damages from the National Blood Authority. Mr Justice Burton, who tried the hepatitis case, made judgment in favour of the claimants, and, in so doing, clarified the relevant aspects of the 1987 Act to such an extent that his judgment is now regarded as an authority on its interpretation. Parts of this interpretation are detailed below.[15]

10.8.1 Interpretation of the Act

The Act requires for the injured party to prove that damage has been caused, that a defect existed, and a link exists between the damage and the defect.

The purpose of the Act was to see whether a defect could have been (previously) discovered or eliminated. Mr Justice Burton ruled that, within the Act, the state of scientific or technical knowledge meant the most advanced knowledge accessible to anyone at the time the product had been put into circulation. The Act is not concerned with the knowledge of the manufacturer or producer in question, or any similar producers. Thus a manufacturer is denied the defence that he did all he could be reasonably expected to.

It is not necessary for the claimant to prove negligence in order to prove his case. Mr Justice Burton considered that liability is defect based and not fault based. In other words, as soon as the possible dangerous nature of the product has been discovered by someone, somewhere, the producer will be liable if he continues to supply those products.

The Act also defines 'that a product is defective when it does not provide the safety which a person is entitled to expect, taking all circumstances into account'. On this point, Mr Justice Burton considered that, with regard to this aspect, a claimant might, in law, be entitled to expect a greater level of safety than that actually expected by the public at large.

Mr H. won his case against the rim manufacturers. Surprisingly, however, even in these days of increasing awareness and litigious escalation, it is understood that Mr H.'s case is only one of three existing product liability cases to have been litigated successfully in England since the introduction of the Consumer Protection Act 1987. Further details of this case can be found in Reference 16.

10.9   The exploding wheel rim (case 2)

A second similar rim failure accident involved Mr B. who, as part of his employment as a youth instructor, was responsible for organising and leading weekly road cycling for a group of apprentices. The bikes used by Mr B. and the apprentices were, to all intents and purposes, identical and, as later became clear, most importantly were all the property of Mr B.'s employer. Mr B.'s description of his accident is as follows:

> It was during an on-road ride that my accident occurred. I was at the rear of a small group of cyclists, we had been riding steadily for about half an hour when we started to descend a hill. I braked lightly to maintain my position. My speed at the time was 20 mph or less. There was a load bang, which was the last thing I remember until I came to some time later. The accident was caused by the failure of the rim of the front wheel, causing a sudden deflation and jam. I was cast head first over the front of the bike onto the road. Luckily, I was wearing a helmet, which was heavily scored. Prior to using the bicycle, I had cleaned and oiled it, checked the tyre

pressures and given it a superficial once-over. During the ride prior to failure, I had used the brakes several times and found no problem. The failure was sudden and unexpected.

Mr B.'s rim failure (shown in Fig. 10.11) was almost identical to Mr H.'s, even to the extent that the rim manufacturer was the same. Unfortunately, Mr B.'s injuries were more far reaching and severe than Mr H.'s, and unfortunately, as a result of his accident, Mr B. is likely to suffer some disablement for the rest of his life. While progressing legal action for damages, however, Mr B. found that his employer's insurers, after receiving barristers' expert opinion, were prepared to meet his claim without going through the courts. The barrister's opinion was that the Mr B.'s case was entirely covered by the precedent of the recent bicycle case of 'Stark v. The Post Office'.[17]

Mr Stark was employed as a postman by the Post Office who provided him with a delivery bicycle. While at work, Mr Stark was riding his bike when the front wheel locked due to a metal stirrup on the front brake breaking into two and one of the pieces lodging in the front wheel. Mr Stark was flung over the handlebars and suffered serious injury. The cause of the break was apparently due to metal fatigue or a manufacturing defect. The case went to the Court of Appeal, where, despite the fact that all parties agreed that the Post Office had maintained Mr Stark's bike as well as it

10.11 Mr B.'s exploding rim.

could have, it was argued that an absolute duty was imposed on employers with regard to equipment used by employees.

The Court of Appeal found in favour of Mr Stark and against the Post Office, and ruled that it did not matter that the fault could not be discovered during a routine inspection, nor did it matter at all that the employer could not have detected the problem before the accident happened. To use the word *luck* is of course entirely inappropriate when considering Mr B.'s accident. However, if Mr B. had any *luck* at all, it must surely be that his accident occurred on a bike provided for him by his employer which, at the time of the accident, he was using during his employment.

10.9.1 National watch-dogs?

Of course, once the possible financial cost and risk of any type of court action is considered, even contemplating taking a bike manufacturer, or retailer, to court would seem to be out of the question for most of us. It is, therefore, with some slight sense of reassurance that occasionally government bodies have taken it upon themselves to censure bike and component manufacturers when they consider an ongoing problem represents an unacceptable risk to the public.

For example, it has been well reported that, in 1997, The Shimano American Corporation voluntarily recalled more than 1 million bicycle cranks after a number of failures had been reported to them. At the same time, after an investigation by the US Consumer Product Safety Commission (CPSC), Shimano agreed to pay a civil penalty of $150 000 to settle allegations that they violated the Consumer Product Safety Act by failing to report, in a timely manner, a defect with three of their bicycle cranks. That The Shimano American Corporation voluntarily paid $150 000 might naively be construed to suggest that there was some foundation in the CPSC's claim that the Shimano cranks were possibly faulty. However, such an inference by the reader should be strongly resisted, since the CPSC clearly state in their published settlement details: 'This settlement agreement and order is agreed to by Shimano to avoid incurring additional legal costs and does not constitute, and is not evidence of, an admission of any liability or wrongdoing by Shimano.'

10.9.2 Product recall

Although the reporting of individual cases such as Shimano's crank recall can provide headline news for specialist bike magazines throughout the world, in reality, items pertaining to bike safety appear only rarely in the national press. Such a lack of reporting might be seen as implying that the incidence of bike component recall seldom occurs. For those of you with time on your hands, however, a quick visit to the CPSC website,[18] and a simple search under the word *bike*, will probably make you wonder which bike components, if any, won't break while you are out cycling. At the time of writing, 230 search results were returned from the CPSC website after searching on the word bike. Typically, the search produced headings such as: 'Diamondback mountain bikes recalled by Sachs bicycle; Technium mountain bike frames recalled by Raleigh; Mountain bikes recalled by Dynacraft Industries; Bike brake and handlebar recalled by Specialized' . . . and so on.

On the other hand, it could be that these results are only the tip of the iceberg. Serious web browsers will soon confirm that, for the same CPSC site, 480 search results are returned for a search on the word bicycle, and a staggering 4500 if you try the combination of bike, bicycle and recall together.

10.10 Conclusions

So, finally, where have we got to in our attempt to discuss bikes, materials and all things relevant?

Certainly, it appears all too easy to design a bike or bike components that will, unfortunately, break unexpectedly during use. It is certainly much more difficult to design bike bits that won't break. Also, for reasons hopefully well argued, most bike failures are likely to be time-dependent fatigue-type failures, which will originate at surface defects, changes of section, places of high stress, etc.

With hindsight, it is often easy to suggest design changes that eliminate particular failures, but most of these design changes would undoubtedly involve making stronger and heavier components, often using steel rather than light alloys, and probably compromising the visual image of components for sound engineering practice . . . all solutions that would be an anathema to most cyclists.

Those of you who have persevered and read to the end of this chapter might feel, if you tend to be slightly pessimistic in nature, that in terms of buying bikes and bike components, you are destined only ever to become an innocent victim of market forces. Regardless of how much you pay for a component, you cannot guarantee its engineering credentials, and may only be purchasing fashion status and image. On the other hand, the optimists among you will point out that there are thousands of cyclists happily pedalling around every day without suffering any injury as a result of component failure. Also, quite correctly these optimists can identify some extremely well-engineered bikes that are available in the market place.

Perhaps, therefore, the moral of this tale (if there is one) is that the more thoughtful among us should pay slightly more attention to the engineering quality of the components we buy, rather than be swayed by the promises of saving a few grams in weight, or of a radical change of image.

As for the most suitable material to choose for your bike or bike components, . . . well, if knotted antelope skin is OK, then almost any material will do, as long as the component design is sympathetic to the material choice.

Finally, I will finish with two thoughts. Even if knotted antelope skin was the perfect material for a bicycle chain, the probability of getting eaten by a lion in order to obtain it is almost certainly a much greater risk to you than using it on your bike. In other words, there have to be far greater dangers that we have to survive in our daily routines other than suffering terminal bike breakage.

. . . And finally, I'm sure that, out there somewhere, there just has to be a bike made from papier-mâché.

10.11 References

1. *BBC Antiques Roadshow*. Also search the web using: Giuseppe Matera, Gota, Tino Sana, Ferdinand Trautmansdorff.
2. *Testing Times* (1999). No 14, Nov.
3. *The Engineer* (1888). Feb 10 p. 118.
4. Roder O., Peters, J.O., Thompson, A.W. and Ritchie, R.O. (2001). Influence of simulated foreign object damage on the high cycle properties of Ti-6Al-4V. University of California, Berkeley.
5. Titanium Metals Corporation – Properties and Processing of TIMETAL 6-4.
6. J. Morgan and D. Wagg (2002). The failure of Bike Cranks. International Standards, Tests and Interpretations. *Sports Engineering*, **5** 1–7.
7. Department of Transport Factsheet. *Pedal Cycle Casualties. Road Accident Statistics Branch*.
8. *Daily Telegraph* (2001). Thursday, 24 May.
9. BS 6102 (1992). Cycles. Part 1. Specification for safety requirements for bicycles. British Standards Institution. London UK.
10. ISO 4210 (1996). Cycles-Safety. *Requirements for Bicycles*, 4th edn. International Organisation for Standardisation. Geneva, Switzerland.
11. JIS D 9415 (1993). Bicycle-Chain-Wheels and Cranks. Japanese Industrial Standard. Japanese Industrial Standards Committee, Tokyo, Japan.
12. Dynamic performance of bicycle cranks. Report, University of Bristol, Department of Mechanical Engineering.
13. Juden C. (1993). Exploding rims. *Cycle Touring and Campaigning*, February/March.
14. Case CK 701879. (2001). H v. Rigida, London County Court, April/May.
15. A v. National Blood Authority (2001). Queen's Bench Division, High Court, London, April.

16. J. Morgan (2002). Product liability – a worked example. *Materials and Design*, **23** 417–421.
17. Stark v. Post Office. (2000). Court of Appeal, 2 March.
18. CPSC website: http://www.cpsc.gov/

11
Materials in mountaineering

J. R. BLACKFORD
University of Edinburgh, Scotland, UK

11.1 Introduction

To gain an understanding of the materials used in mountaineering equipment, it is advantageous to have an appreciation of the activity itself. As you read this chapter, try to imagine situations where the equipment is actually used, e.g. on a rock face or snow slope on a high mountain (Fig. 11.1(*a*), (*b*), (*c*)). This will give you a greater insight into the requirements of the materials that are used in the equipment described.

Over the last 50 years or so there have been considerable advances in materials and equipment. These have created changes in the nature of rock climbing and mountaineering: standards have risen dramatically, ethics and styles have become 'cleaner', e.g. there is a move away from siege tactics, by many alpinists in high mountains, towards a more lightweight, faster approach; and the use of rock protection that damages the rock, e.g. hammered-in pitons, is no longer accepted practice in many climbing locations. The increase in standards is true for both those at the top end of the sport and 'average' climbers because, for example, with modern ropes climbers can fall in relative safety, and with 'sticky' rubber rock boots less skill is required to climb a route (or, for a given amount of skill, a harder route can be climbed). Many 'average' climbers are today climbing routes that were the hardest of the day in the 1950s – in relative safety. The increase in standards is not only due to better equipment but is also a result of improvements in transport, higher economic prosperity, increased leisure time, indoor walls and training.

This chapter will focus on the following rock climbing and mountaineering hardware: ropes, harnesses, karabiners, belay devices, ascenders, rock protection, ice climbing equipment: axes, crampons and ice screws. I have given descriptions of the use of equipment followed by the materials, manufacture and, where appropriate, incidents and failures. Helmets will be considered briefly as, in my view, they are important items of equipment (they are dealt with in more depth in another chapter).

11.1(a) As it was in the late 1950s. Margaret Wellens and Maureen Linton on the Wellenkupe with the Matterhorn in the background (Blackford collection, taken by Alice Palmer).

The interested reader wanting to pursue the topic further will find a wealth of materials science and engineering in: fabrics (windproof, waterproof, breathable, insulating and fast drying) for clothing, rucksacks, tents and sleeping bags; 'sticky' rubber for rock boots; magnetic and electronic

11.1(b) Modern rock climbing. Karl Baker sea-cliff climbing in
Southern Ireland (Blackford collection).

11.1(c) Modern snow and ice climbing. Jon Atkin climbing nearing the top of Dorsal arête in Glencoe, Scotland (Blackford collection).

materials for compasses, global positioning systems, avalanche transceivers and altimeters.

A variety of metallic and polymeric materials are used in rock climbing and mountaineering equipment. In many instances it is not the materials themselves that are particularly advanced but the combination of appropriate choice of material with ingenious design and processing.

11.1.1 Materials selection

Selection of materials is an important aspect of product design. A good understanding of product requirements is needed for both. The main factors to be considered in selection of materials for mountaineering equipment are:

- Low weight
  achieved through a combination of low-density materials and design
- High strength
  static/dynamic loading
  to pass appropriate standard
- Durability
  over a range of temperature (perhaps −40 °C to >50 °C); freeze-thaw
  sunlight – UV resistance
  corrosion resistance in, for example, water or sea water
  abrasion resistance
  degradation with time (seconds/hours/days/years)
  to pass appropriate standard
- Economic
  costs of intrinsic material
  processing infrastructure, e.g. forging press
  processing cost, e.g. power requirements
- Marketing
- Design
  ergonomics
  aesthetics

Many of these factors are interrelated, e.g. the strength of wet ropes has been shown to be much lower than that of dry ropes.

Strength and durability aspects are covered in equipment standards. These standards are often difficult to draw up as it is hard to define precisely how a piece of mountaineering equipment will be used. Making standards too restrictive has the negative effect of potentially stifling innovative design.

When considering materials used in rock climbing and mountaineering, it is worth noting that none have been developed specifically for applications in the pursuits themselves (with the possible exception of some of the modern fabrics). Rather, the more hi-tech materials have been developed in other industries, e.g. aluminium alloys developed in the aircraft industry are used for karabiners. I believe this will continue to be the case, because other industries have more resources for advanced material development and the materials are used in higher volumes. Manufacturers of climbing equipment should be alert to new materials that could be used in future product innovation.

11.1.2 Participation statistics

Market research has shown that climbing/walking/mountaineering-based sports differ from most other UK activities, i.e. they are participation not spectator sports. It is difficult to arrive at an exact figure of how many climbers there are in the UK, but, from British Mountaineering Council (BMC) membership and climbing wall attendance statistics, a figure of 150 000 active climbers in the UK can be estimated (BMC 1997).

A Mintel report (1996) gives more general statistics for walking and mountaineering (see Table 11.1).

It is clear that rock climbing and mountaineering have increased in popularity over the past 40 years from a number of secondary observations:

- increase in number of retail outlets
- pressure for parking spaces at climbing venues
- BMC membership figures (55% increase in individual membership in 1997)
- increased media coverage
- increased fashion interest in outdoor clothing

BMC membership statistics from 1992 are shown in Table 11.2.

Climbing walls

Walls are an increasingly important part of climbing in the UK; this is unsurprising when you consider the typical UK weather. Some 82% of climbers

Table 11.1 Participation statistics for walking and mountaineering

1996 Outdoor market (adults)	Million
Occasional long walks	10
Regular long walks	5
Organised hikers, some climbing	3
Mountaineering/climbing	0.5

Table 11.2 BMC membership statistics from 1992

Participation in mountain activities	(%)
Summer mountain walking	97
Winter mountain walking	92
Rock climbing	90
Indoor climbing	74
Alpine climbing/expeditions	58
Winter climbing	56
Ski mountaineering	21

82% rock climb at grade HVS and below. 66% participate in winter mountaineering up to grade III.

Table 11.3 Growth in the number of climbing walls

Year	Number of walls in UK
1988	40
1995	122
1996	169

Table 11.4 Forecast of estimated value of outdoor markets (£ million) 1996–2000

1996	1997	1998	1999	2000
2000	2200	2500	2800	3000

now use walls, 45% on a weekly basis during the winter months. Many walls now enjoy 30 000 user visits/year, and the Foundry in Sheffield has reported figures of 60 000 (British Mountaineering Council 1997).

There has been a substantial growth in the number of walls as shown in Table 11.3. A further indication of increased participation and spending is provided by Keynote (1996) who estimated the outdoor markets' worth as shown in Table 11.4.

11.1.3 Forces generated in falls

An analysis of the forces generated during a fall on the climber, rope and other parts of the system is valuable in determining the design requirements and developing equipment standards. This analysis is beyond the scope of this chapter; however, good treatments of fall analysis have been given by Smith (1998), Pavier (1996), Blackford and Maycock (2001) and Wexler (1950).

11.2   Ropes

Ropes are one of the most vital pieces of mountaineering equipment and one of the earliest to be used for protecting climbers. They were originally produced from natural fibres such as hemp or manila, twisted and 'hawser laid' (Fig. 11.2(*a*)). The introduction of the synthetic fibre nylon by DuPont during World War II marked a significant advance in providing a rope that gave a safe means of fall arrest. The modern climbing rope came into being in 1951 with the innovation of the 'Kernmantel' construction from *Edelrid* (Fig. 11.2(*b*)). A useful review of modern ropes has been published, in the climbing press, by Bennett (2000).

A rope that is suitable for fall arrest in a climbing environment requires some distinct mechanical and material properties. A rope must both withstand the dynamic force of a falling body without breaking, and must do so without exerting an unacceptably high force on the body. For a given dynamic load, if the elastic deformation of the rope is too low, the peak force experienced by the body will be damagingly high; conversely if the elastic extension is too high (cf. bungee jumping) the peak force will be easily tolerated but the chance of hitting the ground or a ledge is increased. In the case of a falling body, military parachute experiments measuring the impact forces generated during canopy opening have indicated that the body can withstand a maximum load of 12 kN without serious injury.

11.2 Construction details of (*a*) hawser laid rope (*b*) Kernmantel rope.

Reviews of the forces generated in fall situations during climbing are given by Smith (1998) and Blackford and Maycock (2001). A good case study, 'Polymeric ropes for sport activities', has been developed by Nigel Mills for the UK Centre for Materials Education and is available on their website (Mills, 2001).

Ropes may be given 'dry treatments', which coat the sheath, or better still the individual fibres, with a hydrophobic layer. This reduces the moisture uptake of the rope substantially (especially when the rope's new and the coating hasn't worn off), which is an advantage when climbing in the rain in the mountains and particularly when climbing on wet snow when the rope can become soaked and later freeze, making its normal use with belay devices, etc. almost impossible. Knots are important when using ropes for mountaineering: a guide to the types and use of knots for mountaineering is given in the BMC knots publication (British Mountaineering Council, 1997).

Classification of modern climbing ropes

Rock climbing and mountaineering encompass a wide range of activities. Because of this, different types of rope, and accompanying standards, have been developed. Modern climbing ropes tend to be 45, 50, 55, 60 or 70 m in length; 50 m is the most common length currently used in the UK. Their use, properties and standards are summarised in the BMC rope book (British Mountaineering Council, 1998).

Single (full) ropes

Single (full ropes) are used to protect leader falls, e.g. sports climbing. Classically, diameter 11 mm but now available from 9.2 mm upwards.

Half ropes

Two ropes are used together to protect leader falls; rope drag can be reduced by clipping runners selectively with either rope. Very commonly used for traditional climbing in the UK. Having double the length of rope allows abseils twice as long to be made. Classically, diameter 9 mm but now available from 8.1 mm upwards.

Twin ropes

To protect leader falls, both ropes are clipped through each piece of protection. Rarely used in the UK but common in the Alps. Generally smaller diameter than half ropes, 8 mm upwards.

Materials and manufacture

A modern dynamic climbing rope is made from continuous drawn nylon yarns. A consequence of drawing the nylon is strengthening and stiffening, as the molecular chains of this semi-crystalline polymer become orientated. These fibres can be spun to create high-strength, low-stretch 'static' rope – widely used in caving and industrial rope access where shock loading is unlikely and percentage elongation is required to be at a minimum.

To create a yarn with elastic properties suitable for arresting a fall safely, the drawn nylon is heat treated to about 120 °C. Loss of chain orientation and strain-induced crystallinity as a result of the annealing process gives the drawn nylon yarn a reduced Young's modulus. This process increases the energy absorption properties of the yarn and reduces the impact force on the body during a fall. Since the 1960s all ropes have been made from UV-stabilised nylon.

Standards

The EN 892 standard for dynamic ropes specifies the drop test requirements a rope should withstand before it is approved for sale in Europe. In the test, the rope is held statically and passes over a 10 mm diameter edge, 300 mm from the anchor. The falling mass is dropped to give a fall factor of 1.78. To pass the standard, the drop test is repeated five times and the impact force on the first drop must not exceed 12 kN for a single rope (diameter 10–11 mm) and 8 kN for a half rope (diameter 8–9 mm). In addition, other factors such as knotability, sheath slippage and static elongation are measured and have minimum requirements.

Failures and case studies

In the UK 20 incidents of severely damaged or failed ropes have been reported to the BMC over the last 15 years. Two have been because of contamination with corrosive substances, e.g. battery acid (MacNae, 1999), while the remainder were due to abrasion or cutting over a sharp edge (including one fatality) (e.g. Payne, 1994).

In practice, modern climbing ropes will not break at the knot, at a kara-biner on a runner, at the belay device or the free rope (tests have shown that this is also true for ropes 10 to 15 years old) (Schubert, 2000a). But edges or abrasion will lead to rope failure. Climbers should be aware of this and carefully consider the paths their ropes take – or would take if shock loaded.

Improved abrasion resistance of ropes can be achieved by weaving the sheath using a smaller number of bobbins, while increasing the number of bobbins increases with dynamic characteristics of the rope. At present, the standards do not test abrasion or cutting resistance of ropes, but this is an area of much debate among the standards authorities. Creating a realistic and reproducible test is a challenge for the future.

Rope can be damaged from internal abrasion from grit particles, which become embedded in the rope and lead to damage of the fibres if the rope is trodden on or deformed by belay devices, etc. (this is considered in more detail in the section on harnesses and webbing).

The rate of decrease in energy absorption capacity of a rope over an edge decreases with increasing 'metres of use' (metres of use is the sum of the distance the rope is used for climbing and abseiling). Tests have shown the energy absorption capacity to decrease to 50% of the value when the rope is new after only 2000 to 4000 m of use, and to vary from between 50% to less than 10% after 20000 m of use (Fig. 11.3, Schubert, 2000a). These values will be depend strongly on the severity of treatment the rope has experienced, but they are worth bearing in mind as most climbers will climb significantly more than 4000 m with a given rope.

The strength of nylon is reduced significantly when it absorbs moisture, which has implications for mountaineering ropes. A series of tests has recently been reported on the effects of moisture absorption on ropes (Table 11.5). The significant findings include: strength reductions corresponding to 30% of the number of falls being held in drop tests compared with a new dry rope. Thorough drying of the rope leads to a recovery of the dynamic properties. An important conclusion of this work was 'a used rope in good condition, say a rope which can still hold four to five falls in the standard (EN and UIAA) drop tests might only hold one or two falls when soaked after sudden rainfall'.

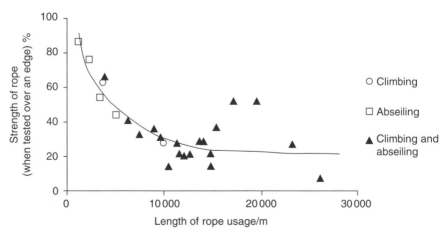

11.3 Strength of rope (when tested over an edge) against distance the rope has been used (redrawn from Schubert, 2000a).

Table 11.5 Test results to examine the effects of moisture absorption on ropes (Signoretti 2001)

Treatment	Test	Rope A Normal New	Rope B Everdry New	Rope C Normal Used
Non-treated (reference)	Number of falls	8	11	4
	Impact force (kN)	8.86	9.46	9.50
Wet in water for 48 h	Number of falls	2.3	3	1.5
	Impact force (kN)	9.26	10.22	10.52
	Falls variation	−71%	−73%	−62%
	Impact force variation	+5%	+8%	+11%
	Weight variation	+45%	+42%	+59%
	Length variation	+4%	+2%	+5%
Wet soaked for 2 h	Number of falls		3	
	Impact force (kN)		9.84	
	Falls variation		−73%	
	Impact force variation		+1%	
Wet with splashes of water	Number of falls		5	
	Impact force (kN)		9.90	
	Falls variation		−55%	
	Impact force variation		+2%	
Wet and dried in normal conditions	Number of falls	6	9.4	
	Impact force (kN)	8.67	8.12	
	Falls variation	−25%	−15%	
	Impact force variation	−2%	+4%	
	Weight variation	−	−1%	
	Length variation	−	−4%	
Wet and dried in 'extra dry' conditions	Number of falls	9	10	3
	Impact force (kN)	7.85	8.26	8.61
	Falls variation	+12%	−9%	−25%
	Impact force variation	−11%	−13%	−9%
	Weight variation	−3%	−3%	−3%
	Length variation	−7%	−8%	−3.5%
Four cycles of soaking and drying under cover	Number of falls		12	
	Impact force (kN)		8.60	
	Falls variation		+9%	
	Impact force variation		−7%	
Four cycles of soaking and drying in sunlight	Number of falls		8	
	Impact force (kN)		8.60	
	Falls variation		−27%	
	Impact force variation		−9%	
Frozen wet and kept at −30°C for 48 h	Number of falls	4	9	3
	Impact force (kN)	8.05	8.98	8.19
	Falls variation	−50%	−64%	−25%
	Impact force variation	−9%	−5%	−14%

Note: These data are the average over three specimens.

Table 11.6 Results of drop rig tests to examine the effects of the anchor or belay system, runner radius and environmental damage on rope performance (Pavier 1999)

Description	Fall factor	Length of rope (m)	Mass (kg)	Number of falls to failure
Effect of belay				
Belay tied off	0.22	4.5	70	18
Sticht plate	0.22	4.5	70	27
Sticht plate	0.22	4.5	70	43
Effect of runner radius				
Runner radius 2.0 mm	0.4	2.5	70	5
Runner radius 4.0 mm	0.4	2.5	70	10
Runner radius 6.0 mm	0.4	2.5	70	18
Environmental damage				
Undamaged	0.9	1.5	70	9
Wet	0.9	1.5	70	7
Grit	0.9	1.5	70	4

A series of drop rig tests have been carried out by Martyn Pavier of Bristol University to determine the effects of the anchor or belay system, runner radius and environmental damage (Pavier, 1999). These results are summarised in Table 11.6. They indicate that the number of falls the rope can withstand before failure is increased by using: (a) a dynamic brake plate rather than tying the rope off, which will reduce the dynamic load on the rope; (b) a larger runner radius, resulting in a lower stress concentration; (c) a dry rope rather than wet rope; (d) clean rope rather than a grit-damaged rope.

Future

Even with the development of improved rope constructions and materials, due to the very nature of climbing, rope degradation from abrasion, cutting or friction melting poses a significant threat, and future development may improve the performance in these areas. Reducing weight is also where future innovation may lie: a comparison between nylon yarn and a spider's drag line, a natural semi-crystalline polymer, has shown that improvements in strength and energy absorption properties appear possible (Blackford and Maycock, 2001). Drop tests on ropes both from test houses and university research have shown a distribution in test results (British Mountaineering

Council Annual Report, 1999; Pavier, 1999). The reasons for this are not fully understood and would be worthy of further investigation.

11.3   Harnesses and slings

Harnesses

Harnesses and slings are made from textile webbing. Before the invention of the sit harness, climbers tied the rope either directly around their waists, or to a swami belt – a few turns of wide tape wrapped around the waist and tied with a tape knot (a very uncomfortable and even dangerous experience if the climber is hanging for any length of time, as crushing of vital organs can occur in just a few minutes).

Harnesses were developed around the late 1960s, with the Whillans sit harness being one of the first. The basic principle of harnesses is that the climber is supported by loops of webbing around the upper thighs and the waist. This makes hanging for prolonged periods relatively comfortable, and, more importantly for sports climbers, makes falling safer. Harnesses are generally fastened with metallic buckles. The rope is tied in directly to the harness. Today, a range of harnesses are produced, which can be used for general aspects of mountaineering and rock climbing; and others designed for more specific uses. Very well-padded harnesses are available for climbers who'll spend substantial periods hanging, for example, on big wall routes, or lightweight harnesses are available for sports climbers who don't want excess weight or bulky padding that restricts movement, or those suitable for use in outdoor centres (it is important for instructors to be able to see quickly and clearly that the harness is on correctly and to be able to check this periodically).

Slings

Slings of some form are carried by all climbers, and provide a link between the climber, protection points and the belay (via karabiners).

Short slings are widely used in the form of 'quick draws' to connect pieces of rock protection to the rope (typically 10–25 cm long). Longer slings are used for constructing belays and for looping spikes of rock or making 'threads' (typically 4, 8 and up to 16 feet long). The first slings were knotted lengths of rope; today, webbing has mostly superseded the use of rope. Webbing for climbing initially was made from nylon with loops made using tape knots but now other polymer fibres with more desirable properties are also used, e.g. *dyneema*, and virtually all slings are made into loops by sewing. Sewn slings are stronger and lighter in weight than knotted slings – a tape knot reduces the strength of a sling by 25% (Perkins, 1991).

The EN standard for sewn slings (EN 566) states a minimum strength of slings of 22 kN, which, considering the loads generated in a fall, is more than adequate.

No incidents of failure of slings during climbing falls have been reported to the BMC in the UK. But, worldwide, many failures of *in situ* slings have been observed. *In situ* slings are slings left in place on a route that are used, for example, as abseil anchors. These failures can be attributed to weathering, for example, freeze/thaw and UV degradation (a limited problem in the UK!) and abrasion.

Quantitative data on the reduction of strength of dyed webbing indicates a loss of 4% or more after 300 h of English summer sun. This loss will increase with longer duration and higher intensities. A 50% loss over 6 months has been predicted for high intensity at altitude. On desert rock, a strength loss of up to 70% in 18 months has been measured (Perkins, 1991). The combined effect with sea water is very severe, so climbers should be especially careful of abseil anchors on sea cliffs.

Another cause of sling failure is from frictional melting. If a rope is threaded through a sling and loaded, the heat generated can be sufficient to melt the sling. This is particularly worth noting for *dyneema* slings, which have melting points even lower than nylon. Repeated use of the same abseil sling, followed by pulling the rope down through the sling, may contribute to the degradation of the slings. One incident has been reported where a knotted sling used as part of a top rope anchor came undone, and another in the Alps where a knot untied when it caught on a rock protrusion.

As with ropes, web will be abraded, both externally and internally. External abrasion is easy to recognise visually, for instance, as a nick in the edge of the web. This leads to a reduction in its strength as there are fewer fibres present to withstand loading and also a stress concentration. If such damage is evident, the sling should be discarded. Damage to stitching can also occur, especially if it is raised proud of the web. Strength losses of 25% have been measured (Perkins, 1991). Internal abrasion is far more difficult to detect (if not impossible); it is a more significant problem for cavers, who spend more of their time crawling through mud than mountaineers do. Internal abrasion occurs when particles of grit penetrate the webbing and their sharp edges abrade the fibres internally. Laboratory tests performed by 'washing' web in a mud slurry indicated a strength reduction of 50% after 100 h, but how this translates to the behaviour of harnesses and slings is not known (Perkins, 1991).

Polyester webbing is woven, but is not used in climbing equipment because of its low elongation under load compared with nylon. However, for use in caving equipment, where absorbing energy in dynamic falls is not an issue, it has advantages in that it retains its strength when wet and is less susceptible to attack from chemicals (e.g. acids from batteries used in head-

Table 11.7 Properties of webbing materials (Perkins 1991)

Property	Nylon 66	Polyester	Dyneema (woven with nylon)
Density (g cm^{-3})	11.4	1.38	0.97
Tenacity (g/dtex)	7.75	7.65	27.0
2% modulus (g/dtex)	38	85	–
% elongation at fracture	13.5	11	2.5
Abrasion resistance	Very high	High	Very high
Strength loss (wet)	10–20%	Negligible	None
Melting point (°C)	250	254	145
Resistance to acid	Low	High	High
Resistance to alkali	High	Low	High
Effect of sea water	Negligible	Negligible	None
Effect of UV	Low	Negligible	Negligible
Effect of lubricating or motor oil	Negligible	–	Negligible
Effect of urine	Negligible	–	Negligible

Table 11.8 Strengths of web (Troll tape book, Perkins, 1991; data marked with *from website of DMM Ltd, 2002)

Web type	Width (mm)	Failure load of web (kN)	Failure or rated load as sewn sling (kN)
Standard	10	5	–
	15	9	–
	19*	–	25*
	25	13.5	20
	26*	–	30*
	50	27	–
Tubular	12.5	8	–
	20	13	22
	25	16	–
Dyneema	12*	–	22*
	15	15.5	25

lamps which may contaminate the webbing; nylon is much more severely degraded by acid attack). Properties of webbing materials are given in Table 11.7, and strengths in Table 11.8.

Web construction

Web is made from weaving yarn and may be followed by a heat treatment process.

Yarn manufacture

The nylon filaments (yarn) are first twisted together; twisting may be followed by cabling with another twisted yarn; this results in a construction like hawser-laid rope. The benefits of this construction are an increase in the web's ultimate tensile strength, and a higher elasticity, hence higher energy-absorbing capacity, while the drawbacks are an increase in mass per unit length (as a twisted fibre shortens) and an increase in web cost as the process is relatively slow and is therefore expensive.

Dyeing

Colour is important in modern outdoor gear! Most web today is therefore dyed, which is actually detrimental to its properties (the strength decreases by about 10%), and it increases the cost by about 30%. However, careful selection of dyes can reduce the detrimental effects of UV degradation on the material.

Weaving

Details of weaving processes and their development can be found in the 'Troll Tape, Slings and Harnesses' booklet (Perkins, 1991). The most important development of weaving techniques in web construction methods means that, if a single strand of the weft or warp is severed (as will happen when the web is abraded), the whole web will not unravel. As such types of web (known as double-weft or knotted edge tapes) are much cheaper to produce, a number of low-cost products from less reputable manufacturers have appeared on the market using this type of web. More sophisticated weaving techniques have appeared and are still under development. Use of coloured web now allows intricately patterned brand-specific webs to be made. Depending on the application of the web, e.g. in slings, harnesses, heavy duty industrial use, it may be heat treated (heat setting). The parameters of these heat treatments are generally proprietary.

Harnesses

Harnesses are constructed from webbing. A review of recent climbing harnesses has been given in *Climber* (2001). There has been a trend for lighter weight and more comfortable harnesses because sufficient strength can be achieved using narrower web combined with polymer foam padding materials. The buckles are typically made from 7075-T6 aluminium alloys.

The harness standard (EN 12277) specifies that a harness must withstand a force of 15 kN for a period of 3 minutes without failure. The test is carried out on a tensile test machine using a specially designed mannequin.

A good review of studies of force distribution in harnesses and an analysis of the forces generated in loading a harness has been reported by Maycock (2001).

No failures of climbing harnesses due to insufficient strength have been reported to the best of my knowledge. However, there have been a number of incidents where accidents have occurred because of incorrect use of harnesses. These include incorrect buckling of the harness. With many web/buckle arrangements, the web should be doubled back on itself to provide sufficient friction to prevent the harness from opening accidentally under load, due to the web slipping through the buckle.

A number of manufacturers now use simple pictures or instructions in words stamped on to the buckle to indicate correct usage.

A couple of harnesses have been reported to the BMC that showed relatively easy buckle slippage, even when threaded correctly (MacNae, 1995; Steele, 1995). This can be attributed to a combination of the shiny surface of the web, resulting in low friction in the buckle and the spacing of the buckle slots. Much of the holding force of webbing through a buckle is generated from the friction of web against web. Climbers should be aware of this and check their harnesses periodically while they are wearing them; repeated loading and unloading (e.g. while ascending ropes using ascenders) may cause the web on susceptible harnesses to slip.

Tying in wrongly has led to a number of incidents and probably many more potential incidents, which have been avoided as the climber didn't load the system. This is a real problem with sports climbing and climbing at indoor walls, and a number of incidents – some fatal – have occurred (Schubert, 2000b). The nature of these pursuits often leads to a false sense of safety compared with the more obvious dangers in many mountaineering situations, e.g. loose rock or the high potential of hitting ledges if you fall off. This, combined with the large number of times you tie into a harness in an indoor wall environment and the fact that you're loading the harness every time you 'lower off', means climbers should be vigilant.

Tying the rope through the leg loops only results in the upright climber being supported below their centre of gravity, thus increasing the likelihood of them hanging upside down and suffering head injuries or falling out of the harness. A potentially disastrous incident in the USA was averted by luck when a very good and experienced climber top-roped a route and, for training purposes, decided to down-climb the route. When he was safely back on the ground, he realised he'd threaded the rope through his harness but not tied a knot in it.

Other incidents have occurred when climbers have clipped abseil devices to gear loops or other non-load bearing parts of a harness (gear loops are loops designed for carrying climbing equipment: rock protection, quick draws etc., but are not designed to be load bearing).

11.4   Karabiners

Description and use of the equipment

Karabiners are connectors that are extensively used in linking components of the mountaineering system, for example, linking belay devices, runners, belay anchors attaching the rope to the system. Karabiners are alloy loops with a sprung gate in the circumference. The word karabiner originates from the name of the connectors used by German and Italian soldiers around the 1900s for securing a carrying strap to their rifles – *Karabinerharken* in German and *carabiner* in Italian. They are classified by their shape and by whether the gate is a simple 'snap' gate or has a more elaborate gate locking mechanism. A selection of modern karabiners is shown in Fig. 11.4.

The snap gates can be operated quickly; the advantage of being able to do this on steep or overhanging rock is clear. They are also lighter as there

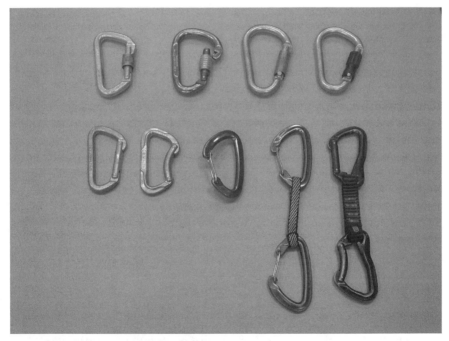

11.4 A selection of karabiners. Top row, locking karabiners, from left to right: standard off-set D screw gate karabiner; HMS screw gate karabiner, gate deliberately open to show nose of karabiner; HMS twistlock karabiner; HMS twistlock (nylon) karabiner. Lower row, snap gate karabiners: standard off-set D karabiner; off-set D with bent gate; off-set D with wire gate; a 'quick-draw'; two wire gate karabiners joined by a textile sling; a 'quick-draw' with two semi-captive eye karabiners.

is no locking mechanism on the device. However, the very fact that they open easily during correct operation makes them susceptible to accidental opening during use, which has caused a number of incidents, and has led to a change in the karabiner standard.

The likelihood of accidental opening of the karabiner gate can be almost eliminated by using a locking karabiner. The most common type of locking karabiner is a 'screw gate', which has a threaded gate and an internally threaded sleeve that screws on the gate. The karabiner operates as a snap gate when the sleeve is undone, but the gate can be locked by screwing the sleeve up the gate and securing it against the nose of the karabiner.

More sophisticated closure mechanisms have been developed, for example, the 'twistlock'. When the gate is in the closed position, it is always locked as the locking sleeve is sprung (in a torsion axis around the long axis of the gate); to unlock the gate, the sleeve is swivelled then the gate is pulled open. The twistlock has advantages over conventional screw gate karabiners as it can be operated more rapidly in many circumstances (in others its spring loaded sleeve can be a real pain), and, when the gate is closed, it is always locked whereas a screw gate has to be screwed shut. A potential disadvantage is the ease of opening of the sleeve: for instance, if it is caught and twisted by the rope it could open accidentally, as the sleeve only needs to be rotated through 90° compared with a screw gate, which often requires several turns to allow the gate to be opened. Recently, a number of accidents have been reported in Germany, where locking karabiners have opened accidentally while being used to secure climbing harnesses to the rope. The likelihood of this happening seems to be higher with twistlock karabiners (Schubert, 2000b). Recommendations of using two locking karabiners back to back were made, or to use the rope to tie in directly to the harness, which is standard practice in the UK.

More elaborate closure mechanisms have been developed. These involve an additional action, e.g. pressing a securing button (Petzl 'ball-lock karabiner').

Standards

The first standard for karabiners was produced by the UIAA Safety Commission in 1965. The most recent standard is EN 12275 'Mountaineering equipment – connectors – safety requirements and test methods' (1998). The standard details seven types of connector (the information given above covers all types except type Q). The standard specifies tensile testing procedures for karabiners. They must be tested along the major axis closed gate, open gate and along the minor axis with the gate closed; minimum breaking loads must be attained, which are given in Table 11.9. The karabiners are tensile tested at a crosshead velocity of 20–50 mm per minute,

Table 11.9 Karabiners: types of connector and minimum values from EN 12275

Type	Description	Major axis gate closed (kN)	Major axis gate open (kN)	Minor axis (kN)
B	Basic, general purpose	20	7*	7
H	HMS, for dynamic belaying	20	6*	7
K	Klettersteig, higher strength for 'via ferrata' use	25	–	7
D	Directional, with captive sling, e.g. DMM Mamba	20	7*	–
A	Specific anchor, for use with a specific bolt hanger	20	7*	–
Q	Screwed-closure, Quicklink or Maillon Rapide	25	–	10
X	Oval, for aid climbing use	18	5*	7

*No requirement if gate fitted with an automatic gate-locking device.

with the karabiner loaded using steel bars of diameter 12 mm. A gate opening pressure between 5 and 15 N is specified in the standard (if it is too low, the gate will open too easily; too high, and the climber won't be able to operate it). The standard specifies the minimum clearance between the gate and the nose when the gate is fully open; and karabiners should have surfaces free from burrs. It is important to inspect karabiners periodically for burrs as they can be very detrimental to ropes.

Materials and manufacture

Today, the majority of karabiners (c. >95%) used for mountaineering are made from 7000 series aluminium alloys. These are wrought age-hardened alloys, based on the Al–Cu,Zn,Mg, Cr system; typically 7075-T6 is used for karabiner bodies and gates. The gate is hollow and sprung loaded using a spring pusher (generally pressed from stainless steel strip) with a spring mounted on the end. The gate is attached to the karabiner using a stainless steel rivet, and the gate generally locates on the nose of the karabiner with a second rivet pin (although some designs use a bayonet contact at the end of the free end of the gate/nose). The closure systems on locking karabiners are metallic, again typically 7075, or polymeric, typically injection-moulded nylon.

Karabiners made from austenitic stainless or alloy steels, e.g. micro-alloyed steels containing B and Mn, are used in certain situations where weight saving is not of primary concern (a rarity in mountaineering!), in applications where the higher corrosion resistance, or higher wear resistance makes the materials more suitable, for example, in caving, top roping or in industrial use.

Most simply shaped karabiners are made from circular cross-section wrought wire; the C-shape of the karabiner is bent while the material is in a soft (annealed) condition; relatively minor modifications in cross-section are achieved by cold forging in press tools. The nose profile is punched out and rivet hole(s) drilled. More dramatic and innovative karabiner shapes are produced by hot forging as more of the alloy can be 'moved /deformed' cf. cold forging (e.g. the DMM Mamba). After forming, the karabiner is heat treated (by solution treatment, quenching into water, followed by an ageing heat treatment). This results in a fine dispersion of hardening precipitates. The karabiner is ground and polished to remove sharp edges. Gates and metallic sleeves are made by turning, milling and drilling, and similarly ground and polished. Soft colour anodising is used for 'colour coding', cosmetic or brand recognition; knowing that a certain piece of equipment is on a blue karabiner can be helpful for selecting it quickly, or distinguishing your equipment from that of your climbing partners! Further details regarding the processing, properties and microstructure of aluminium alloy karabiners have been published by Blackford (1996).

Wire gates

Wire gates are basic non-locking karabiners with a gate made from a martensitic, precipitation hardenable 17Cr–4Ni stainless steel (diameter c. 2.3 mm). They emerged in the early 1990s, pioneered by Black Diamond (1999), to the generally sceptical climbing public – the initial appearance of a gate resembling a paperclip didn't inspire confidence. These types of karabiners are now widely accepted, as they are lighter in weight, the gate itself is stronger, and they are less liable to open gate failures (which are due to the gates flicking open) because of their low inertia in shock loading.

Failures and case studies

Karabiners are one of the most extensively used items of climbing equipment. There have been multiple reports of failures especially 'gate open' and misalignment during loading. These will be considered below.

As with all climbing equipment, knowledge of its correct usage and thinking about what you're doing, coupled with some knowledge of incidents, reduces the likelihood of failures. Climbers should be aware of the limitations of snap gates. In critical situations, for example, on belays or on your last piece of protection before a long runout, it makes sense to use two snap gates back to back (or with opposing gates) or to use locking karabiners.

Faults on karabiners discovered before or during their use (one of each of these incidents has been seen in the UK by the BMC over the past 20 years):

- displaced spring-pusher in the gate – causing the gate not to close.
- loose hinge pin.
- loose latch pin.

The most common incidents reported to the BMC involve failure in use because of:

- loading the karabiner with the gate open.
- abnormal loading of the karabiner forcing the gate to open under load.
- loading with a sling (rather than a dynamic rope) – one instance only. Using dynamic rope, the load on a karabiner in a fall is likely to be <10 kN, but, if a karabiner is connected only with static slings, the load generated can be considerably higher than this.

The first of these has been a cause of considerable concern for accident investigators and manufacturers since the 1980s.

Open gate failures

To be able to confirm a karabiner failure as *open gate*, all the parts of the karabiner must be found and the gate and the nose of the karabiner should not show any gross deformation – if the accident occurs high in the mountains or on a sea cliff, this is not always possible.

Before the emergence of sports climbing, the phenomenon of open gate failures was hardly noticeable as climbers didn't fall so often. It is always possible to break a karabiner in a fall with the gate open, as forces exceeding 6 kN can be generated in severe falls.

So, how frequent are open gate failures of karabiners?

In the UK, five open gate incidents have been reported to the BMC since 1990, e.g. Dickens (1990), Grandison (1993), Huyton (1994). In France, over a 3-year period, 20 open gate failures were reported; 18 of these were on artificial walls, or sports climbing crags and 2 were in alpine/traditional climbing areas (Schubert, 1998). In Germany, around the early 1990s, 4–6 incidents were reported to the DAV Sicherheitskreis annually; in the first 6 months of 1993 four incidents had been reported (Schubert, 1998). Fortunately, none of these incidents was fatal – typical injuries were broken arms, torn tendons, and a head injury (to a climber not wearing a helmet!).

How do gates come open?

The gate needn't open far, or for very long, for a karabiner to fail gate open. During a fall, the gate can open accidentally, owing to the inertia of the gate during shock loading. It can be snagged open on a rock protrusion, or by the rope running rapidly through the karabiner, generating vibrations that cause the gate to flick open. The first evidence of a gate opening during a 4 m fall (simulated in a laboratory) was reported by Charlet and Lassia (1989). They caught the slight opening of the gate on high speed video (400 frames/s); shortly after the gate opened, the karabiner failed.

These incidents led to a change in the standard for gate open strength from 6 kN to 7 kN, although some believe this value should be increased. Karabiners of higher gate open strengths are available, which give an increased margin of safety. Some inspiration for designs of higher open gate strengths was found from the design of crane hooks, first reported in the 1930s.

Alignment of karabiners during loading

Karabiners are designed to be loaded along their major axis and, because of their shape, they tend to align correctly. Incidents have occurred when they rotated in use and had been loaded abnormally.

An incident occurred when a good German climber fell unexpectedly on relatively easy ground 6 m above a bolt. The karabiner failed and she fell 40 m to her death. Investigations showed there to be no material or manufacturing faults with the karabiner and it was determined that it had been incorrectly aligned and loaded across its minor axis during the fall (Schubert, 1998).

Related incidents have occurred when karabiners have been loaded badly because of the use of wide webbing; narrow webbing concentrates the forces along the spine of the karabiner, while a wider web transmits the force further away from the spine, creating a higher moment and increasing the likelihood of failure. Most modern web products are relatively narrow (12–15 mm cf. 25 mm in the past).

The problems of incorrect alignment of karabiners have been addressed by: designing a steeper nose angle – forcing the quick draw and rope to be adjacent to the karabiner spine; making a smaller loop in the quick draw so it cannot move about as easily; by fixing the quick draw to the karabiner using an elastic band (or similar proprietary version); by designing a captive eye to retain the quick draw in the correct position, e.g. DMM Mamba. In addition, once a climber is aware of this problem, looking at the karabiners and making sure they are in the correct orientation substantially reduces the problem.

Figure-of-eight failures

Failure of the locking sleeve of screw gate karabiners during abseiling with a figure-of-eight (due to abnormal loading) has been reported (McMillan, 2000). Essentially, these incidents occurred because of user error.

Corrosion

There have been no reported failures of karabiners under load that have been influenced by corrosion. It is likely that any severely corroded karabiners will have been discarded by most climbers, because of common sense. Severely corroded karabiners have been reported, e.g. one used by a canoeist, kept in a lifejacket pocket and periodically exposed to sea water. There have been some concerns about the susceptibility of 7075 to stress corrosion cracking (scc). Pitting and cracking of an aluminium alloy karabiner has been reported (Riley and Maddock, 1983); however, the history of this karabiner was not known as it was found lying in a cave. Whether scc plays a role for karabiners under normal operating conditions is still an open question. It is likely to be more significant for cavers than climbers, but there is scope for a systematic study to investigate this.

Future

In the future new alloys with higher specific strength are likely to be developed and used in karabiners. New designs and shapes will emerge and these are likely to be strongly aided by computer simulation.

11.5   Belay, descending and ascending devices

11.5.1 Belay and descending devices

Belay and descending devices work by bending the rope around a radius/radii creating friction to brake a fall or to descend a rope in a controlled manner (Fig. 11.5). The earliest belays – known as direct belays – were made simply by wrapping the rope around a spike of rock (if a suitable one could be found!) and using the friction generated to stop a fall. Large falls sometimes led to the belayer being pulled off, or the forces generated were high enough to break the rope itself. Then came indirect belays, where the climber is part of the system. The climber is secured to the rock and then belays from his body – initially by wrapping the rope around the body to create friction (shoulder or waist belay). Descending was carried out in a similar manner – both of these procedures are rather painful! A development came with the Italian or Munter hitch – a knot that creates friction by looping the rope around itself. This knot is still widely used in some parts of Europe today.

Modern belay devices are generally metallic and depend on creating friction that enables a high force on the exit side to be controlled by a lower force on the entry side (it's very easy for a lighter climber to hold a fall of a significantly heavier climber with a modern device). The operation of a device for belaying or descending is essentially the same – in belaying, the

11.5 Belaying, abseiling and ascending devices attached to rope. From left to right: a *Petzl* ascender (for climbing up ropes); a *Petzl* shunt (designed as an abseiling safety device but also used for ascending ropes); a *Black Diamond ATC-air traffic controller* (for belaying and abseiling); a *Chouinard* figure-of-eight (for abseiling).

device remains stationary while the rope moves, and in descending the device moves while the rope stays stationary. Because of this, devices are often used for both belaying and descending. Devices can be divided into three classes.

Brake plate

The classic version of a brake plate is the Sticht plate, which is a circular disc with two holes through which the rope is looped, and wrapped around a karabiner to provide the friction for the device. When the rope is braked, it forms an 's' shape through the device. Many variants on this type of device are available today, which operate in the same manner but slip more or less easily.

Figure-of-eight

An '8'-shaped piece of alloy. The rope is wrapped around it to create friction. A karabiner is only used for connecting it to the climber's harness.

Auto-locking devices

Auto-locking devices have gained popularity over recent years. They are particularly used for sports climbing as a climber can be held on the rope without the belayer having to maintain tension on the rope. These devices should lock up automatically in a fall but users should be aware of their correct operation and read manufacturers' instructions carefully. They are generally heavier and more complicated than the other two types of device.

Materials and processing

The brake plate devices are made from high strength aluminium alloys – either die cast to shape or extruded (and machined) from wrought material. Figure-of-eights are die cast. The materials and processing of the auto-locking devices are similar to those used for ascenders (and, not surprisingly, they are often made by the same manufacturers with the appropriate processing infrastructure); see the section below on ascenders.

Standards

Currently, there are no standards for mountaineering belaying and descending devices. There is an EN standard (EN 341) for descending devices that are defined as 'rescue devices' – designed for lowering people to safety using a rope. Unlike most other items of metallic climbing equipment, these devices are not classified as Personal Protective Equipment (PPE) and so do not have a CE mark. The UIAA is currently working to produce a standard.

Failures

No failures of the device itself (i.e. material failure) have been reported. However, numerous incidents involving incorrect use of these devices have been reported, which include (Lyon, 2001):

- inserting the rope(s) wrongly into the device
- using the wrong diameter rope
- incorrect alignment of the device and connecting karabiner
- operator error – inattentive holding technique, or distraction
- clothing or hair being jammed in the device
- relying totally on the 'auto-lock' mechanism of such devices, or using it 'hands free'. A good report on the correct use of these devices has been published in *Summit* (Ingram, 2001).

11.5.2 Ascending devices

Ascenders are mechanical devices used for climbing up ropes. The simplest 'device' for climbing rope is using hands and feet – but this is very strenuous, especially given the diameter of climbing ropes. Prusik knots are widely used for ascending ropes in emergency situations or for very occasional use. These knots use a length of cord (of diameter less than that of the rope) looped at least twice around the rope. They can be pushed up the rope but grip when loaded. If climbers wish to ascend a significant distance up a rope or rope lengths, then using an ascender makes sense (Fig. 11.5).

Ascenders work by trapping the rope between a cam and a housing. The device is loaded via the housing with a toothed cam (termed *body loaded ascenders*), or via the cam itself with a smoother cam surface (termed *cam loaded ascenders*). Body loaded ascenders are more efficient and are used for climbing clean ropes, e.g. in big wall climbing, or rope access work. The toothed cam can damage the rope sheath, and extensive use of these devices on a rope will certainly reduce the rope's lifetime. Ascending devices are not designed for dynamic loading, so care should be taken when using them. Shock loading can lead to severe sheath damage of the rope. These devices are often made from bent plate, which creates a channel for the rope to run through; under excessive loads it is possible that they could begin to open. The cam loaded ascenders are more suitable for situations where high loads may occur, e.g. tensioning ropes in a rescue situation. They enclose the rope completely so are stronger compared with body loaded ascenders. They exert a lever effect on the rope, distorting its passage; the load on the rope is distributed over a larger area (with no stress concentrations created from the teeth cf. body loaded ascenders). Cam loaded ascenders provide better grip on icy or muddy ropes and they wear the ropes less.

The most recent ascending devices on the market are small 'emergency' use devices. These are intended for 'emergency' or occasional use. They are not as effective as the larger mechanical camming devices but they are more effective compared with prusik knots, and their relatively light weight makes them suitable for mountaineering use.

Materials and processing

Ascenders are generally made from high strength aluminium alloys, depending on their design they may be die cast, made from bent plate, and extruded bar material. The devices are often sprung loaded and assembled using rivets. Nylon or synthetic rubber may be used for safety catches and grip coverings.

Standards

The devices are covered by the standard EN 567 'rope clamps'. The main requirements for the device are:

- to be marked with the rope diameters that may be used
- to hold 4 kN without significant deformation to the rope and without it breaking
- to have a locking device to prevent it becoming detached from the rope.

Failures

Ascending devices are used less extensively than many other items of climbing equipment and they should not be used for dynamic loading. This has probably contributed to the very limited number of failures observed. Only one failure of an ascender has been reported to the BMC in the UK, over 20 years ago. This happened when the device was either dropped or crushed and this caused the cast alloy body to break. However, a number of potential failure mechanisms have been suggested (Lyon, 2001):

- failure of the body of cast alloy (as above)
- failure of the cam springs, or of the cam retaining (locking device) springs.
- slippage due to wear of the toothed cam
- slippage due to mud or ice on the rope
- breakage of the rope by the ascender under high load.

11.6   Rock protection

There is a bewildering array of rock protection equipment available. Modern traditional climbers often carry extensive racks of gear (Fig. 11.6). Slings tied in rope and looped over suitable rock spikes were the first rock protection used. Knots tied in lengths of ropes were jammed into suitable cracks – using different diameter rope it was possible to protect different sized cracks. This form of protection is still used exclusively today in sandstone areas of the Czech Republic, because the soft rock would be severely eroded by harder rock protection (although drilled holes with glued in bolts are used for belays). Then came the idea of wedging pebbles or rocks into cracks, threading a sling over them and connecting it to the rope via a karabiner. Prior to climbing their routes, climbers would scour the ground looking for suitably sized pebbles. The obvious extension of this idea appeared in the late 1950s and 1960s in the form of drilled-out machine nuts that were threaded with cord and could be carried by the climber and placed in a suitable fissure in the rock. In the early 1970s the

11.6 Rock protection (attached to karabiners). From left to right:
pitons, brass microwires, aluminium alloy chocks on wire,
camming devices (note flexible and rigid stems), hexentrics.

first chocks designed specifically for climbing appeared – the *Moac*, a tapered aluminium alloy wedge, and the hex from *Clog*. Since this time, great efforts and innovation have gone into creating a huge range of rock protection, which can be classified as 'passive', devices having no moving parts, e.g. wired chocks, or 'active', devices with moving parts, e.g. camming devices.

Today, chocks are available in sizes ranging from about 1 mm to 100–200+ mm, although the most commonly used sizes tend to be those from several mm up to about 50 mm – as the very small micro-chocks are often not too reliable because of their intrinsically small size and the size of the placement in the rock. The larger sizes are relatively heavy and awkward to carry and it is often possible to place a smaller chock in an alternative placement. The sides on many chocks are curved – this allows them to fit better in many placements and to cam into certain positions.

Camming devices were first developed in the 1970s. Chocks require a tapered crack or constriction for them to jam in the rock, whereas camming devices can be placed in parallel sided or flared cracks. They have revolutionised rock climbing, enabling runners to be placed on previously unprotectable sections of rock. The first devices to be marketed were *Friends* designed by Ray Jardine. They have four spring-loaded cams attached to an axle, which is attached to a stem. The cams are activated with a trigger bar. They work by the friction generated by the cams on the rock. Today, several designs are marketed, and many now use a flexible wire stem as opposed to the original rigid stem, which makes them more suitable for use in horizontal placements.

Standards

Chocks are covered to standard EN 12270, which specifies a tensile test with, at first glance, a low force requirement of only 2 kN. This value was agreed upon so that the micro-chocks could conform to the standard and still be marketed. This is an aspect climbers should be aware of and realise that chocks with ratings of 10 kN will provide substantially more protection in the event of a fall.

The standard for spring-loaded camming devices (SLCDs) is EN 12276. These devices rely on friction so the standard specifies a tensile test pulling the devices via their stem with the device located between two parallel steel plates. A minimum force of 5 kN is specified. The steel plates do not replicate use in a real climbing situation. Most rock surfaces will generate significantly more friction than these plates. However, certain types of rock are relatively smooth, and on other occasions the rock will be wet or icy – in these circumstances the friction provided will be reduced substantially.

Materials and manufacture

Chocks and the cams of camming devices are made from high strength wrought aluminium alloys. Often, the raw stock is purchased in extruded form close to the shape of the final product. The extruded material is cut to size and may be forged (hot or cold) or machined to shape. The components are finished by drilling, grinding, polishing and possibly by anodising. The components of the chocks or camming devices are assembled with galvanised steel wires, which are joined by swaging.

Some micro-chocks are now manufactured from aluminium alloys as described above, but some use stainless steel wire, e.g. Black Diamond micro-stoppers. The smaller ones, however (HBs or RPs), tend to be made from investment cast brass alloys. These alloys are softer than the Al alloys, so they are less likely to rip through a placement in the rock; in fact they will often deform to rugosities in the rock surface. However, this means that they cannot be attached to a swaged loop of wire as it reduces the strength – the wire will rip through the chock itself, so they are joined to the wire by soldering.

Incidents

In the UK no incidents have been reported to the BMC that could be attributed to the failure of chocks during use. Rock protection pulls out now and then – but this can be almost entirely attributed to the placement – learning to place good protection is an important art in climbing. A number of cases have been observed of the presence of pre-existing defects in the extruded alloy, and several of severe scratching of the chock in use. Inspection showed these flaws did not reduce the stated strength of the unit. In use, the wires can become broken from repeated bending or snagging on the rock, but this tends to be clear on inspection of the unit, which should be retired. Corroded wires have also been observed on wired nuts – there is a high potential for galvanic corrosion in sea cliff environments as the chock, wire and swage are different alloys. Since 1988 seven incidents involving camming units have been reported to the BMC. None of the failures were attributed to poor materials or manufacturing. In cases where the unit broke, it could be attributed to a poor placement.

Pegs and pitons

Pegs and pitons – lengths of angle or tapered alloy which are placed in a crack in the rock by hammering – were used extensively in the 1950s and 1960s. The placement of pegs causes scars in the rock and has inadvertently created some excellent free climbing routes, e.g. the peg-scarred cracks on Peak District gritstone. Ethics limit their use today in the UK because of the damage they do to the rock, and because of the limited amount of rock available and the view that it should be climbed free – or left for someone who can climb it free. Today, pegs are still widely used for aid climbing, big wall climbing, winter climbing, and they are often found *in situ* on crags and sea cliffs. A good review of pegs found on sea cliffs was published in *Summit* (Hillebrandt, 2001), which demonstrates the corrosive nature of the sea cliff environment and how care should be taken when using *in situ* gear. Stainless steel pegs are more suitable for extended use in such environments. In the late 1950s the use of relatively hard steel pegs made possible the ascent of El Capitan, a big wall in Yosemite, California. Up until this time softer 'iron' pegs (probably rather low carbon steel) were used, but these deformed so much when they were hammered into the cracks that they couldn't be removed and reused for later pitches – the number of pegs that would have to be carried for a complete ascent of such a wall (today it's about 34 pitches) would have made it virtually impossible.

11.7   Ice climbing equipment

The main items of equipment used for ice climbing are ice axes and hammers and crampons. These will be considered in detail below, and are shown in Fig. 11.7.

On mixed terrain the same rock protection equipment is used as for rock climbing (although less reliance on camming units in potentially icy cracks makes sense!). For climbing pure ice routes, ice screws are used for protection. These screws are generally tubular and machined to give a coarse thread on the outside of the tube. Most ice screws are now made from steel, e.g. 4130 (a CrMo steel; also used as aircraft industry hydraulic pipe), with a 'hanger', for clipping a karabiner into the ice screw, made from 4340 plate (Fig. 11.7). Titanium alloy ice screws were common in the early 1990s; a few years ago (after the fall of the Iron Curtain) Eastern European climbers notoriously funded their climbing trips to the Alps by selling titanium alloy ice screws around the car parks and climbing venues. This, no doubt was connected with the abundance of titanium alloys in these countries.

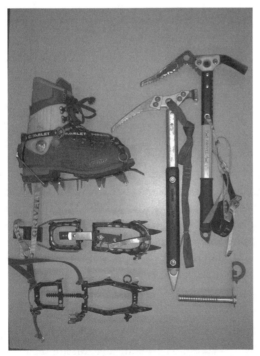

11.7 Ice-climbing equipment: crampons, ice tools, ice screws. Left side from top: rigid plastic mountaineering boot with crampon; step-in style crampon; strap-on style crampon. Right side from top: modular drooped-pick technical ice axe; modular alpine-pick ice hammer; ice screw.

Ice tools

The first ice axes were long wooden shafts with a pick and adze fixed to the top of the shaft. These were documented in the nineteenth century. Initially, the picks were straight, but curved picks were found to be more effective and soon developed – termed alpine picks. The adze is generally a triangular-shaped plate that is used for cutting steps or clearing areas of snow or ice. Many first ascents of ice routes were made impressively by cutting steps before the advent of front point crampons. Today, ice axes are used for winter walking, mountaineering and steep technical ice climbing. As the climbing becomes steeper and more technical, two ice tools are used – an axe and a hammer. Ice hammers are similar to axes but have a hammer head opposite the pick instead of an adze. Shorter tools are used for more technical climbing and they are held close to the end of the shaft near the spike, while for walking on snow an axe is held at the head – more like a walking stick – which means it is in the correct position for use for self-arrest (an important technique to learn).

Modern technical axes have drooped or reverse curved picks. These originated in the early 1970s with Hamish MacInnes in Scotland (the *Terror-dactyl*) and Chouinard in the US. They are better for 'hooking' on mixed ground and are certainly easier to remove from the ice than more conventional alpine picks. Most technical tools are modular – that is the pick, adze or hammer can be replaced – generally by bolting in a replacement. Axes designed more for walking generally have a pick and adze constructed as a single integral part. Today, there is a large variety of axes available and a range of specialist tools, e.g. narrow picks designed specifically for waterfall ice climbing, and thicker picks designed for mixed climbing, which are more suitable for torquing as they are less likely to snap or bend, but are more likely to shatter waterfall ice. Modern axes intended specifically for steep climbing have bent shafts, which offer a number of advantages including a reduced likelihood of bashing your knuckles on the ice and giving more clearance when reaching over bulges. Recently, some very lightweight axes made from aluminium alloys have come on the market; these are considered suitable for ski mountaineering or for approaches on snow following a rock climb.

Standards

Ice tools are covered by EN 13089. The standard distinguishes between technical and basic ice tools – both are subjected to similar tests, but the technical tool is required to have a stronger shaft, head and pick than the basic tool. Additionally, the picks on technical tools are fatigue tested.

Materials and manufacture

The pick, adze, hammer are generally made from stamped plate or forged steel (type 4340, a 1.5% NiCrMo steel) and heat treated by quenching and tempering (to 40–44 Rockwell C hardness), although there are now axes with heads made using wrought 7000 series alloys. Early shafts were made of wood, e.g. ash. Today, most are manufactured from extruded 7075-T6 aluminium tube (it may be bought in a softer T3 condition to enable the tube to be bent), but carbon fibre shafts are becoming more common and, because of the construction techniques, they allow for more flexible innovative designs. The spikes on the lower end of the axe shaft are generally steel, e.g. 4340. The head of the axe and spike are secured to the shaft by riveting (stainless steel rivets) or by gluing with toughened epoxies. Injection-moulded nylon-66 inserts are often used as spacers inside the shaft. Synthetic rubber is used to cover the shaft for grip.

Crampons

Crampons are spiked frames that are fixed to mountaineering boots to provide better grip on hard snow or ice compared with the boot sole itself. In the 1930s the 12-point crampon was developed. This revolutionised ice climbing by allowing front pointing, enabling vertical ice to be climbed using horizontal spikes that protrude in front of the boot. Developments since then have made crampons lighter, stronger and more adjustable. Different types of crampons are available for particular activities, e.g. steep ice climbing or walking on snow. The fit of a crampon to the mountaineering boot is very important and many hours are spent in retailers sorting out the correct combination of boot and crampon. Ski boots and bindings are made to standard dimensions, e.g. the distance from the binding to the top of the toe welt is specified, which makes a ski boot for one manufacturer suitable for use with a binding from another. There are moves in the outdoor trade to follow suit with certain types of crampons and boots. A good recent review of technical aspects of crampons has been published by the BMC (British Mountaineering Council, 2002).

Standards

Crampons are covered by European Standard EN 893. This specifies design requirements, material hardness, and strength of the spikes, attachment points and frame. It states crampons must have a minimum of eight points, at least six of which must point downwards.

Materials and manufacture

The majority of crampons are made from quenched and tempered steels, e.g. 4340 (Mountain Technology) or 4130 (Black Diamond). There have been a very limited number of models made from aluminium alloys, but these have not caught on so far, as for crampons to function correctly, it is important that they are sharp and thin enough. The lower stiffness of aluminium alloys compared with steels makes meeting this design criterion difficult. The crampon blank is pressed, or laser cut, from plate, may be drilled, bent to shape while hot, and heat treated. The crampon fittings (straps or bail bar and heel system) are attached to the crampon using rivets or alloy loops or a combination of both. The stainless steel bail bar is generally fitted through drilled holes and secured by flattening the protruding ends. Straps are made from nylon tape or neoprene-coated fabric. Over the last couple of years injection-moulded polyurethane toe cradles have become popular on crampons. These allow the crampons to be fitted quite rapidly and they will fit a wide range of shapes of boots. The heel clamp is generally injection moulded nylon-66. Crampons are made to be adjustable. There are several designs available that can be adjusted by hand in the field, while others use a system of nuts and bolts (generally *nyloc* nuts).

Incidents

Manufacturers of ice axes and crampons can learn a lot from the renowned Liberty ship incidents of the early twentieth century. Making equipment from bcc structured metal alloys (steels) that will be subjected to impact loading at low temperatures requires careful alloy selection and heat treatment (or alternatively the selection of different materials). Because of the inherent problems of the ductile brittle transition temperature (DBTT) in bcc alloys, combined with the weight constraints of mountaineering gear (it's no good having a pair of crampons or axes you can't lift off the ground), along with the abuse equipment receives in winter mountaineering, it's inevitable that numerous failures of ice climbing equipment have been observed.

To take an extreme view of the behaviour of axes and crampons in use – if they are hit hard enough (which is quite possible with the abuse winter mountaineering equipment sees), they will either bend or break.

In the UK, from 1985 to 1999, the BMC technical committee investigated nearly 30 incidents involving ice tools and 31 incidents involving broken crampons.

Incidents reported concerning ice axes include:

- bending of the pick from high impact forces, for example, when accidentally hitting rock.

- bending or breaking of the pick when torquing.
- fracture of the pick originating from stress concentrations from sharp radii on part of the pick. One such incident has been documented by Simon and Taylor (1991). The pick fractured, with the fracture originating from a sharp radius where the pick enters the shaft of the axe. Repeated loading resulted in a fatigue failure, which was confirmed by examination of the fracture surface. This showed a crack initiation site at the sharp radius and stage I growth of the crack, then a large area of brittle fracture (stage II), before a final zone of ductile fracture.
- fracture of modular heads, for example, of a cast stainless steel alloy. The design of this component also showed sharp radii (Blackford, 1994).
- failure of the pick retaining bolts/system resulting in the pick rotating in the modular head and becoming unusable (Blackford, 1995a).

Crampon incidents

- Brittle fracture or fatigue (19 cases). Walking in crampons generates significant fatigue loading on them (with each step being one cycle), which will be worse with poorly fitting crampons. It is not uncommon to observe fatigue cracks in crampons, which is something climbers should be aware of and check periodically.

  An example of brittle fracture of a front point has been reported (Blackford, 1995b). There were no signs of long-term fatigue (as no 'old' dark brown rust was observed). The fracture surface exhibited a flat region on the bottom of the crampon and a rougher region over the remaining surface covered in 'new' orange rust. The final crack was thought to have originated on the flat underside of the crampon, due to a sudden high load, and is consistent with front point usage. It would have propagated rapidly possibly later during that climb. Examination of the composition and hardness of the material suggested it was tempered at too low a temperature. Tempering at a higher temperature would increase the toughness of the material but increase the likelihood of failure by bending.
- Crampons becoming detached from boots because of poor fitting.
- A few failures due to poor design or manufacture. The manufacturers have either withdrawn or modified the particular models.

Grossly deformed front points were observed on a pair of crampons after a winter climb (Hellen, 1996). The dimensions of the crampons were correct compared with other crampons of the same model. Inspection of the crampons indicated the hardness of the points to be too low, which was attributed to incorrect heat treatment.

11.8 Helmets

Helmets are good. There have been numerous incidents in climbing and mountaineering where helmets have saved lives and others where falls have been too long or falling rocks too large and helmets unfortunately haven't helped. In mountaineering, helmets are used to protect the wearer if they fall but additionally, and often more commonly, to provide protection from falling rocks, etc. Helmets are considered in detail in another chapter in this book. Reports on a comprehensive helmet testing programme, conducted at Leeds University, have been published in *Summit* (MacNae and Taylor, 2000, 2001; Taylor, 2002).

Mountaineering helmets are covered by standard EN 12492. This specifies the maximum force that should be transmitted to the wearer for a given impact, and tests for penetration of the helmet to simulate the impact of sharp rocks. The strap retention system is also tested.

Incidents

In the UK over the last ten years, six incidents involving helmets have been reported to the BMC. Of these, five people were protected by their helmets and survived; the other case involved a long fall during which the helmet came off the wearer's head, leaving him exposed to further impacts. He died from the injuries sustained. A subsequent investigation showed the helmet was manufactured to the industrial helmet standard, which requires the chinstrap to release on high/impact loading – this is to prevent strangulation, for instance, if the helmet became caught on scaffolding (McMillan, 1995).

11.9 Future trends

Future trends for specific equipment have been considered in sections of this chapter. In the following section I will cover some more general aspects. Before considering the future of materials in mountaineering equipment, I'd like to look at future directions that rock climbing and mountaineering will take. This is a personal view.

Future directions of rock climbing and mountaineering

Rock climbing standards will continue to increase in technical difficulty, i.e. moves with smaller and fewer holds and on steeper terrain. Longer climbs will be made in better 'style', and although the highest walls on the planet have been climbed, there remain possibilities to climb them in better style, e.g. by using less equipment to directly aid upwards progress.

Climbing involves risk. Leading climbers of every era have pushed boundaries of boldness in climbing, i.e. climbing routes where a fall potentially has very serious consequences. Often, this can entail climbing relatively blank sections of rock (or loose rock), where it is impossible to place protection – in essence, this form of climbing needs less (or no) equipment. That said, some of the confidence to tackle such routes comes from a belief in the minimal rock protection placed and from knowing that your rope will withstand a fall.

The trend among many of today's top mountaineers is towards faster, higher, lighter approaches. Because of the more varied environment (great extremes of temperature and terrain and often the extended length of excursions) compared with rock climbing, mountaineering has a higher requirement for equipment. In addition to climbing hardware, mountaineers may require: ice tools, specialised clothing, rucksacks, tents, stoves, sleeping bags and navigational equipment.

To enhance a faster, more minimalist approach to mountaineering, a number of generalisations can be made about materials and equipment.

Future trends in materials in mountaineering equipment

Materials and equipment with higher performance are desirable, e.g. lower weight, increased strength (or the same strength at lower weight), and improved durability.

This chapter has considered materials but has related them closely to the application and equipment design. The application, materials and design are intimately linked – in this context materials should not be considered in isolation. Innovative designs will continue to emerge. I believe these will come from people involved in the sport because they have more insight and interest into what's needed than people outside.

As mentioned at the start of this chapter, advanced materials developed in other industries will be used in mountaineering equipment in the future. A current trend in the materials and engineering communities is towards smart materials and systems – these effectively react or interact with their environment; examples could be helmets or ropes that change colour if damaged.

Much mountaineering equipment is simple and uncomplicated in design. This works well especially when you consider the environment it's being used in – you may be in extreme environments and days away from a replacement or a workshop. I believe there are opportunities to simplify equipment. Another direction in equipment design that has proved popular is gear that serves multiple purposes, e.g. Swiss Army knife, and *Leatherman* multi-purpose tool; there have been more bizarre examples like a rucksack and tent incorporated in a single item – but these haven't stood the test of time.

Computer modelling is playing an increasingly important role in most aspects of materials and engineering, and I believe it will be utilised substantially more in mountaineering equipment design.

Environmental issues are likely to play an increasingly important role in materials for mountaineering equipment. 'Cradle–grave product lifecycle' is becoming more important and legislation is likely to make recycling issues more prominent. Some fleece fabrics are made from recycled polymer bottles, which also gives the marketing departments good 'material'. Claims have been made by some boot manufacturers that their soles cause less erosion and so are more environmentally friendly.

Marketing plays a huge role in much of the sporting equipment industry today. Being able to tag equipment with fancy labels, with hi-tech sounding materials, is a bonus for marketing departments. This will no doubt continue in the future.

In my view mountaineering is about being outside, doing things that you, mostly, enjoy doing. Equipment and materials are important, but good equipment tends to be unobtrusive and should add to your experience rather than cocoon you from it or get in the way. In the end anything goes – gear is designed with specific uses in mind but when needs must – use what you have for whatever . . . use your imagination . . . belay from clumps of grass, hammer nut keys in to fissures for protection, abseil off gloves buried in the snow or boot laces . . .

11.10 Sources of further information

Technical books

The references in the following section give full details of a number of key technical references for mountaineering equipment. Pit Schubert's book *Sicherheit und Risiko in Fels und Eis* (Safety and risk on crags and ice), written in German, is a most authoritative book covering aspects of equipment use, and misuse, and failure investigations over a 25-year period. It makes interesting reading for materials scientists and engineers, and contains a wealth of valuable and horrifying information for climbers.

The British Mountaineering Council (BMC) has published a number of information leaflets, on specific equipment, e.g. ropes and crampons, and a recent one on care and maintenance of equipment. The BMC, over the past 30 years, has conducted numerous incident investigations; reports of specific incidents are available from the BMC.

Manufacturers' catalogues and leaflets contain some very good information, but readers should be aware of the inevitable bias and marketing hype.

Andy Perkins has written a good guide to the properties and use of web (Perkins, 1991). This contains a wealth of information that is hard to find elsewhere.

The UIAA published a special issue of their *Journal* dedicated to 'Equipment and its application' (2000, no. 3). This has several excellent contemporary articles on problem issues with modern equipment.

Journals from various eminent climbing and mountaineering clubs contain a wealth of articles predominantly dealing with climbing exploits but a few classic older technical articles, e.g. Wexler's 1950 article on theory of belaying, can also be found in volumes of the Alpine Club, American Alpine Club, Climbers' Club and Scottish Mountaineering Club.

'How to' books

The following books are aimed at imparting information about mountaineering practices to climbers and mountaineers, but many contain details about equipment design and use.

Fyffe, A. and Peter, I. (1997). *The Handbook of Climbing*. Pelham Books. Includes a good glossary.
Langmuir, E. (1995). *Mountain Craft and Leadership*, MLTB, Scottish Sports Council, 3rd edn.
March, W. (1985). *Modern Rope Techniques in Mountaineering*, Cicerone Press, 3rd edn. Now quite dated, but has sound information about: ropes, knots, belaying, prusiks, and rescue techniques.
Schubert, P. (1991). *Modern Alpine Climbing, Equipment and Techniques* (translated into English by George Steele and M. Vapenikova), Cicerone Press. (First published as *Die Anwendung des Seiles in Fels und Eis* by Bergverlag Rudolf Rother, 1985.) Comprehensive coverage of equipment and its use for rock and ice climbing.
Dill, J. (1998). *Staying Alive* (National Park Service Search and Rescue), *Rock Climbing – Yosemite Free Climbs*, Don Reid, Chockstone Press, 2nd Edn, pp. 8–21. This is an excellent account of accidents and near misses in Yosemite National Park, but the lessons are applicable everywhere. Highly recommended if you climb.
Twight, M. and Martin, J. (2001). 'Extreme alpinism – climbing light, fast and high', *The Mountaineers*. American, modern, driven.
Raleigh, D. (1995). *Ice Tools and Technique*, Elk Mountain Press. Tips on equipment, snow anchors, etc.
MacInnes, H. (1999). *International Mountain Rescue Handbook*, Constable, 3rd edn.
Moran, M. (1988). *Scotland's Winter Mountains – The Challenge and The Skills*, David and Charles Publishers.

How to rock climb series: clearly written and illustrated examples are:
Mountaineering Anchors, John Long, Chockstone Press, 1993
– details placing and configuring anchors in a variety of situations.
Big Walls, John Long, John Middendorf, Chockstone Press, 1994
– details gear, procedures, suppliers, etc.

Guide books and coffee table books

Guide books give descriptions of routes and their difficulty and are available for many – but not all – regions of the world. There are plenty of 'coffee table books' with amazing pictures of wild, inspiring action and scenes from bouldering to big mountains.

Conferences

The British Mountaineering Council runs a technical conference (biannually). There are no formal proceedings and the main aim is to transfer technical information to the 'general' public, through further dissemination via retailers, instructors, mountain guides and journalists from the mainstream outdoor press.

A conference series on 'The science and technology in climbing and mountaineering' has been initiated at Leeds University; proceedings are available on CD-rom. The first conference was 7–9 April 1999, and the second 3–5 April 2002. The conference does include papers on equipment, but in addition covers physiological, psychological, risk and ethical issues.

Magazines

Numerous glossy outdoor magazines exist, and there are generally articles on equipment; many specialise in aspects of climbing and mountaineering, e.g. UK-based: *High Mountain Sports*, *Climber*, *On the Edge*; and US-based: *Climbing*, *Rock and Ice*, etc. *Summit* is the magazine of the British Mountaineering Council and often contains good technical articles; similarly, most national mountaineering organisations have journals or magazines.

Trade organisations and shows

Outdoors Industries Association
Morrit House
58 Station Approach
South Ruislip
Ruislip HA 6SA

UK
Tel: 020 8842 1111
Fax: 020 8842 0090
email: info@go-outdoors.org.uk

The Outdoors Industries Association (OIA) represents companies in the UK and Republic of Ireland providing clothing, equipment and services for recreational outdoor activities. The OIA runs 'Go-outdoors', a trade show, annually in the UK. European trade shows (ISPO) are organised in Munich, Germany, twice a year to showcase summer and winter equipment.

Governing bodies and national clubs

British Mountaineering Council (BMC)
177–179 Burton Road
Manchester
M20 2BB
UK
Tel: 0870 010 4878
Fax: 0161 445 4500
email: office@thebmc.co.uk

Mountaineering Council of Scotland (MCoS)
The Old Granary
West Mill St
Perth, PH1 5QP
Scotland
Tel: 01738 638227
Fax: 01738 442095
email: info@mountaineering-scotland.org.uk

Mountaineering Council of Ireland (MCI)
House of Sport,
Longmile Road,
Dublin 12,
Republic of Ireland
Tel: +353-1-4507376
Fax: +353-1-4502805
email: mci@eircom.net

Union Internationale des Associations d'Alpinisme (UIAA)
Street address:
UIAA

Monbijoustrasse 61
CH – 3007 Bern
Switzerland

PO Box address:
UIAA
Postfach
CH – 3000 Bern 23
Switzerland
Tel: +41 (0)31/3701828
Fax: +41 (0)31/3701838
email: office@uiaa.ch

National centres

Plas y Brenin
National Mountain Centre
Capel Curig, Gwynedd
LL24 0ET
Wales
Tel: 01690 720 214
Fax: 01690 720 394

Glenmore Lodge
Sportscotland National Centre
Aviemore
Inverness-shire PH22 1QU
Scotland
Tel: 01479 861256
Fax: 01479 861212

Tollymore Mountain Centre
Bryansford
Newcastle
County Down BT33 0PT
Northern Ireland
Tel: 028 4372 2158
Fax: 028 4372 6155

Standards

Union Internationale des Associations d'Alpinisme (UIAA) standards
were set up as standards of excellence – with only the best equipment

meeting the standards, other equipment could still be sold because the standards were not mandatory. In 1990 the European Directive on Personal Protective Equipment (PPE) was applied to mountaineering equipment. The requirements of the PPE directive are set out in the European Norm (EN) standards. Now equipment must conform to the EN standard to be sold in the EU; EN standards should be all-inclusive 'minimum standards'. For example, the standard for ice axes has sections referring to basic axes (to which all ice axes must conform) and technical axes which are designed for more extreme use and the standard requires stronger shafts and picks. The UIAA standards are based on the EN standards and frequently have additional safety requirements.

The UIAA standards can be changed more easily than the EN standards as less bureaucracy is involved with the UIAA Safety Commission than CEN and the national standards bodies. It is believed that this is the way standards will be altered in the future, with changes being incorporated into EN standards at the next revision (McMillan 1999). A list of EN and UIAA standards for mountaineering and climbing equipment is given in Table 11.10.

Table 11.10 List of EN and UIAA standards for mountaineering and climbing equipment (*UIAA Journal*, 2000)

Title of Standard	EN Standard	UIAA Standard
Dynamic ropes	EN892	101
Accessory cord	EN564	102
Tape	EN565	103
Slings	EN566	104
Harnesses	EN12277	105
Helmets	EN12492	106
Low stretch ropes	EN1891	107
Connectors (karabiners)	EN12275	121
Pitons	EN569	122
Bolts	EN959	123
Chocks	EN12270	124
Frictional anchors (e.g. friends)	EN12276	125
Rope clamps (ascenders)	EN567	126
Pulleys	EN12278	127
Energy absorbing systems (EAS) for Via Ferrata	EN958	128
Abseil devices (in preparation)	–	129
Belaying devices (in preparation)	–	130
Ice anchors	EN568	151
Ice tools (axes and hammers)	EN13089	152
Crampons	EN893	153
Snow anchors (in preparation)	–	154

11.11 Acknowledgements

Thanks to Stuart Ingram at the BMC, Hugh McNicholl of Mountain Technology, Ed Maycock, Chris Hall, Vicky Pugsley and Andy Perkins of Troll, and Jeff Maudin at Black Diamond.

11.12 References

Bennett, F. (2000). Learning the ropes, *Summit*, **20**, 28–29.
Blackford, J.R. (1994). Camp Hypercouloir ice axe, fracture of head, British Mountaineering Council Technical Report TCMD 94/4.
Blackford, J.R. (1995a). Collapse of ice axe pick, British Mountaineering Council Technical Report TCMD 95/2.
Blackford, J.R. (1995b). Fracture of front point, British Mountaineering Council Technical Report TCMR 95/8.
Blackford, J.R. (1996). Materials in mountaineering equipment: A look at how processing and heat treatment influences the structure and properties of aluminium alloy karabiners, in *The Engineering of Sport, Proceedings of the First International Conference on the Engineering of Sport*, Ed. S Haake, Publ. Balkema, Sheffield, UK 2–4 July, pp. 161–167.
Blackford, J.R. and Maycock, E.C. (2001). Climbing safety reaches new heights in Europe, *Materials World*, European supplement **9**, (8), pp. 8–12.
Black Diamond (1999). Changing the way climbers think about gear. The story of wiregates, information leaflet published by Black Diamond Equipment Limited.
British Mountaineering Council (1997). Participation statistics.
British Mountaineering Council (1997). Technical Series *Knots*.
British Mountaineering Council (1998). Technical Series *Ropes*.
British Mountaineering Council (1999). Annual Report, p 12.
British Mountaineering Council (2001). Care and maintenance.
British Mountaineering Council (2002). Crampons.
Charlet, J. (2000). Karabiners: be aware, *Journal of the UIAA*, **3**, 14–15.
Charlet, J. and Lassia, R. (1989). Etude Dynamique du Mousqueton, Annex 5 in UIAA Safety Commission papers, Chamonix 28 June–1 July 1989; published in British Mountaineering Council Technical Report 90/1, 1990.
Dickens, P.M. (1990). Examination of DMM featherlite karabiner which broke – gate open, British Mountaineering Council Technical Report TCMC 90.
Grandison, N. (1993). Camp lightweight (6 kN) UIAA Karabiner broken – fall factor 0.3 – keeper open, British Mountaineering Council Technical Report TCMC 93/11.
Hellen, T. K. (1996). Stubai bent front points, British Mountaineering Council Technical Report TCMR 96/5.
Hillebrandt, D. (2001). Hidden secrets corrosion of sea cliff pegs, *Summit*, **22**, 52
Huyton, A. (1994). Faders karabiner broken – fall factor 0.13, keeper open, British Mountaineering Council Technical Report TCMC 94/2.
Ingram, S. (2001). Grigri – unmasking the myths, *Summit*, **23**, 26–27.
Keynote UK sports market (1996). Reported in British Mountaineering Council Participation statistics, 1997.
Lyon, B. (2001). *Care and Maintenance*, British Mountaineering Council, 29–32.

Maycock, E. C. (2001). Troll Attachment Loop Development, MSc Thesis, Centre for Materials Science and Engineering, University of Edinburgh.

McMillan, N. (1995). Industrial helmet – chinstrap release, British Mountaineering Council Technical Report TCMF 95/13.

McMillan, N. (1999). The BMC, technology, and the safety of climbers, *The Science and Technology in Climbing and Mountaineering* Conference Proceedings, Leeds University, Chapter 14.

McMillian, N. (2000). Karabiner breakings when using a figure-of eight, *Journal of the UIAA*, **3**, 5–6.

MacNae, A. (1995). Harness buckle slippage, British Mountaineering Council Technical Report TCME 95/15.

MacNae, A. (1999). Acid rope, British Mountaineering Council Technical Report TCMB 99/01.

MacNae, A. and Taylor, M. (2000). Head games: the BMC helmet testing programme, part 1, *Summit*, **19**, 30–32.

MacNae, A. and Taylor, M. (2001). Heads up: the BMC helmet testing programme, part 2, *Summit*, **20**, 30–36.

Mills, N. (2001). Materials Learning and Teaching Case Studies: Polymeric ropes for sport activities. http://www.materials.ac.uk/resources/casestudies/ropes.html

Mintel leisure intelligence, Sporting activities in the great outdoors, August 1996; reported in British Mountaineering Council Participation statistics 1997.

Pavier, M. (1996). Derivation of a rope behaviour model for the analysis of forces developed during a rock climbing leader fall, *The Engineering of Sport, Proceedings of the First International Conference on the Engineering of Sport*, Ed. S Haake, Publ. Balkema, Sheffield, UK 2–4 July 1996, pp. 271–279.

Pavier, M. (1999). Presentation at British Mountaineering Council Technical Conference, Plas y Brenin, UK, November.

Payne, R. (1994). Mercury rope broken in fatal abseil accident, British Mountaineering Council Technical Report TCMB 94/10.

Perkins, A. (1991). Troll tapes, slings and harnesses booklet, Troll, UK.

Riley, A.M. and Maddock, P. (1983). Pitting corrosion of alloy karabiners, British Mountaineering Council Technical Report TCMC 83/11.

Schubert, P. (1998). *Sicherheit und Risiko in Fels und Eis* [Safety and risk on crags and ice], Bergverlag Rudolf Rother GmbH.

Schubert, P. (2000a). About ageing of climbing ropes, *Journal of the UIAA*, **3**, 12–13.

Schubert, P. (2000b). Attaching to the rope by karabiner, *Journal of the UIAA*, **3**, 18–19. Also see www.klettertraining.de/divers/unfaelle.html

Signoretti, G. (2001). Wet and icy ropes may be dangerous! *Journal of the UIAA*, **2**, 25–28.

Simon, M. and Taylor, I. (1991). *Ice axe/crampon failure investigations*, British Mountaineering Council Technical Memorandum, TCM 92/2.0.

Smith, R.A. (1998). Risk reduction in rock climbing, *Sports Engineering*, **1**(1), 27–39.

Steele, G. (1995). Climbers belt buckle slippage, British Mountaineering Council Technical Report TCME 95/4.

Taylor, M. (2002). Helmets off centre impacts, *Summit*, **26**, 50.

UIAA Journal (2000). EN and UIAA standards for mountaineering and climbing equipment, *UIAA Journal*, **3**, 23.

Wexler, A. (1950). The theory of belaying, *American Alpine Club Journal*, **7**, (4), 379–405.

12
Materials in skiing

H. CASEY

Los Alamos National Laboratory, USA

12.1 Introduction

Many observers of the 2002 Winter Olympic Games in Utah will be aware of the increasing influence of advanced technology on the character and quality of athletic performance. Technology has, for many years, played a dominant role in sports such as motor racing or yacht racing, but in recent years advanced materials and sophisticated manufacturing technology have produced revolutionary changes in equipment used for track and field, golf, tennis, skiing, biking and many other sports. The effects of modern equipment are apparent, not only at the professional and competition levels, but also at the recreational level, where the progress is reflected by shorter learning periods, increased comfort and safety, and a more rewarding experience. These benefits are stimulating interest and growth in sports as well as making a positive contribution to the consumer market in sporting goods.

The alpine ski is representative of sports equipment that has evolved through the years in a fairly predictable manner, but has undergone more radical changes in recent years as a result of improved design and manufacturing technology. The proceeding review tracks ski design and manufacturing development with emphasis on the contribution from materials technology. Topics include the impact of technology on the ski industry, the contribution of advanced materials in equipment performance, performance trends in competition and recreational skiing, and speculation regarding future trends and advanced concepts in ski design.

The link between equipment and technique was emphasized many years ago by the great 'ski teacher' Georges Joubert[1]: 'The objective of the (ski) technique of any given period is to obtain the best results possible with the equipment available at the time.' My intent is to provide information which reinforces this premise based on personal observations stemming from my background as both a materials engineer and a ski instructor.

12.2 The impact of technology on the ski industry

Public awareness and participation in winter sports is on the increase, and includes competitive and recreational skiing (Alpine and Nordic) as well as snowboarding and similar related pursuits. The status of the industry is reflected by the equipment sales and the growth and development of ski towns and recreational communities, particularly in Europe and the United States, but also in many other parts of the world. Advances in equipment have contributed to the popularity of skiing as a recreational pursuit, and the market has grown in response to the demand for equipment designed to meet the requirements of racing and recreational skiing. Equipment includes skis, poles, bindings, boots, as well as special clothing and other accessories. This text deals primarily with alpine skis, although the complete range of equipment has benefited from the use of modern technology.

The earliest skis used for downhill skiing were derived from cross-country skis and were long (>230 cm) and very stiff. The first competition downhill races took place in Switzerland during the period 1911–1925 and the focus on speed and maneuverability on steep terrain resulted in attempts to produce a shorter, more flexible tool. Designs were refined in the 1930s and 1940s, driven by the quest for success of the national teams, although the innovations in design and manufacturing were often inspired by individual racers and entrepreneurs. The most significant improvements during this period were in the bindings and boots, which were modified to provide a direct linkage between skier and ski, and increase stability and control. Binding technology continued to evolve with the development of sophisticated release mechanisms, which would ensure foot stability under the most severe operating conditions, but release instantly when the impact or twisting force reached a level which constituted a threat to the skier. The developments in boots and binding paralleled alpine ski development, but will be included only for reference in this review, which is focused on the ski. Figure 12.1 indicates the extent of the changes in appearance and design during the last century.

Equally important is the progress in the bindings, i.e. the attachment between boot and ski. A typical binding from the 1940s is contrasted with a modern (1990s) release system in Fig. 12.2.

Developments in materials and in processing and manufacturing technology are important factors responsible for the progress in ski design and production. The materials used in ski construction are selected to meet established specifications and are therefore a critical component of the overall technology enabling the high performance of the equipment. The terminology 'high performance' is normally associated with competitive-level skiing, inferring a high degree of performance under extreme conditions including high speeds and demanding terrain. However, the

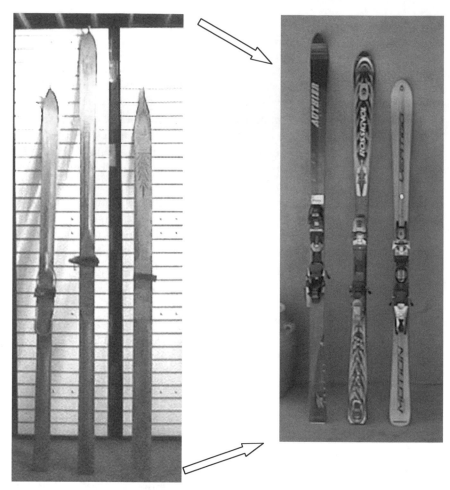

12.1 Skis dating from 1900–1940 contrasted with modern equipment from the period 1980–2001. Long (>230 cm) hickory skis are compared with shorter (160–200 cm) modern equipment built with multiple materials.

performance of the ski can also be rated on the ease of operation at the beginner level and the contribution to the learning process for the developing skier. For example, using today's modern ski, the novice skier can advance to intermediate level with just a few weeks of instruction; intermediate skiers are much better equipped to attain expert status, a level once reserved for only a small faction of recreational skiers. New ski equipment has not only revolutionized Olympic and professional-level racing, but it has opened up the sport of skiing to thousands of less gifted athletes. Advanced terrain and extreme snow conditions are now within the reach

12.2 A 1940s ski represents the technology of that era, namely steel edges (fastened with screws), grooved and lacquered base, and cable bindings. On the right is a modern release binding with a combination of titanium and plastic components.

of the average skier who is sufficiently motivated to learn contemporary ski technique.

The improvements in equipment performance have had an enormous impact on the entire ski industry, influencing essentially every aspect of skiing. For example, consider the fundamental mechanics of skiing. Ski technique has undergone significant changes over the years, and as will be detailed later, the changes can be directly linked to the performance of the equipment. Skis, bindings and boots all benefit from the use of sophisticated designs made possible by new materials and advanced manufacturing technology. As a result, revolutionary changes have occurred in ski technique, initially at the competitive level, and subsequently influencing all levels of the recreational skiing.

The change is also apparent in the accessibility of the terrain. Difficult slopes and variable snow conditions are no longer a barrier to a moderately competent skier. New equipment has resulted in increased numbers of skiers venturing off-piste into demanding terrain. In addition, modern skis and groomed conditions provide the opportunity for the average skier to ski faster and to 'carve turns', using technique previously within the capability of only racers and very advanced skiers.

Finally, the market for ski equipment has continued to expand in response to public demand. The increase in the popularity of skiing is due in part to the modern ski, which is shorter, lightweight, durable, easier to use, and has been fitted with highly efficient safety release bindings.

Compare these to the early skis, which were long, heavy, easily damaged, and had to be firmly strapped to the boot with no release available in the event of a fall. The 'user-friendly' aspects of the modern ski attract many potential skiers who would shy clear of its predecessor's cumbersome appearance and limited performance characteristics. Improvements in safety, comfort and performance are all key factors identified with contemporary ski equipment, and each has contributed to the increased popularity of Alpine skiing. The market has been expanded to include 'specialty skis' such as 'fat skis' for powder (off-piste), 'ballet' and 'acrobat' for terrain gardens, and 'blades' (very short skates) for training and limited terrain use. Each season, new 'designs' are introduced to meet (or create) the specialty markets and popular trends of the period.

12.3　Contribution from materials and manufacturing

In the early days of skiing, design and construction concepts were limited primarily by availability of material. Wood was the obvious choice; it was readily available, its properties and characteristics were reasonably well understood, and woodworking was an established practice. Wood could be hand crafted or machined and formed to shape (Fig. 12.3).

Wood has a cellular structure providing (in the absence of moisture) low density, high strength properties. The small 'cells' and larger 'sap' passages are quite recognizable in the macrograph. The specific modulus (modulus/density) of wood is comparable to steel, aluminum and titanium. For this reason, wood was (and continues to be) an important option for ski construction, where weight, strength and flexibility are important criteria.

A ski requires strength and flexibility along its primary axis as well as torsion rigidity to provide a stable platform underfoot. This was achieved by making the skis long (>220 cm) and relatively thin except for a thickened platform under the binding position. The ski was built with an upturned front tip to ride over obstacles, and 'camber' was introduced to increase

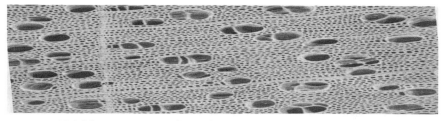

12.3 Macrostructure of wood.

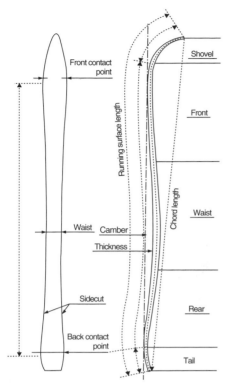

Adjacent figure illustrates typical alpine ski profile indicating key design aspects: tip, tail, waist, camber, sidecut and overall shape.

12.4 Alpine ski profile, indication key design aspects.

ski/snow contact area and improve glide. Various forms of grooves were cut into the base to improve tracking, and many concepts were developed to bind the skier's feet to the skis. The early bindings (derived from cross-country skis) lacked lateral support and heel restraint, and control was reduced accordingly. The basic design of the ski is as shown in Fig. 12.4.

Despite its obvious advantages, wood has some serious negative characteristics. Because of its fibrous structure, it is highly anisotropic and has poor strength in a torsion mode. It is sensitive to humidity and absorbs moisture, which increases the weight and results in warping and distortion. The early skis were shaped using steam heating, and warping was a constant problem. A dry heat method for shaping hickory skis was developed, enabling production of a thinner, more flexible ski. Wood bases are soft and easily damaged and lacquer and wax treatment was added to improve durability and to prevent sticking and buildup of snow on the running surface. The early hickory wood skis were heavy and difficult to maneuver; glide and durability was limited and warping and structural failures were common. It took many years and a persistent and determined effort of a few ski

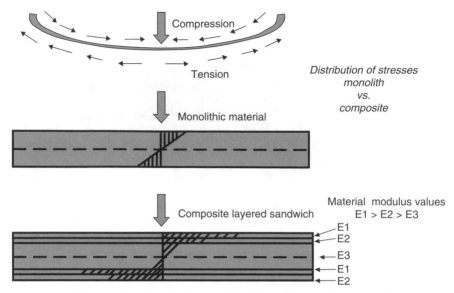

12.5 (Two-dimensional) stress distribution in simple monolith, versus multilayer dissimilar material lay-up.

pioneers to move from the traditional monolithic wood structures and incorporate plastic bases, steel edges, and aluminum alloy structural members. Flexibility in a monolithic structure is a function of the thickness. In a multi-component design, different materials and sub-structures are used to tailor the flex and torsion strength according to the performance requirements. This concept is illustrated in Fig. 12.5.

In a single material the stress resulting from flexing is a function of the distance from the neutral axis. In a sandwich multilayer, the stress distribution is a function of the different materials properties, e.g. modulus; high modulus materials will carry the bulk of the stress. In the figure above, three different materials are combined in a layered sandwich configuration. For any given level of strain, the high modulus (E) material will carry a corresponding higher level of stress. This simple diagram does not attempt to explain the complex dynamic tension, compression and torsion stress/strain conditions prevailing during operation of the ski. It merely attempts to illustrate the significance of selecting the appropriate engineering materials to meet the design requirements, and the potential benefits of a multicomponent structure.

Although European racers used steel edges as early as the 1930s, the development of the 'metal' ski, introduced in the 1950s, is attributed to Howard Head,[2] who is also acknowledged as the designer (decades later) of the large body tennis racquet. Head's original skis emerged as alu-

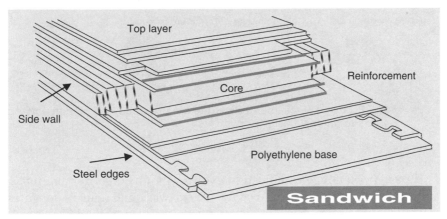

12.6 Typical multilayer design from the 1970s.

minum/wood-bonded structures, and were further developed with plastic base material and steel edges. Ski testing took place on the slopes. Head's 'revolutionary' metal skis were provided to ski instructors and colleagues to test and many prototypes were subject to premature failure before a successful manufacturing process was developed. Combining different materials (wood/metals/fiberglass, etc.) into a single structure presented many problems. It was difficult to reproduce the specific dimensions in a consistent manner. Bonding and joining of the dissimilar materials was a major difficulty. Adhesives were under development for use in aircraft and other critical structures in the 1950s but the technology was complicated and equipment for assembling and curing complex material combinations was expensive. Common failure modes included rupture of the bonds between the dissimilar materials causing 'de-lamination' of the structural members, and deformation due to overstress in the course of normal operation. As significant advances were made in the manufacturing technology, particularly the bonding and curing, the durability of the product improved.

Wood laminates, metal, fiberglass and plastics were incorporated into ski structures as the materials became available and commercial interest in skiing grew during the 1950s and 1960s. The design goal was to combine good strength and flexibility along the length of the ski with adequate torsion strength to sustain a platform for the skier. In addition, as skis were designed for ever-increasing speed and more demanding terrain, it became necessary to dampen the structures to absorb dynamic impact loading and to suppress vibration at variable frequencies. Ski construction evolved to complex designs involving a base, core, sidewalls, top layer, reinforcing and damping layers. A typical 'sandwich' design from the 1970s is shown in Fig. 12.6.

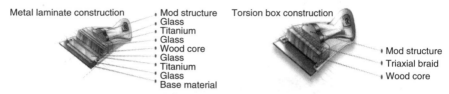

Metal laminate construction
Torsion box construction

· Mod structure
· Glass
· Titanium
· Glass
· Wood core
· Glass
· Titanium
· Glass
· Base material

· Mod structure
· Triaxial braid
· Wood core

12.7 Product line examples illustrate the use of advanced materials
and original designs.

The sandwich design continued to evolve with torsion boxes and tubes, and using glass materials of choice for the sidewalls, and foamed polymers were used for the core. Foamed polyurethane is used in many skis, although lower density acrylic foams or metal honeycomb are alternate choices for core material for lightweight skis. The higher density plastics are more effective in reducing vibrations and creating a smoother ride, but the ski is less responsive and more suited to long radius turns and high speed cruising. Some designers prefer damping systems external to the main structure of the ski in an attempt to create the 'universal' ski, smooth as well as quick and responsive. Damping technology has dominated the design of racing skis because of the prevailing snow conditions (ice, ruts) and the need for precise control at high speed. By contrast, skis used for 'soft' and 'powder' snow conditions need only moderate damping since the snow is a most effective 'cushion', and performance is enhanced by a 'lively' ski.

Two designs from a popular ski range for 2001/02 season are shown in Fig. 12.7. Note that the traditional wood core has been preserved but alongside titanium, fiberglass, and an 'elastomeric' secondary core to dampen shock and vibration. In the torsion box structure, the wood core is wrapped with fiberglass in a proprietary 'triaxial braid' weave, a technique aimed at providing strength in torsion in a lightweight, responsive ski.

In some instances materials have been highlighted primarily for marketing purposes, and advertising has overstated the effects of the specific 'advanced materials' and 'space-age' alloys. This was particularly evident during the 1980s, but the knowledge gained from ski testing resulted in remarkable improvements in equipment beginning in the mid-1990s. As designers became more aware of the engineering characteristics of the materials, the performance was greatly increased, culminating in a dramatic reduction in the length, and a corresponding improvement in the response, stability and maneuverability of the ski.

Prior to 1990, an 'all-mountain' recreational ski would range from 195 to 210 cm in length. By 1995, the length had been reduced to 198 cm or lower (with improved performance), and to 180 cm by the year 2000. Olympic slalom skiers were using 155 cm length skis at the Utah Games. The performance of the shorter, compact ski is the result of modifications to the design (shape and contours) plus better use of engineering materials to tailor strength, flex and stability.

Design and processing technology are closely guarded secrets in what has become an intensely competitive industry. The ski industry appears willing to explore the most advanced concepts, including the use of electro-active polymers and piezo-ceramics to achieve optimum performance at either the competitive or recreational level. Piezo-electric materials are materials that develop an electric charge when deformed, and conversely, deform in the presence of an electric field. Piezo-based 'smart' materials both actively sense and respond to the surrounding environment. Ski designers have already incorporated such systems into the ski, although the success of such an experimental concept has yet to be confirmed. There is a fine line between appearing to improve the performance as a marketing ploy, and actually delivering a significant improvement by application of complex new technology. The ski industry has been quick to exploit the 'high tech' feature of the period in its advertising. It takes a skilled professional skier/racer to distinguish the characteristics of the numerous models of skis on the market from many different manufacturers. However, it is obvious that the modern skis are significantly improved over their predecessors, and the improvements can be attributed to an increased knowledge of engineering design and use of appropriate materials.

12.4   Development of competitive and recreational skiing

The first downhill ski race (the Kandahar) took place in 1911. Alpine skiing emerged as a competitive and recreational sport in the 1930s. Although ski technique has been continuously modified and refined over the years, the changes have been evolutionary and derived from competition skiing. Speed and precision tracking are the goals of ski racing, and the athlete is required to maintain dynamic balance on two moving platforms (the skis) as they slide over continuously changing terrain. The racer must direct the skis down a predetermined course at the maximum speed possible, countering the gravity and momentum forces with angular body movements.

The size and general characteristics of the early racing equipment is illustrated in the photograph of the 1936 USA Olympic team (Fig. 12.8).

Early ski equipment could best be described as primitive, and ski technique was determined based on the exploits and 'style' of a few outstanding racers. Alpine skis were derived from cross-country skiing, and the attachment device (lateral binding) allowed the heel to 'float'. This factor, plus soft flexing boots, provided very little stability when pressuring or pivoting the ski, and the main goals of the period involved controlling speed

12.8 1936 US Olympic ski team.

and remaining upright. Downhill ski technique improved as bindings were developed to stabilize the foot and allow the skier to control pressure on the skis. As lateral stability increased (with improved boots and bindings) the skier could change direction with long skidded turns.

The 'Arlberg' method relied on snowplow and stem turns and the skier's weight primarily to the back of the ski. As early as the 1930s, ski racers were experimenting with flexion and extension of the legs to lighten the skis and allow them to be redirected by rotary twisting movements. The Austrians refined these movements by skiing with their legs locked together and employing a series of upper body rotary motions; this was the predecessor to the 'Kruckenhauser' Austrian technique, which dominated in the 1950s. Rapid vertical body movements were used to decrease ski/snow pressure (un-weighting), allowing the ski to be redirected by rotary twisting of the upper body or by more subtle rotary leg and foot movements (steering). However, the ultimate goal was to limit the amount of lateral slippage or skidding such that the skis would track or 'carve' precisely the fastest path down a race course. Ski technique has at various times emphasized 'pressuring and edging' versus 'pivoting and rotating' the ski to execute a change of direction. In contemporary skiing, speed control is determined by the shape of the turns as opposed to skidding and braking maneuvers.

As the equipment improved, ski technique would progress through many changes. The trends in technique were heavily influenced by the competition at the international level. The central European countries struggled for domination in alpine skiing. Northern European countries such as Norway and Sweden also had a significant impact due to the influence of skiers such as Stein Ericksen and Ingemar Stenmark. As the knowledge of the physics of skiing grew, the ski design and equipment performance became increasingly important.

Skis were developed to be lighter, stronger, responsive, and more stable as the designs and manufacturing technology improved. Most of the innovative designs were aimed at improved racing performance. The flex pattern of the ski was tailored to meet the racing performance criteria (downhill, giant slalom, slalom).

Recreational equipment also benefited; for example, skis were more durable and less prone to breakage or de-lamination, but many of the skis produced for the recreational skier were marketed as 'de-tuned' racing designs with misleading information, particularly with regard to the materials and their benefits. Starting in the 1970s, the graduated length method (GLM) of ski teaching used short skis to introduce beginner skiers to the ski slopes. The novice skier graduated to increasing lengths in accordance with the skill development. These skis were not suitable for high performance skiing and lacked the stability and response characteristics of the modern ski. Many 'high performance' short skis produced during this period were designed for twisting and skidded turns, for braking and resisting gravity forces. This contrasted sharply with the racing technique which emphasized gliding, carving and working with gravity to 'shape' the turn and maintain control during a high speed descent.

The techniques being developed on the race course during this period were not easily transferable to the recreational skier. Carving with a traditional length ski requires the ski to be displaced well outside the skier's body and is possible only at high speeds where centrifugal force is sufficient to maintain balance (analogous to a bicycle turn). Only expert skiers were capable of employing the subtleties of angulation, inclination, edging and pressure control, as well as skiing at the speed necessary to carve their turns. By contrast, recreational skiing relied on completing each turn by twisting and skidding to slow down or come to a stop. On the more demanding terrain, recreational skiing could frequently be described as 'a series of linked recoveries'.

This dichotomy between racing and recreational technique created many problems for the ski schools in Europe and the United States and was (is) the subject of many heated discussions among professional ski instructors and racers.

The changes in technique (as reflected by the skier's stance) during the period 1950–1990 are captured in the photo series (Fig. 12.9).

Stein Ericksen's classic narrow stance (legs close together) and counter-rotated shoulders were the primary influence throughout the 1950s. Warren Witherall illustrates a narrow stance, with the upper body in a countered

Stein Ericksen / 1950s ..... Warren Witherall / 1960s .... Phil Mayer / 1980s

12.9 Three well-known racers demonstrate the style and technique of the specific era.

(open) position, characteristic of the technique of the 1960s. Phil Mayer in the 1980s shows a wider, square stance and demonstrates independent leg action.

The trend over five decades has been a reduction in vertical and rotary movements, thus stabilizing the upper body, and developing a natural athletic stance, a progression which can be linked to improvements in equipment performance. Significant changes have occurred during the last decade as the 'user-friendly' aspects of short skis have been combined with the characteristics of the high performance carving skis. This was accomplished with the development of the 'shaped ski'. Although the original 'shapes' were designed as race trainers, and the new designs have dominated international ski racing, the concept was quickly adapted for recreational skis. To appreciate the advantage of the new designs, consider the principal forces operating during high speed skiing, as shown in Fig. 12.10.

As long as the resultant force is aligned inside the ski/snow contact points, the skier is in balance. Carving with a traditional length ski required the ski to be displaced well outside the skier's body and is only possible at high speeds where centrifugal force is sufficient to maintain balance (analogous to a bicycle turn). Increasing the 'shape' of the ski produces more ski/snow contact at a lower edge angle, with less displacement, and consequently lower speed. The advantage is that novice skiers can experience the 'user-friendly' aspects of shorter skis and can develop good posture (stance) which emphasizes gliding and edging. Intermediate skiers can refine edging

Centrifugal force

Center of mass

Resultant force

12.10 Simple two-dimensional illustration of the principal forces during the control phase of the turn.

12.11 US Olympic champion Bodi Miller exemplifies ski racing technique.

skills and experience the thrill of carving turns at moderate speeds. Experts can expand their horizons to exciting new levels.

Strong lateral motion and a wide square stance is the trademark of contemporary ski technique, allowing for more consistent ski/snow contact and better tracking (carving) of the skis. The photograph of the US Olympic champion Bodi Miller (Fig. 12.11) illustrates precise carving of the skis and the wide track square stance typical of today's professional racers.

The technique is directly attributable to the improvements in equipment, and particularly the skis. Compared to previous decades, they are shorter and wider, with more sidecut (wider tip and tail, narrow waist), increased longitudinal flexion, greater resistance to twisting, and improved resistance to shock and vibration (damping).

These characteristics benefit the recreational skier as well as the racer. The shorter length is more maneuverable, and the platform is more stable due to the use of stronger materials that also transmit less shock and vibration forces. The skier can balance on an edged ski at the lower speeds associated with intermediate level skiing. The result has been a dramatic change in ski technique at both the racing and the recreational level.

12.5   Future trends

It is interesting to note the similarity in choice of materials for skis, golf clubs, tennis racquets, bikes, etc. The 'common thread' is the requirement for the athlete to remain in balance and control in a situation involving dynamic motion and impact. He/she needs to be aware of the prevailing dynamic conditions in order to react and exert control. The novice needs equipment designed to absorb and insulate her/him from external impact forces, whereas the experienced athlete can profit from equipment that is more responsive. 'Sensitivity' and 'feel' are juxtaposed to shock and vibration, in much the same way as an automobile suspension system affords some level of isolation, but provides appropriate sense of the road conditions. The problem can be compared to signal conditioning, i.e. using a filter to remove 'noise', and clarifying/amplifying the desired components of the feedback. The approach in skis relied on the use of complex multilayered designs and/or external damping systems, together with the use of sophisticated advanced materials. It is likely that this trend will continue, and will include 'smart materials' and 'smart systems'.

Earlier we referred to piezo and electro-active materials which have already featured in some skis as well as in boots and clothing. These multifunctional materials are being refined for automotive, aircraft and other commercial use, and we anticipate additional exploitation of this technology in ski equipment.

Reference was made earlier to developments in binding technology. The latest bindings appear to offer the ultimate in safety and control, but still allow the ski to flex undisturbed by the attachment device. This is accomplished by innovative designs such as a rail attachment concept (Vokl Skis) or an internal free gliding axle (Atomic Skis). Details can be accessed over the appropriate websites. Binding design and development has paralleled ski design and future trends will be influenced by the changes in ski boot design, which is likely to be the next significant change.

As recreational skis respond to more subtle control forces, the boots should become softer, lighter and generally more comfortable. Again, we might anticipate the use of smart materials to provide an orthotic foot-bed that will meet the conflicting requirements for comfort and control. Textiles fabricated from spun electro-active polymers could have some application in future developments.

Finally, it is worth noting the increasing sophistication of ski training systems. The current equipment is likely to benefit from sensing and control technology, opening the door for simulators that could be used to prepare novice or expert skiers for real experiences on the slopes. Virtual reality (VR) systems for sports would be a predictable trend for the next decade. The term VR is applied to a collection of technologies which together form an interface between humans and computers. It is reasonable to contemplate appropriate computer generated images linked to a ski trainer system equipped with adaptive sensing and control technology. Such a system could be used for teaching and coaching as well as marketing of equipment and ski areas.

12.6   Acknowledgements

The author wishes to acknowledge the knowledge and information gained over many years of interactions with friends and colleagues including the following organizations:

- Professional Ski Instructors of America/Rocky Mountain Region
- Ernie Blake Ski School, Taos Ski Valley, New Mexico
- Pajarito Mountain Ski School, Los Alamos, New Mexico
- Alpine Sports, Santa Fe, New Mexico
- Los Alamos National Laboratory
- Office of Naval Research (International Field Office/London)

12.7   References and sources of further information

1 *Journal of Professional Ski Instructors*, September 1982.
2 *Ski Pioneers* (Ernie Blake, Friends and the Making of Taos Ski Valley), Rick Richards, 1992.
3 *The Professional Skier* (journals through Winter 2002).
4 *Tomorrow's Materials*, Ken Easterling, Institute of Metals, 1999.
5 *Ski Moderne*, Joubert and Vuarnet, 1957/58.
6 *Teach Yourself to Ski*, G. Joubert, 1970.
7 Skis Dynastar (personal communications).
8 Vokl Skis (personal communications).
9 Atomic Skis (personal communications).
10 Rossignal Skis (personal communications).

13
Materials in cricket

A. J. SUBIC
RMIT University, Melbourne, Australia

A. J. COOKE
Cooke Associates, Cambridge, UK

13.1 Introduction

The exact origins of the sport of cricket are unknown – although there is mention of a game resembling cricket *c.* AD 1300 in southern England. Indeed, much of the history of cricket development has been lost or never recorded. However, it is known that, by the second half of the eighteenth century, crowds of up to 20 000 were watching certain games and high stakes gambling was involved (Green, 1988). The evident popularity of the game necessitated some consistent rules, and a code of laws was laid down by the Marylebone Cricket Club (MCC) in 1788 and adopted throughout the game. Prior to this code, the rules varied depending upon in which part of the British Isles the game was being played. The MCC is still the custodian of laws relating to cricket around the world.

The international cricket game is run by the International Cricket Council (ICC) and each country has its own cricket board or governing body. The ICC facilitates two types of international cricket: test cricket (5 days) and limited overs (1 day, 50 overs per side). There are ten test-playing nations – Australia, New Zealand, South Africa, England, West Indies, India, Pakistan, Sri Lanka, Zimbabwe, Bangladesh – and many more countries play limited overs cricket. Each major cricketing country (all test countries and a few of the others) has its own cricket board or governing body that defines the exact regulations for play in that region. For example, in the United Kingdom, the regulations are laid down by the England and Wales Cricket Board (ECB). The cricket board for each country cannot impose rules that contravene the official MCC Laws of Cricket. However, they do define the structure of domestic competitions as well as regulating aspects of the equipment used, e.g. which brand of cricket balls is to be used for competitions (both domestic and international) held in their area.

This chapter examines the rules of the game and their correlation with the design of cricket equipment: in particular, the materials used and construction of balls, bats and protective equipment. In all cases, a literature review was carried out to identify published work. General engineering databases such as Bath Information and Data Services (BIDS), Compendex and OCLC FirstSearch were examined along with the more specialised sports engineering resources provided by the International Sports Engineering Association. Also, an in-depth review of leading cricket bat/ball manufacturers and their practices has been conducted. A summary of the literature can be found within each section together with a more detailed discussion of the equipment, materials and construction. The chapter concludes with thoughts on future trends for cricket and its associated equipment.

13.2   Cricket balls

This section examines the manufacture of cricket balls and how aspects of their construction can influence their performance. There are three main areas where the ball can affect the game. The first is the aerodynamics of the ball and how it affects flight (for example, spin bowling versus seam bowling). However, the aerodynamics of sports balls is covered in Chapter 5 and so will not be discussed here. Second is the issue of how the ball's makeup influences the major impacts (ball/pitch and ball/bat) of the game. The nature of the ball will dramatically affect how the ball bounces off the pitch and how the ball leaves the bat when struck. Thirdly, cricket balls must maintain their shape and not drastically soften over long periods of time and after repeated impacts; for example in test match cricket a new ball is only issued after 90 overs, which is a minimum of 540 deliveries. Both of the last two areas are considered in this section.

13.2.1 Historic background

Of the 42 Laws of Cricket only one concerns the ball: Law 5. This Law specifies the tolerances for circumference and mass, but does not specify any other characteristic. Since 1927 the requirements for First Class Cricket balls have remained constant, with Law 5 stating: 'The ball, when new, shall weigh not less than $5\frac{1}{2}$ oz/155.9 g nor more than $5\frac{3}{4}$ oz/163 g; and shall measure not less than $8\frac{13}{16}$ in/22.4 cm nor more than 9 in/22.9 cm in circumference' (MCC, 2000b).

However, in the late 1970s, the Test and County Cricket Board (TCCB), the Cricket Council and leading ball manufacturers joined forces, along with the British Standards Institute (BSi), to develop a standard for cricket balls used in the British Isles to ensure consistent ball quality. The latest version of the standard was published in 1995 (BSi, 1995). Each company is visited two to four times a year and four samples representative of current production are selected for testing. The experimental process consists of a sequence of tests that determine various aspects of the ball, the first two of which cover the requirements of Law 5:

1. Circumference
2. Mass
3. Seam width
4. Seam height
5. Shape
6. Height of bounce
7. Hardness
8. Impact resistance
9. Wear resistance.

For use in first-class cricket in the UK, a ball must pass the British Standard. The only other cricket board to have a standard is the Australian Cricket Board (ACB), which differs significantly from the UK in that only traditionally produced cores can be used, but machines can stitch the leather (ACB, 1999).

It is interesting to compare cricket with the situation in professional baseball. All the balls used in US Major League Baseball are made by one company. At the time of writing, four companies have had balls pass the British Standard and have thus been allowed to supply balls for first-class cricket. Bowlers sometimes have a preference for which brand they use. The reasons can be quite vague, but often the bowler has an innate feeling that one type of ball will do more for him than any other make. This issue was demonstrated in the 1992 England vs. Pakistan series, where each side preferred a different brand. Before each match, a coin was tossed and the team that won could choose the make of ball (Oslear and Mosey, 1993).

Since the introduction of the standard, there have still been incidents of balls being used in first-class cricket that do not meet the specifications. In 1984, the TCCB ordered that each country must supply scales to weigh cricket balls. The results were surprising, for although a few were under weight, many were more than 163 g (Oslear and Mosey, 1993). More recently, there were allegations that certain brands of balls licensed for first-class cricket were underweight and it was claimed that the slightly underweight balls enabled bowlers to generate more swing on their deliveries (Dean, 2000).

In baseball any increase in the number of home runs being scored is usually greeted with claims of the balls being 'juiced' (Kaat, 2000). Over the past century there have often been claims that the manufacture of balls has been subtly altered in order to change its performance (Wright, 1999), but after rigorous testing nothing conclusive has ever been proved (Kaat, 2000).

In 1999 there was some unsubstantiated speculation that the white ball selected for use by the ECB for all limited overs matches held in the UK gave a great advantage to the bowler. It was alleged that the ball swung more in the air and moved more off the pitch. Some testing was carried out, but nothing was proven. In 2002, the ECB decided to adopt the same brand of white ball as used in Australia and South Africa. The main reason given was to let English players get used to a ball they invariably use whilst on tour and in World Cup competition, but it also removed the possibilities of unsubstantiated speculation.

13.2.2 Literature review

A review was carried out to identify published work on cricket ball manufacture and construction. An indication of how cricket research is directed can be gained by reviewing the papers presented at the first three International Conferences on the Engineering of Sport (Haake (ed.), 1996; Haake (ed.), 1998; Subic and Haake (eds), 2000). In total 14 papers were presented about cricket, but only six were on cricket balls. Of these, three papers concerned aerodynamics, two concentrated on ball/pitch impacts and one examined ball/bat impacts. Thus, the majority of research into cricket balls has traditionally been directed at understanding their aerodynamic properties, with issues such as 'swing' and 'reverse swing' featuring heavily.

The work carried out on ball/pitch impacts has concentrated on two performance characteristics: the coefficients of friction and restitution. Most of the research has been aimed at understanding how the pitch (soil types, etc.) affects the bounce of the ball (Carre *et al.*, 2000). It is interesting to note that, as in almost all cricket papers, the type of ball used for testing purposes is not stated; most researchers seem to make the tacit assumption that the ball is a constant.

At a more general level, several pieces of research have been published that compare the dynamic properties of a range of sports balls (Giacobbe *et al.*, 1995; Cross, 1999). These studies discuss coefficient of restitution, along with compression testing and provide experimental frameworks for dynamically evaluating balls. Unfortunately, the authors concentrate on baseballs, golf balls and tennis balls, with only the study by Giacobbe *et al.* examining a cricket ball. However, the ball was unspecified and of an old design.

The results of this literature review indicate that limited research has been carried out into the material properties and construction of cricket balls, and the effect of the cores on overall ball performance. One paper considers these areas in more depth (Jarratt and Cooke, 2001) and this is the basis for the following section.

13.2.3 Materials and construction of cricket balls

Traditional cricket balls are produced using a 'quilted' or 'layered' core. A moulded cork centre (approximately 2 cm in diameter) is alternately layered with wool yarn and more cork. The wool is wound wet and under

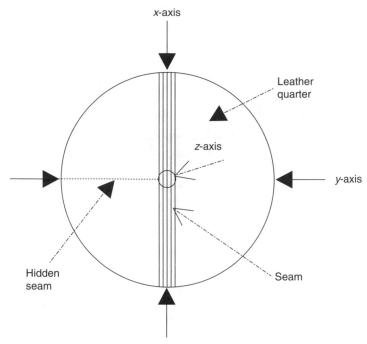

13.1 Diagram of a cricket ball showing the principal axes and location of the seam.

tension to compress each of the layers of cork – as the yarn dries it contracts and so the compressive force on the cork layers increases. Typically, five layers of cork and six of yarn are put down to form the core or centre.

The leather cover consists of four pieces (or quarters) of hide, which is normally dyed red. On the equator of the ball there is a visible seam (the 'primary seam') that consists of six rows of stitching; all other stitching to join the quarters together (the 'quarter seams') is hidden. Traditionally, the quarter seams are set at right angles to each other. Figure 13.1 shows the positioning of the seams and the three principal axes of the cricket ball. The nature of thread used and the number of stitches used are closely specified by both the British Standard and the ACB specification as the seam has a critical affect on the ball's aerodynamic properties (BSi, 1995; ACB, 1999).

Traditional cricket balls have many similarities in construction with baseballs and the balls used in the game of Real Tennis. A baseball core consists of a composite cork/rubber pill, which is surrounded by two layers of rubber. Around this are wound three layers of wool and a finishing layer of cotton string. The leather cover consists of two 'figure-of-eight' pieces that are double stitched using distinctive red thread. To this day, Real Tennis balls have been made by hand and have changed very little – a core or pill is constructed from cut-up cork pieces and then wrapped with cotton webbing before being finished off with a hand-stitched felt cover. As individual club professionals who make the balls will testify, this lack of regulation results in a wide variety in performance (Knowles *et al.*, 2000).

Of the four companies licensed to produce balls for first-class cricket, only one uses a traditionally produced core. Two use a rolled core and the other a composite centre. When the British Standard first came into force, it was felt that non-traditional cores were the only centres capable of meeting the requirements, especially those governing shape after repeated impacts, but the manufacture of quilted cores has become much more consistent. All quilted cores are machine wound to a high level of precision and repeatability. Indeed, when examination was made of traditional and non-traditional cores, the quilted cores were one of the most consistent types in terms of mass and all types were comparable in terms of diameter and sphericity (Jarratt and Cooke, 2000). Cost has become an important issue: composite and rolled cores are cheaper to fabricate.

Composite moulded cores are made by mixing cork and rubber particles together, which are then placed into a spherical mould. Pressure and heat are then used to bond the particles of the composite together. Rolled cores are fabricated by layering alternate sheets of cork/rubber mix and rubber. The sheets are cut to size, rolled up (often likened to a 'Swiss Roll') and then heated in a hydraulic press. Schematic diagrams of all three types of core are shown in Figs 13.2–13.4.

13.2.4 Analysis of cores/balls

During the summer of 2000, the England and Wales Cricket Board (ECB), which was formed from a merger of the TCCB and the National Cricket Association, commissioned a limited research project to examine the properties of the cores of cricket balls licensed for use in UK first-class cricket (Jarratt and Cooke, 2001). At the time, only the three manufacturers using non-traditional cores were producing such balls.

Five different batches of cores (at least one from each licensed manufacturer) were tested: two were composite (one with coarse cork particles and the other with fine grains), two were rolled and one was produced in the traditional manner. Four complete, brand new balls were also

13.2 Cross-section of a traditional core (direction of cut is not important). The darker areas represent cork whilst the mottled areas show the layers of wound flax. (Courtesy of Rowan Huppert.)

13.3 Cross-section of a composite core (direction of cut is not important). The mottled region represents the cork/rubber particles bonded together. (Courtesy of Rowan Huppert.)

examined: one was made with a composite core, two with rolled cores and one with a traditional core. For comparative purposes a baseball was also used in the experiments.

All the samples were weighed and had their volume calculated. Callipers

13.4(a) Cross-section of rolled core cut in the plane of the seam (*x–z* plane). The darker regions show the rubber sheets and the lighter stripes represent the cork/rubber sheets. (Courtesy of Rowan Huppert.)

13.4(b) Cross-section of a rolled core cut perpendicularly to the plane of the seam (*y–x* plane). The darker regions show the rubber sheets and the lighter stripes represent the cork/rubber sheets. (Courtesy of Rowan Huppert.)

were used to measure the diameter along the principal axes and at several points in between. After these initial measurements, two tests were carried out to investigate the hardness of the samples along the three principal axes: a drop test and a compression test. Samples of each type of core were tested on all of the three principal axes. The whole cricket balls were also tested in this way along with the baseball.

The diameter along the three main axes (x-, y- and z-), along with several points in between, were measured using callipers for each core type. Within experimental error, all types of core were spherical and all samples within each batch were consistent (Jarratt and Cooke, 2000). This indicated that all the manufacturers supplying cores for testing, be they traditional or composite, were producing consistent centres in terms of diameter. The manufacturers use carefully machined measuring rings or hoops (similar to those issued to umpires for use on complete balls) to check the diameter of their product. We tried to use an adapted Talisurf™ machine (measures surface roughness and roundness – made by Taylor Hobson), but owing to the rough nature of the various cores' surfaces, the results were inconclusive.

13.2.5 Findings

The diameter measurements showed that all the cores were close to spherical with very little recorded variation between values. This was also true of the whole balls. In terms of mass and volume, there was a significant difference between types of core. In the case of the composite cores, there was a significant spread of masses within the two core types (over 7%). This raised questions about the quality control of the particular process used to fabricate them. Table 13.1 shows the data gathered for average mass, volume and density. The rolled cores are, on average, 16% heavier than the traditional cores and the traditionally produced cores were the least dense. It must be noted that all the complete balls met the mass and diameter requirements as stipulated by the British Standard.

During the drop test the peak loads and decelerations were recorded and averaged. Both sets of data showed similar trends. Due to space limi-

Table 13.1 Mass, volume and density measurements for the different core types

Core type	Mean mass (g)	Mean volume (cm³)	Mean density (gcm⁻³)
Composite 1	93.3	127	0.73
Composite 2	96.9	131	0.74
Rolled 1	105.3	146	0.72
Rolled 2	101.9	130	0.78
Traditional	90.4	131	0.69

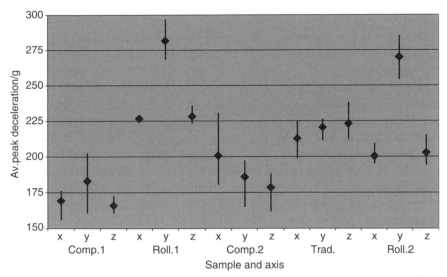

13.5 Mean peak deceleration by sample and axis. The data point is
the mean recorded value and the error bars show the spread of
recorded values.

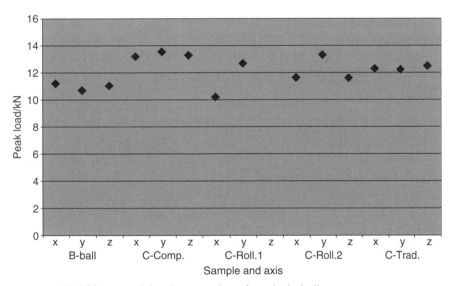

13.6 Mean peak load comparison for whole balls.

tations only data from the drop test can be shown. Figure 13.5 shows the
values gathered for peak deceleration for each sample and axis. The
results of the tests on whole balls (including the baseball) are shown in
Fig. 13.6.

13.2.6 Discussion

The results of the drop test indicate that the traditional core and the composite cores did not show any marked variation in peak load between the three principal axes – examining the deceleration data shows that the spread of results for each of the three main axes lie within those of the other two axes. In the case of the rolled cores, the values show that there is no significant variation between the x- and z-axes. However, the y-axis data is significantly different from that of the x- and z-axes: 16% and 17% for the two rolled core types.

When Fig. 13.6 is examined, it must immediately be noted that due to the limited nature of the study only one ball of each type was available for testing. Hence there are no error bars. The general trend appears to be that the values for peak load recorded on the three principal axes for the baseball, the composite-cored cricket ball and the traditional-cored cricket ball are similar. The values obtained for the two rolled-core balls show a marked increase in peak load for the y-axis.

The results for the compression test support these observations. Values recorded for the traditional and two composite cores show that there are no striking differences between the three axes. The compressive load required for the y-axis for the two rolled cores is significantly greater than that required for the x- and z-axes: well over 50% in both cases. This axial variance was also apparent in the whole balls constructed with rolled centres, whereas the composite-centred ball showed no dramatic variation between the three axes.

13.2.7 Conclusions

Rolled cores are constructed in a way that gives obvious axial variation when the cores are sectioned. Cricket balls produced with rolled cores exhibit large discrepancies in hardness between the plane of the seam and the axis perpendicular to the seam. This variation will have an effect on many of the main dynamic interactions of the game depending upon which part of the ball is involved, e.g. ball with ground and ball with bat. A further important area of note must be when the ball strikes the body (with associated health risks), which occurs both intentionally and unintentionally throughout the game of cricket.

Cricket is a game that has many areas of natural variation ranging from the composition of the pitch to the overhead weather conditions. It is an issue for governing bodies to address in conjunction with the manufacturers as to whether there should be noticeable variation between different types of first-class cricket balls, all of which have passed the British Standard. One possible course of action is to undertake an extensive review of the requirements of the British Standard, a result of which could be the introduction of regulations governing core construction. To examine this issue further, tests on more complete balls are required. In terms of other possible tests, a series of experiments to investigate the coefficient of restitution would show how the different ball types perform at speeds similar to those seen in the actual game (bowling speeds up to 100 mph) – examining the ball/pitch and ball/bat interactions. It is interesting to note that rolled cores were phased out subsequent to this research.

13.3   Cricket bats

This section examines the design and manufacture of cricket bats and how these aspects can influence their performance. Although there are no accepted scientific measures of cricket bat performance, the following bat properties are considered relevant: mass and moment of inertia, coefficient of restitution (COR), centre of percussion (COP), node of the fundamental mode of oscillation and the maximum power point. These parameters have a significant effect on how a bat 'feels' and performs during impact with the ball. The bat exhibits both rigid body and elastic behaviour during impact, which must be considered when evaluating bat performance. How fast and how far a cricket ball will go after impact with the bat will depend on how well the bat has been designed, manufactured and conditioned to achieve the most suitable balance between the parameters described above. These issues, in particular, are considered in this section.

13.3.1 Historic background

Law 6 of the Laws of Cricket concerns the bat. It states that the bat shall not be more than 38 inches/96.5 cm in length and that the blade of the bat shall be made solely of wood and shall not exceed $4\frac{1}{4}$ inches/10.8 cm at the widest part. Law 6 allows for the blade to be covered by another material for protection, strengthening or repair. It specifies that such material shall not exceed $\frac{1}{16}$ inches/1.56 mm in thickness, and shall not be likely to cause unacceptable damage to the ball. Within the Laws of Cricket, there is no restriction on the type of material and geometry used for the handle of the bat. Also, there is no prescribed maximum weight of the cricket bat.

The requirement that the blade must be made of wood came about soon after an incident in 1979 when the Australian player, Dennis Lillee, used an aluminium bat in a test match against England. After only a few deliveries, Mike Brearley complained that the bat was damaging the ball, which resulted in the umpires instructing Lillee to replace the bat (Laver, 2002). This has triggered the MCC to amend Law 6 in order to stop similar attempts in the future. Over the years manufacturers have experimented with different types of wood but English Willow or the so-called Cricket Bat Willow (*Salix alba caerulea*) has remained the most suitable material to date.

Despite the common perception that the design of cricket bats has not changed much over the years, it is nevertheless important to note here those changes that have occurred and that have had a profound effect on the game of cricket. Perhaps the most significant change is in the mass of the bat. Cricket bats used in the past were much lighter than those used by international cricketers today. For example, the weight of bats in the days of Sir Donald Bradman and Sir Jack Hobbs was in the region of 2 lb 2 ozs – 2 lb 4 ozs, while it is not uncommon for international cricketers today to use bats of 3 lb and more (Laver, 2002; Grant, 1998). Bats today have longer blades with larger profiles and thicker edges, and with the mass concentrated lower down the blade. Many cricketers today playing high level cricket use heavier bats as this enables them to hit the ball further with more power. This style of play has evolved in particular with the introduction of the 1-day matches. But more power comes at a cost, as players today do not use the full range of shots as often as they did in the past with lighter and faster bats. In addition, heavier bats cause players to tire more quickly, especially during test matches.

The design of the handle has also changed over the years. While early bats were made out of one piece of wood, modern bats are made out of two parts (the blade and the handle), which are fitted together during manufacturing. Solid wooden handles have been replaced by lighter cane handles interlaced with strips of rubber (typically 2–4 strips of rubber), which aim to reduce the transmission of shock and vibrations from the blade of the bat to the player's hands. High quality handles are typically made from Asian cane (Manau or Sarawak cane). As there is no restriction on the material of the handle, recent research has recommended that other more advanced materials such as composites be explored in the future. For example, John and Li (2002) have applied carbon fibre-reinforced rubber strips in the handle in order to stiffen the bat and expand the 'sweet spot' area on the blade of the bat, thus increasing the amount of energy submitted to the ball during impact.

13.3.2 Literature review

A review was carried out to identify published work on cricket bat design and manufacture. The majority of research into cricket bats has traditionally been directed at the problem of maximising the speed of the cricket ball after impact with the bat. In relation to this problem most of the published work is aimed at describing the mechanics of impact between the bat and the ball, and the effects of parameters such as the coefficient of restitution, centre of percussion, mass and mass moment of inertia on the post-impact speed of the ball. Considerable work has been done on the physical interpretation of the 'sweet spot' and its location on the cricket bat using the research on baseball bats as a basis. It is possible to establish such correlations as the mechanics of swinging the bat is similar for both games. The length and weight of the cricket bat and baseball bat are also similar. In recent years research has concentrated on vibration analysis of cricket bats and more specifically on the ways inherent structural dynamic characteristics (such as natural frequencies and mode shapes) can be tailored to improve cricket bat performance.

The greatest post-impact or rebound velocity of the ball will be achieved when the ball hits the bat at the 'sweet spot' or the so-called 'middle'. This is the impact location on the bat where the transfer of energy from the bat to the ball is maximal while the transfer of energy to the hands is minimal, which provides a comfortable or 'sweet' sensation in the hands of the hitter. Based on earlier rigid body models of the bat, it was believed that the 'sweet spot' coincides with the centre of percussion of the bat (Wood and Dawson, 1977; Noble, 1983; Noble and Eck, 1985), where the centre of percussion represents an impact point, which generates a minimum reaction force at the grip. Some manufacturers still identify the centre of percussion as the 'middle' of the bat, which may be true in some cases especially when a stationary bat is tested for maximum post-impact or rebound velocity of the ball. Based primarily on extensive research on tennis racquets and baseball bats, today it is widely accepted that there are other impact locations on the bat that are capable of producing the greatest post-impact ball velocity, including the node of the fundamental mode of vibration and the so-called maximum power point (Brody, 1986, 1996; Grant, 1998; Noble, 1998; Watts and Bahill, 2000). The interaction between various locations and the rating of their respective effectiveness are still not fully understood.

More recent research has focused on the elastic properties of cricket bats and their dynamic behaviour during impact. This has led to identifying the particular effects flexural vibrations have on bat performance. Based on this type of approach, Grant and Nixon (1996) have developed a parametric model of the dynamic performance of a cricket bat, which was used primarily to compare different design solutions with respect to the relocation of the third flexural mode of vibrations outside the excitation range. Although this work reinforces the importance of 'tuning' the structural dynamic properties of cricket bats it does not offer a solution to this problem. Subic (1998) and Subic and Vethecan (2002) introduced some more elaborate techniques for structural dynamic analysis and local structural dynamic modification precisely for this purpose. In particular, they have developed a novel technique for evaluating the most effective locations for vibration attenuation in structures such as tennis racquet frames, cricket bats and baseball bats based on the dynamic coupling principles. Penrose and Hose (1998) used the FEM to investigate different design strategies for shaping the bat to control its flexural and vibrational properties. This work investigated in particular the effect of the first mode of oscillation on the impact by establishing the relationship between the location of the node along the bat and the post-impact velocity of the ball. The examples of using FEA and modal analysis for design sensitivity studies described above offer new opportunities for more systematic design optimisation of cricket bats in the future.

13.3.3 Performance of cricket bats

Although there are no accepted scientific measures of bat performance available, research to date has identified a range of parameters that have the greatest effect on maximising the post-impact velocity of the ball and which therefore may be used for this purpose. These parameters include mass and moment of inertia, coefficient of restitution, location of the node of the fundamental mode of oscillation and the location of the centre of percussion.

Coefficient of restitution is typically considered as an appropriate measure of bat performance as it has a significant effect on the post-impact velocity of the ball. It represents the ratio of the relative velocity between the ball and bat immediately after impact and immediately before impact. Coefficients of restitution of the ball and the bat are found separately through simplified impact tests. Manufacturers use such tests, for example, to compare the performance of different bats. Typically, they drop a steel ball vertically from a set height onto the clamped blade of the bat and measure the rebound height of the ball at different points. For this test condition, the coefficient of restitution is expressed as the square root of the ratio between the rebound and drop height. Although the value of the coefficient of restitution varies from bat to bat, according to the authors' research a value of around 0.6 is expected for cricket bat and ball collision.

The duration of impact between the cricket bat and ball is typically around 0.6–0.7 ms, which is consistent with the estimate of around 1 ms provided by Daish (1972). This corresponds to an excitation in the frequency range of 0–1000 Hz. The damping ratio in this range is considered insignificant, which implies that the modes of oscillation excited in this range may have a significant effect on the performance of the bat. Figure 13.7 shows an example of modes of oscillation identified in this frequency range using experimental modal analysis on a competitive cricket bat ($m = 1196.8$ g, $L = 835$ mm, $W = 108$ mm). During impact, the dynamic behaviour of the bat corresponds to that of a free, non-supported bat, irrespective of the firmness of the grip, which is the main reason the bat is tested in a free–free condition.

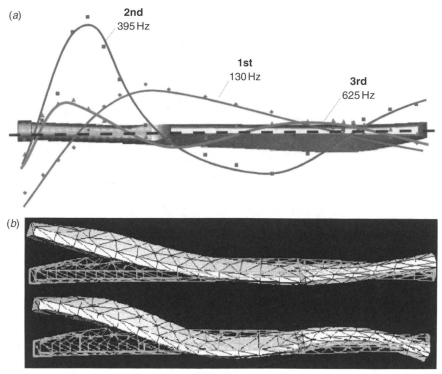

13.7 Modes of oscillation of a cricket bat: (*a*) experimental modal analysis results; (*b*) finite element analysis results.

In acknowledging the importance of the flexural modes of oscillation on the performance of the cricket bat, Grant and Nixon (1996) and Knowles *et al.* (1996) suggested that the 'middle' of the bat may in fact be located at the node of the first flexural mode. If the ball strikes the bat at the node, the mode would not be excited and therefore would not absorb any impact energy, which means that more could be returned to the ball in the form of kinetic energy generating a higher post-impact velocity of the ball. There is much evidence that the first mode does in fact dominate the impact (Noble and Walker, 1994; Hatze, 1994). In addition, Noble and Dzewaltowski (1994) found that the node–COP distance represents one of the most powerful predictors of player's perception of bat performance.

When selecting cricket bats, players tend to use the weight and feel (balance) of the bat as the main criteria. But what the ideal or optimal weight of a bat is for each player is not fully understood (although there is much anecdotal and empirical evidence that the ball will go the fastest when

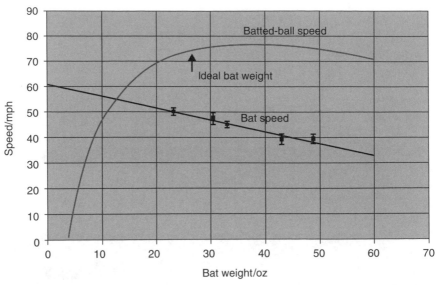

13.8 Ideal bat weight for baseball (with permission from Terry Bahill).

hit by a bat of weight that is well suited to a particular player). In the absence of similar studies for cricket, a graph of ideal bat weight for baseball and softball players developed by Bahill (Watts and Bahill, 2000), shown in Fig. 13.8, can be used as a good indication of the type of relationships that need to be established for each player when determining their ideal bat weight. The effect of the mass moment of inertia of the bat will also have a profound effect on the performance of the cricket bat and on how a particular bat feels during the swing. The greater the mass and mass moment of inertia the greater will be the momentum and power of the bat. But this will also require a greater impulse to produce the change in the velocity or direction of the bat, which will in turn compromise the velocity of the swing. In baseball there is a trend today towards lighter bats (Adair, 1990; Noble, 1998; Watts and Bahill, 2000) while in cricket, as discussed earlier, the trend is towards heavier bats. Clearly, the relationships describing the ideal mass and mass moment of inertia of cricket bats for each player have yet to be determined.

13.3.4 Materials and construction of cricket bats

High performance cricket bats are invariably made from English Willow or the so-called Cricket Bat Willow (*Salix alba caerulea*). English Willow is very light and durable with good shock resistance properties, which makes

Table 13.2 Physical properties of English Willow and
Cane (after Lavers, 1983)

Material	Young's modulus, E (MPa)	Density, ρ (kg/m^3)
Willow	6600	417

this wood particularly suitable for cricket bat blades. This wood is best when its moisture content is 12–14% (Laver, 2002). Although it has very high compressive strength, it needs to retain some moisture to enable the fibres to stretch, rather than crack, under the strain from the impact of the ball. The density of the blade varies spatially depending on the moisture content and the compression achieved during the rolling or knocking-in process. Laver and Wood specify a density of 417 kg/m^3 when at 12% moisture content (Laver, 2002). This is consistent with the physical properties of English Willow provided in Table 13.2. The density will be different for the surface of the blade and for the centre (core) due to the rolling/pressing and knocking-in process.

The value of the Poisson's ratio for Willow is not readily available in literature. References typically quote general Poisson's ratios for orthotropic materials such as hardwoods and softwoods. There are six potential values for any given type of wood, depending on which way the grain runs. However, we are normally concerned only with wood in which the grain runs longitudinally (e.g. along the long axis of the piece) as is the case with the cricket blade, which is always cut with the long axis parallel to the axis of the tree. This enables the bat to utilise the maximum strength and stiffness available in the wood. The significant Poisson's ratios are then v_{LR} and v_{LT}, arising from tension or compression in the longitudinal direction with a passive strain in the radial or tangential directions. Therefore, Poisson's ratio can be expressed as follows,

$$\text{Poisson's ratio } (v) = \frac{\text{passive strain (normal to force)}}{\text{active strain (in direction of force)}}$$

$$v_{LR} = \frac{\gamma_R}{\gamma_L} \qquad [13.1]$$

Based on the average values of Poisson's ratios for hardwoods and softwoods obtained from Bodig and Jayne (1982), Possion's ratio for Willow is determined as 0.35.

Manufacturing of cricket bats involves the following main stages: harvesting of Willow, splitting and cutting the clefts with each cleft forming one blade, seasoning of clefts to reduce the moisture content in the wood,

13.9 Shaping of the cricket bat blade (www.gray-nicolls.co.uk).

13.10 Pressing/rolling and knocking-in of the cricket bat (www.gray-nicolls.co.uk).

pressing or rolling the seasoned blades, handle production and fitting, final shaping, balancing and finishing (see Figs 13.9 and 13.10). Grading of blade quality is a qualitative and subjective process involving visual inspection of the face of the blade for knots and blemishes, colour of the wood, straightness and number of growth lines on the face of the blade. In particular, quality is dependent on the number of latewood rings present in the blade. According to Sayers *et al.* (2000), most manufacturers recommend that ideal bats have between five and eight latewood rings.

Pressing or rolling will compress the bat blade and reduce the size of the Willow by around 4–5 mm, which will alter its mechanical properties slightly, producing an in-homogenous structure. This process will yield a hardened surface layer of 1–2 mm on the bat (1.4 mm, according to Grant and Nixon, 1996), which will enable the bat to resist damage during the rest of the manufacturing process or when striking a cricket ball. According to Grant and Nixon (1996), the maximum theoretical compression ratio for Willow is 3.6 with an average ratio of around 3. Pressing or rolling does not preclude the need for knocking-in of the bat after manufacturing – it just makes the process easier. Knocking-in involves final conditioning of the blades surface prior to use. The main purpose of the knocking-in is to com-

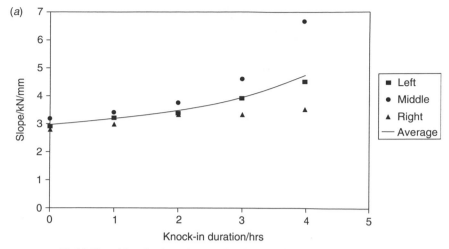

13.11 Knocking-in duration and related effects (Sayers *et al.*, 2000).
(*a*) Slope of load penetration curve vs. knock-in duration. (*b*)
Scanning electron microscope studies of the bat face (*cont'd over*).

press the wood fibres, which will protect the bat from cracking during impact and increase its useful life. According to Sayers *et al.* (2000), the duration of knocking-in is around 4–6 hours, during which the hardness of the bat face can be doubled. Figure 13.11 shows the knocking-in duration in relation to the load or slope penetration curve. During knocking-in the core of the bat is left unhardened. This will result in a more elastic response of the bat during impact and in higher rebound velocities of the ball. The coefficient of restitution of the bat will increase significantly as a result of the surface hardening and stiffening of the bat achieved during the knocking-in process.

13.3.5 Conclusions

Cricket bats have experienced very little change over the years in terms of the geometry and materials used. This is mainly attributed to stringent design constraints imposed by the Laws of Cricket. The little change that has occurred over the years was mainly initiated by individual players and implemented by the local craftsmen to suit player's personal preferences in terms of weight and feel. There is very little scientific research done to substantiate and validate such design decisions. Recent research on tennis rackets and baseball bats in particular points to a number of directions of relevance to cricket bat design. For example, we now understand that both rigid body and elastic behaviour affect the performance of the bat and that

(b)

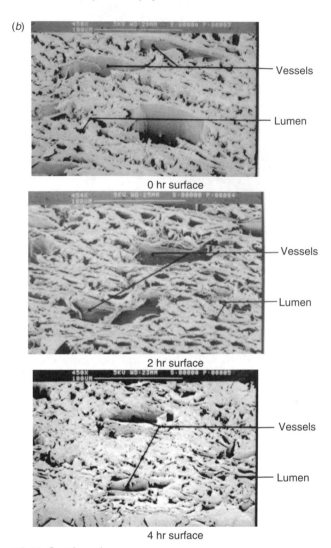

Vessels

Lumen

0 hr surface

Vessels

Lumen

2 hr surface

Vessels

Lumen

4 hr surface

13.11 Continued

how fast the ball will go when impacted depends on the parameters such as the coefficient of restitution, the relative locations of the 'sweet spots' on the bat, the inertial and vibrational characteristics of the bat. It is important to define these measures more precisely for the cricket bat in the coming years. Underpinning relationships and measures of ideal weight and mass moment of inertia of the bat for individual players must be determined if we are to understand and quantify more accurately players' ability to swing the bat faster or to hit the ball harder. Although English Willow has been the preferred material for cricket bat blades, there is much scope for design optimisation of the handle using advanced engineering materials in conjunction with substantially different geometry.

13.4   Protective equipment in cricket

The revised MCC Codes of Laws of Cricket define protective equipment (e.g. Appendix D) and its legitimate use during the game (e.g. Law 40, The Wicketkeeper). Specifications for protective equipment, e.g. equivalent to the British Standard for cricket balls, are in place (BSi, 1998; BSi, 2000). Although early forms of protective equipment (such as helmets, gloves and leg pads) have been used in cricket since the 1800s, it is only in the past 20 years that serious injuries in cricket have declined dramatically with the advancement of materials available for manufacture of this type of equipment.

In cricket, batsmen are most prone to injuries primarily from the direct impact of the ball, which at test level weighs 156 g and can reach velocities of 160 km/h. The most common injuries occur to the head/face, hands/fingers, followed by arms, legs and chest. An illustration of the main injury areas of the batsmen with the respective average rates of injury occurrence is given in Fig. 13.12 (based on a comprehensive sports injury survey published in the *Herald Sun* on 10 May 1994).

As the risk of severe injuries to the batsmen is high, the protection of the batsmen in cricket is therefore of primary concern. Protective equipment introduced over the years for this purpose includes:

- helmet
- batting gloves
- chest guard
- arm guards
- thigh guards
- protector
- leg guards.

The rules of the game have also evolved over the years in an attempt to protect the batsmen from injury. As fast deliveries are one of the most common threats to the batsmen some limitations have been introduced to reduce the associated risks. For example, a fast delivery in the form of a full toss above waist height is called a 'no ball'. Also, a limit on the number of 'bouncers' that can be delivered per over has been introduced (a maximum of two allowed), whereby the term 'bouncer' is used for a ball that bounced above shoulder height endangering the head of the batsmen. Other amendments to the rules can be expected in the future if the game becomes more dynamic and powerful.

The main role of the protective equipment is to help prevent and reduce the severity of injuries. Protective equipment for cricket has advanced

26%

15%

21%

7%

12%

14%

Other: 5%

13.12 Injuries in cricket.

significantly over the years, perhaps more then any cricket technology. The technological advancement of this equipment is mainly attributed to new and advanced materials and manufacturing processes that have been made available over the past decade, such as fibreglass and other impact resistant polymers, high and low density foams. Protective equipment can provide protection in a range of ways, such as by increasing the impact area and distributing the load over a greater area, transferring or dispersing the impact area to another part of the body, adding mass to the body part thus limiting deformation and impact, reducing friction between contacting surfaces and by absorbing the impact energy (Hyde and Gengenbach, 1997; Anderson and Hall, 1997). The ability to absorb impact energy depends primarily on the density of the material and the ability of the material to regain its initial shape. The discussion that follows in this section will focus on the properties and design of the two most representative and complex pieces of protective equipment used in cricket today – the helmet and the gloves.

13.4.1 Cricket helmet

The main objective of equipment used for head protection is to reduce the acceleration of the head during impact, which can act to squash the brain against the skull and to spread the applied impact load over a greater surface area in order to reduce the pressure and likelihood of skull fracture. Modern cricket helmets are designed to achieve precisely this by combining a stiff shell, which provides an impact face that spreads the impact force across a larger surface area, and a softer internal padding that cushions the blow and absorbs the energy of impact. Most helmets on the market use a fibreglass or ABS shell and low density polyethylene as the padding. In addition, heat-formed liners are included with the padding that have the ability to reshape to personal fit under body heat generated during play. Cricket helmets are also fitted with a faceguard or 'cage' typically made of powder coated stainless steel instead of transparent plastics that in the past had a tendency to fog-up during play. New materials, such as polycarbonate can replace the cage structure as they do not fog-up and can withstand high impact forces. Polycarbonate can be moulded in any shape as required. Cricket helmets used for example in Australia comply with the Australian and New Zealand Standard AS/NZS 4499.1 for cricket helmets.

The selection and combination of the materials used for the shell and the padding is critical for the performance of the helmet. Knowles *et al.* (1998) have explored various materials and combinations of materials with respect to the effectiveness of the helmet in attenuating the shock of a cricket ball impact and reducing the strain transmitted from the helmet to the head. They have tested a range of padding types in conjunction with a variety of plates made of reinforced, un-reinforced and impact modified plastic materials using separate tests and tests involving helmets fitted on an instrumented head-form. The properties of tested closed cell polyethylene (PE) foams of the low density (LD) recoverable type used for padding in cricket helmets are given in Table 13.3. The comprehensive experimental investigation reported by Knowles *et al.* (1998) found that the stiffness of the shell

Table 13.3 Properties of closed cell polyethylene (PE) foams (after Knowles *et al.*, 1998)

Type	Density (kg/m³)	Force at compression of			
		25%	40%	50%	60%
LD33	33	40	75	115	175
LD45	45	50	90	135	210
LD60	60	70	115	170	255
HD60	60				

and the thickness of the padding have the most significant effect on the performance of the helmet whereby the effect of shell flexural rigidity above $0.05 \, N/m^2$ (for prototype plate made of 2.5 mm ABS) is very small. The thickness of the foam pads for design and aesthetic reasons was limited to 15 mm.

Recent research on protective helmets has focused on the development of appropriate computational models typically based on the FEM (Mitrovic and Subic, 2000; Ko *et al.*, 2000). The main purpose of such models is to simulate the impact and determine the energy absorption capability of helmets. This approach enables more efficient development of new helmet designs instead of producing and testing a large number of prototypes, which can be very costly and time consuming. Mitrovic and Subic (2000), in particular, have developed FE models of protective helmets that allow for the material, geometric and contact non-linearity during frontal impact to the 'face', which is of particular importance to cricket helmet design. They have used their model to simulate and determine the impact force and displacement relationships for different helmet designs, which enabled them to select designs that have the best energy absorption capability under impact (see Fig. 13.13 for characteristic results involving two different helmet designs). This has enabled them to experiment with new materials for the helmet shell and to develop more appropriate composite designs that have better stiffness to weight ratio than more traditional ABS shells. Ko *et al.* (2000) have used the FE method to simulate the impact between helmet and anvil during a drop test. They have included in their model the helmet (shell and padding) and the head-form. Although this model does not include the non-linearity of materials and the viscoelasticity of the padding they have found that the Young's modulus of the padding represents one of the most important properties affecting the head-form acceleration. The approaches described above offer new opportunities for design optimisation of cricket helmets, which is currently under way.

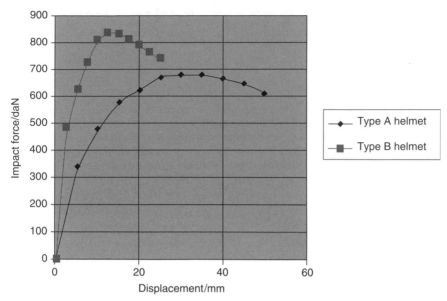

13.13 Simulation of impact force vs. displacement for two composite helmet designs.

13.4.2 Cricket gloves

Batsmen have suffered for a long time from injuries to their hands as a result of being hit by cricket balls travelling at high speeds. There are still no batting gloves available on the market that totally prevent the batsmen's hands from being broken when struck by the ball. Alexander *et al.* (1998) reported two common designs of gloves available on the market: a square finger glove and a 'sausage' (or pre-bent) finger glove. The glove consists of three main components: a finger insert, the protective foam and the leather outer. The simplest material construction for a protective layer of the cricket glove is homogeneous foam. Knowing the energy absorption of the foam, kinetic energy of the ball and the approximate impact area on the finger, the thickness of foam required for maximum impacts at 90 mph can be calculated using the following equation:

$$\text{Densification strain} \times \text{plateau stress} = \frac{\text{kinetic energy of cricket ball}}{\text{area of impact} \times \text{thickness of foam}} \quad [13.2]$$

The foams used in the manufacture of these designs of cricket gloves are high density polyethylene foams, which recover their shape after impact and

Table 13.4 Specification of insert tested for cricket gloves (Alexander *et al.*, 1998)

Insert NO.	Type	Material
1	Hinged insert	Polypropylene material (Vivac)
2	Hinged insert	2 : 1 Kevlar/glass fibre mix: polypropylene hinge
3	Hinged insert	2 : 1 Kevlar/glass fibre mix: Kevlar fibres, resin hinge
4	Hinged insert	Duro aluminium alloy: 1.5 mm section
5	Hinged insert	Duro aluminium alloy: 2.5 mm section
6	Pre-bent insert	Woven glass fibre and polyester resin
7	Pre-bent insert	Kevlar/glass fibre mix, glass fibre and polyester resin
8	Pre-bent insert	Kevlar/glass fibre mix, glass fibre and epoxy resin (epoxy resin is far stronger than polyester resin)
9	Pre-bent insert	Kevlar/glass fibre mix, epoxy resin

are flexible. In case of existing foams the thickness required to safely withstand the impact would be 2.5–3 cm, which exceeds the acceptable level. Therefore, many foams used are inadequate for resisting cricket ball impact at the highest bowling speeds of around 160 km/h at current thickness. The foams with higher energy absorption characteristics, although able to resist impact, are unsuitable due to their rigidity and permanent deformation. To overcome this problem Alexander *et al.* (1998) explored different insert types for more appropriate cricket glove design that would satisfy the requirements described above. Table 13.4 shows a summary of the inserts tested. All inserts were tested using a ball cannon at Dunlop Slazenger. Tape was wrapped around the inserts to simulate a leather pouch and Plasticene fingers were inserted before fixing the insert to a metal bar, simulating a bat handle. A 5.5 oz cricket ball was then fired at the inserts at speeds ranging from 40 mph to 90 mph (approximately 18–40 m/s). The speed at which the insert broke was noted. The diameter of the fingers was measured before and after the impact and recorded as percentage deformation. The ball speed was recorded using two light beam speed sensors.

Results showed that the hinged insert designs did not provide adequate protection. Pre-bent inserts made from Kevlar and glass fibre mix woven cloth proved to be far stronger and provided adequate protection during the tests. It was also found that epoxy resin with Kevlar/glass mix was the strongest option. A final insert was designed and manufactured, based on the test results. A final glove design was proposed which incorporated the insert, extra padding in certain positions and a Kevlar sheet on the back of the hand. The improved cricket glove design proposed by Alexander *et al.* (1998) included the following features:

- A pre-bent double finger Kevlar insert (with either epoxy or polyester resin) to protect the first two fingers of the bottom hand.
- Sausage-style fingers to maximise the protection and eliminate the problem of hinged fingers opening up under impact and exposing the joints.
- A Kevlar sheet required over the back of the hand as the current use of foam is inadequate.
- Foam padding around the web between the thumb and first finger of the bottom hand and along the outside of the top hand. Further work required to choose optimum foam for flexibility, good energy absorption and no permanent deformation.

These recommendations have been featured in subsequent designs of cricket gloves that are now commercially available.

13.5 Conclusions

This chapter has provided a comprehensive overview of the key design and manufacturing issues relating to the cricket balls, cricket bats and protective equipment in cricket. Each section included a comprehensive literature review of work to date, a discussion of the governing rules and the historical background of the particular piece of equipment describing the evolution of both the game and the equipment. The main objective of this section was to provide a comprehensive overview of current knowledge of equipment design, and the materials and processes used in the manufacture of cricket equipment. The chapter ends by discussing some future trends and opportunities for improving this technology.

The section on cricket balls did not include a discussion on the aerodynamics of the ball and how it affects flight (for example, spin bowling versus seam bowling) as this was covered in greater detail in Chapter 5. Instead, the focus here was on the materials and manufacturing processes used. In particular, the section on the materials and manufacturing of cricket balls examined the properties of the ball core and related those properties to ball performance.

The section on cricket bats examined the parameters that affect bat performance the most and identified strategies for identifying particular measures of performance. It provided a discussion on both rigid body and elastic properties of cricket bats. Various models and approaches to assessing cricket bat performance have been identified, primarily based on tennis racket and baseball bat research involving both experimental and FEM techniques. A discussion on the materials and manufacturing processes used for production of cricket bats indicated that very little change has occurred over the years in this regard apart from the re-distribution of mass along the blade and the inclusion of rubber strips in the handle. There is considerable scope for work in this area (e.g. new materials and designs of the handle, shape and weight optimisation of the bat, etc.). Properties of the English Willow and of the knocking-in process have been described in detail.

The section on the protective equipment in cricket focused on the equipment used for the protection of the batsmen, and especially on the design of cricket helmets and gloves. A comprehensive overview of the materials used in the manufacture of the helmet shell and padding has been provided and the main parameters (shell stiffness and padding thickness and modulus of elasticity) that affect the performance of the helmet identified. It has been found that many cricket gloves do not provide adequate protection at the extreme bowling velocities in cricket and new designs have been suggested to overcome this problem.

13.6   Future trends

Since the inception of the game of cricket technological improvements have been limited compared to other bat and ball sports. There are many reasons for this, including conservative governing rules and the lack of scientific research in the design and manufacture of cricket bats and balls. For example, most design improvements have been initiated by players and carried out by local craftsmen rather than as a result of scientific research. Clearly, there are gaps in knowledge that must be overcome in the future if this technology is to advance. The game of cricket also faces a challenge of maintaining and improving its popularity and participation in the sport. This may initiate new amendments to the Laws of Cricket in the future allowing for a slightly more liberal approach to the technology used.

Since the research on cricket balls reported in this chapter was concluded, there has been an appreciation of the fact that rolled cores have different material properties depending upon which axis is examined. At the time of writing, the ball with the traditional core is being used exclusively for limited overs (1-day) cricket to prepare players for the forthcoming World Cup in South Africa where that brand of ball will be used. It is also recognised that an international sport such as cricket must be supplied with high-quality equipment. One of the companies producing the rolled core balls has been taken over by a specialised rubber company and has moved to using high-quality composite cores. Extensive testing has been conducted to produce the new core and the exact details are shrouded in secrecy. Generally in the industry there is ongoing consolidation, and one of the larger ball manufacturers has recently bought one of the smaller ones. As the manufacturers find new ways to increase the coefficient of restitution of cricket balls, it is expected that the Laws of Cricket will specify their upper limits as in case of baseball.

It is evident from previous discussions that cricket bat manufacturers will need to integrate and apply engineering sciences in the design and manufacture of cricket bats to a greater extent if they are to achieve more significant improvements in the performance of the existing technology. The discussions given in the section on cricket bats open a number of strategic directions, including:

1. Maximising the post-impact velocity of the ball, by:
   - Increasing the coefficient of restitution of the bat and/or ball by using advanced wood treatment techniques, modifying the design and materials within the given rules, etc.
   - Bringing the node of the fundamental mode of oscillation of the bat closer to its centre of percussion and grouping them lower down the blade through structural dynamic modification and by combining Willow with other treated timber
   - Reducing the anti-nodes of the handle and relocating the third mode of oscillation outside the excitation spectrum by changing the design and material of the handle
   - Establishing the ideal mass and mass moment of inertia of the bat for individual players, which will improve the feel and velocity of the swing
   - Increasing the 'sweet spot' area by combining the approaches described above.
2. Minimising the shock and vibrations generated and transmitted from the blade to the handle, by:
   - Modifying the design of the handle using advanced materials (such as composites and fibre reinforced rubber strips) and vibration attenuation devices
   - Increasing the 'sweet spot' area using the approaches described above
   - Establishing the ideal mass and mass moment of inertia of the bat for individual players
   - Re-distributing the mass of the bat by reshaping the bat using structural dynamic modification methods and by combining Willow with other treated timber.

Clearly, the strategies described above will need to be implemented within the governing rules. As the performance in cricket improves more drastically with new designs of bats and balls it is expected that the rules will be amended, as in case of baseball, to maintain the nature and integrity of the game while trying to increase popularity and participation. For example, in baseball the coefficients of restitution of the ball and bat are limited and standard tests have been developed to evaluate these parameters. A bat performance factor (BPF) is specified as the ratio of bat and ball COR to the ball COR, and most levels of play have adopted the maximum bat performance standard. Similar approaches are expected in cricket in the future despite the current lack of initiative. Further research in the sport of cricket is necessary to support this transition.

The dependence on protective equipment in cricket has increased dramatically with advances in the materials and manufacturing methods used. It is most likely that protective equipment will gain increased importance as batting and balling performance improves with new technologies and training methods. In particular, existing equipment will need to become more thermally comfortable as cricket is basically a summer sport. The continuous improvement of the protective equipment and the perceived improvement in safety should have a positive impact on increasing the participation in this sport. Already, in Britain protective equipment is compulsory in schools, and it is expected that this will happen at other levels and in other countries participating in this sport. In recent years, the British governing bodies of cricket facilitated the introduction of protective equipment in British schools. This has cost well in excess of £250 000 but is surely a necessary investment for any cricket organisation.

13.7 Acknowledgements

The authors would like to thank Tim Jarratt, Cambridge University Engineering Department, for his invaluable contribution to the cricket ball research section; John Carr and Andy Smith, the England and Wales Cricket Board, for their comments on the game; Rowan Huppert for photography assistance; Peter Llewellyn, Timber Development Association of South Australia for his comments on the Willow properties; Terry Bahill, University of Arizona, for the provision of the ideal bat weight graph; cricket equipment manufacturers worldwide.

13.8 References

ACB (1999). *Specification for Manufacture of a Cricket Ball Suitable for First Class Cricket in Australia*. Sydney, Australian Cricket Board.

Adair R K (1990). *The Physics of Baseball*. New York, Harper and Row Publishers.

Alexander S, Underwood D and Cooke A J (1998). Cricket glove design, in *Proceedings of 2nd International Conference on the Engineering of Sport*. Oxford, Blackwell Science, 15–22.

Anderson M K and Hall S J (1997). *The Fundamentals of Sports Injury Management*. Pennsylvania, Williams and Wilkins.

Bodig H and Jayne (1982). *Mechanics of Wood and Wood Composites.* New York, Van Nostrand Reinhold, 117.

Brody H (1986). The sweet spot of a baseball bat. *American Journal of Physics,* **54**(7), 640–642.

Brody H (1996). The modern tennis racket, in *Proceedings of 1st International Conference on the Engineering of Sport.* Oxford, Blackwell Science, 79–82.

BSi (1995). *BS 5993 – Specification for Cricket Balls.* London, British Standards Institution.

BSi (1998). *BS 7928 – Specification for Head Protectors for Cricketers.* London, British Standards Institution.

BSi (2000). *BS 6183 – Specification for Protective Equipment for Cricketers.* London, British Standards Institution.

Carre M J, Haake S J, Baker S W and Newell A J (2000). Predicting the dynamic behaviour of cricket balls after impact with a deformable pitch, in *Proceedings of 3rd International Conference on the Engineering of Sport.* Oxford, Blackwell Science, 177–184.

Cross R (1999). The bounce of a ball. *American Journal of Physics,* **67**(3), 222–227.

Daish C B (1972). *The Physics of Ball Games.* London, English University Press.

Dean G (2000). Balls put to the test in light of faulty scale, *The Times,* 24 August, London.

Giacobbe P A, Scarton H A and Lee Y-S (1995). Dynamic hardness (SDH) of baseballs and softballs, in *Safety in Baseball/Softball.* Atlanta, ASTM, 47–66.

Grant C (1998). The role of materials in the design of an improved cricket bat. *MRS Bulletin,* 50–53.

Grant C and Nixon S A (1996). Parametric modelling of the dynamic performance of a cricket bat, in *Proceedings of 1st International Conference on the Engineering of Sport.* Oxford, Blackwell Science, 245–249.

Green B (1988). *A History of Cricket.* London, Barrie & Jenkins.

Haake S J (ed.) (1996). *The Engineering of Sport.* Rotterdam, Balkema.

Haake S J (ed.) (1998). *The Engineering of Sport – Design and Development.* Oxford, Blackwell Science.

Hatze H (1994). Impact probability distribution, sweet spot, and the concept of an effective power region in tennis rackets. *Journal of Applied Biomechanics,* **10**, 43–50.

Herald Sun (1994). Sports injuries and possible prevention. Melbourne, *Herald Sun,* 10 May.

Hyde T E and Gengenbach M S (1997). *Conservative Management of Sports Injuries.* Pennsylvania, Williams and Wilkins.

Jarratt T A W and Cooke A J (2000). *A Comparison of Cricket Ball Cores.* Report commissioned by ECB, Cambridge University Engineering Department.

Jarratt T A W and Cooke A J (2001). A comparison of cricket ball cores, in *Materials and Science in Sports,* Coronado, CA, TMS.

John S and Li Z B (2002). Multi-directional vibration analysis of cricket bats, in *Proceedings of 4th International Conference on the Engineering of Sport.* Oxford, Blackwell Science, 96–104.

Kaat J (2000). Baseball's new baseball. *Popular Mechanics,* **177**, 62–67.

Knowles K M, Cooke A J, Lennox T and Mastropietro S (2000). Design of real tennis balls, in *Proceedings of 3rd International Conference on the Engineering of Sport.* Oxford, Blackwell Science, 43–50.

Knowles S, Mather J S B and Brooks R (1996). Cricket bat design and analysis through impact vibration modelling, in *Proceedings of 1st International Conference on the Engineering of Sport.* Oxford, Blackwell Science, 245–249.

Knowles S, Fletcher G, Brooks R and Mather J S B (1998). The development of a superior performance cricket helmet, in *Proceedings of 2nd International Conference on the Engineering of Sport.* Oxford, Blackwell Science, 405–413.

Ko C W, Ujihashi S, Inou N, Takakuda K, Ono K, Mitsuishi H, Nash D (2000). Dynamic responses of helmets for sports in falling impact onto playing surfaces, in *Proceedings of 3rd International Conference on the Engineering of Sport.* Oxford, Blackwell Science, 399–406.

Laver J (2002). www.laverwood.co.nz, Laver & Wood Ltd.

Lavers G M (1983). *The Strength Properties of Timber.* London, HMSO.

MCC (2000a). *MCC History*, http://www.lords.org.

MCC (2000b). *The Laws of Cricket*, London, Marylebone Cricket Club.

Mitrovic C and Subic A (2000). Simulation of energy absorption effects during collision between helmet and hard obstacles, in *Proceedings of 3rd International Conference on the Engineering of Sport.* Oxford, Blackwell Science, 389–397.

Noble L (1983). Empirical determination of the centre of percussion axis of softball and baseball bats, in Winter D A, Norman R W, Wells R P, Hayes K C and Patla A E (Eds), *International Series on Biomechanics*, Champaign, Human Kinetics Publishers, 516–520.

Noble L (1998). Inertial and vibrational characteristics of softball and baseball bats: research and design implications, in *Proceedings of 16th Symposium of the International Society of Biomechanics in Sports*, Konstanz, University of Konstanz, topic 5 – keynote lecture.

Noble L and Dzewaltowski D (1994). A field test to determine the attributes of aluminium softball bats that influence perception and preference: a pilot study, Unpublished research report submitted to Kansas State University Research Foundation, Manhattan.

Noble L and Eck J (1985). Effects of selected softball bat loading strategies on impact reaction impulse. *Medicine and Science in Sports and Exercise*, **18**(1), 50–59.

Noble L and Walker H (1994). Baseball bat inertial and vibrational characteristics and discomfort following ball-bat impacts. *Journal of Applied Biomechanics*, **10**, 132–144.

Oslear D and Mosey D (1993). *The Wisden Book of Cricket Laws.* London, Stanley Paul.

Penrose J M T and Hose D R (1998). Finite element impact analysis of a flexible cricket bat for design optimisation, in *Proceedings of 2nd International Conference on the Engineering of Sport.* Oxford, Blackwell Science, 531–539.

Sayers A T, Koumbarakis M. and Sobey S (2000). Surface hardness of cricket bats following knocking-in, in *Proceedings of 3rd International Conference on the Engineering of Sport.* Oxford, Blackwell Science, 87–94.

Subic A (1998). Evaluation and design optimisation of sports technology based on structural dynamic modelling, in *Proceedings of 2nd International Conference on the Engineering of Sport.* Oxford, Blackwell Science, 79–91.

Subic A J and Haake S J (eds) (2000). *The Engineering of Sport – Research, Development and Innovation*. Oxford, Blackwell Science.

Subic A J and Vethecan J K (2002). Evaluating the effectiveness of dynamic vibration absorbers on high performance tennis racquets, in *Proceedings of 4th International Conference on the Engineering of Sport*. Oxford, Blackwell Science, 262–273.

Watts R G and Bahill A T (2000). *Keep Your Eye on the Ball: Curve Balls, Knuckleballs and Fallacies of Baseball*. W H Freeman.

Wood G A and Dawson M (1977). Physical properties of cricket bats and their implications for hitting. *Sports Coach*, **1**(3), 28–34.

Wright R O (1999). *A Tale of Two Leagues: How Baseball Changed as the Rules, Ball, Franchises, Stadiums and Players Changed, 1900–1998*. Jefferson, NC, McFarland & Company Inc.

www.abcofcricket.com
www.cricket.org
www.gray-nicolls.co.uk
www.mmat.ubc.ca
www.nafi.com.au
www.newbery.co.uk
www.rfs.org.uk/totm/cricket.htm
www.sgcricket.com
www-uk.cricket.org

14
Materials in Paralympic sports

J. MACARI PALLIS

Cislunar Aerospace, Inc., USA

14.1 Introduction

The word 'Paralympics' literally means 'alongside the Olympics'. Although commonly thought of as relating to the word 'paraplegia', which refers to paralysis of the lower half of the body, the words are unrelated. Paralympians may be paraplegics, quadriplegics, amputees, the hearing or visually impaired, and other individuals with motor disabilities. In the past, Paralympic athletes only included individuals with physical disabilities, while athletes with intellectual disabilities were included within the Special Olympics. Eligibility requirements for inclusion of events for individuals with intellectual disabilities into the Paralympics have been initiated with the International Paralympic Committee, although events for athletes with intellectual disabilities will not be a part of the Athens 2004 Paralympics.

As its name defines, the Paralympics are conducted in parallel with the Olympic games. Currently, 22 sports fall under the Paralympic games – 18 summer sports and 4 winter sports. The summer events include: archery, athletics, boccia, cycling, equestrian, goalball, judo, powerlifting, sailing, shooting, soccer (football), swimming, table tennis, wheelchair basketball, wheelchair fencing, wheelchair rugby, wheelchair tennis, and volleyball. The winter games are alpine skiing, cross-country (Nordic) skiing, ice sledge hockey and wheelchair dance.

To recognise the impact, contributions and advancements material science has made, and can make, in sports for individuals with physically disabilities, it is important to understand and define some common athlete disabilities, applicable sport regulations, adaptive equipment and prosthetic devices used.

As with any aspect of sports engineering, the key objectives of material science are to contribute to the athlete's safety, prevent or reduce the likelihood of injury and increase comfort and accessibility. For sports equipment, such as an ice sled or wheelchair racer, performance enhancement is an important goal within the allowable limits of the sports' regulations.

Assistive and adaptive devices may utilise high technology, state-of-the-art production and manufacturing techniques with carbon fibre, titanium or aerospace materials or may use the simplest individualised solution the athlete or colleagues develop with local fabric or hardware store supplies. Clearly, there are tremendous opportunities and ways for engineering and material science to contribute to sports for individuals with physical disabilities, not just at the elite Paralympic level but for all categories of competition and recreation. It is important to remember that these same innovations many times contribute solutions for an active ageing able-body population and others (such as women and children), who may simply lack the height or physical strength of the 'average' man.

14.2   Physical disabilities

Physical impairments vary greatly by sport. Athletes qualify within specific classifications for each sport. Classification is determined by physical examinations by a sanctioned sports body and examiner. For example, wheelchair tennis players must have a medically diagnosed mobility-related disability, which results in a significant or complete loss of use of at least one extremity (arms or legs).

Figure 14.1 is a chart of the neurological levels of injury. The letters in the diagram refer to the neurological level: C for cervical, T for thoracic, L for lumbar and S for sacral. In turn, these correspond to specific motor, sensory and autonomic nervous system functions. Based on the level of neurological injury, an individual with a spinal cord injury (SCI) in general will not have motor function or sensation below that level.

While different athletes may have a variety of medical conditions (cerebral palsy, post-polio paralysis) most Paralympic regulations specify a neurological level of disability versus medical condition to compete in an athletic event. For example, although the pathology of spina bifida and spinal cord injuries is completely different, in wheelchair tennis, athletes would compete with other athletes of similar neurological level of injury regardless of the medical condition. Players without limitations in the upper body would qualify for paraplegic wheelchair tennis, while those with both upper and lower body strength limitations would qualify for the quadriplegic field regardless of the disease or injury.

Each individual's injury and injury level is unique. Consequently, it remains important to understand the disability, the extent and how it manifests limitations. Understanding a player's disability enhances the athlete's, coach's and trainer's ability to maximise performance, training and conditioning, adapt for limitations, minimise discomfort and prevent serious injury or illness.

The major medical conditions include amputation, spinal cord injury,

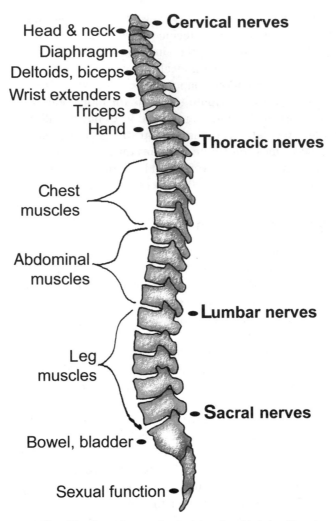

Head & neck
Diaphragm
Deltoids, biceps
Wrist extenders
Triceps
Hand

Cervical nerves

Thoracic nerves

Chest
muscles

Abdominal
muscles

Lumbar nerves

Leg
muscles

Sacral nerves

Bowel, bladder

Sexual function

14.1 Classification of neurological levels of injury. (Courtesy Cislunar Aerospace, Inc.)

spina bifida, polio and cerebral palsy. In turn, these medical conditions have characteristic impairments, which may affect voluntary motor functions, sensation, body temperature regulation, fatigue, cardiovascular, bowel and bladder function, respiratory, spasms and skin breakdowns. Some are unique to the specific medical condition, others common to all.

Injuries are often differentiated as 'complete' or 'incomplete'. An individual with an incomplete injury can move and have sensation far below the neurological area of injury. In contrast an injury is called 'complete' if there is no voluntary motion or sensation below (within three segments) the area of neurological injury.

In Paralympic and recreational sports for individuals with physical disabilities, material selection takes an extra dimension of importance in safety and injury prevention. Lack of stability, balance, strength of materials or friction between the body and equipment, prosthetic or assist devices can spell disaster to the athlete or recreational player with physical disabilities. An assistive device may be compensating for lack of motor control or provide balance. Instability or a structural failure could easily and obviously defeat the purpose of the equipment.

As well, athletes with paralysis may not be able to feel rubbing, chafing and pain. Friction may cause pressure sores (also called bed sores, decubiti or decubitus ulcers), which undetected and uncared for can cause infection. Often, these sores develop where bones are close to the skin's surface. Sores can form deep craters that can infect all the way to muscle and bone. A severe skin sore can mean several weeks of hospitalisation, bed rest, surgery or skin grafting.

14.2.1 Amputation

Amputation of a portion of the body may have been needed due to an accident, related to a disease such as cancer or due to circulation problems such as diabetes. Individuals may choose to compete in a wheelchair versus using a prosthetic device (like an artificial limb). Amputees usually have normal voluntary muscle movement and sensation, cardiovascular, respiratory, bowel and bladder function and body temperature regulation.

It is important to note that, in many sports, amputees dominate the top rankings. Although missing a limb, typically the amputee has full function of their body with no paralysis or other weakness. Thus, they are able to condition their entire body. In tennis they can rotate their full body to generate power in a stroke or bend to place themselves in an aerodynamically correct position in wheelchair racing. Individuals with paralysis and muscular weakness are often unable to condition and maintain muscle strength in those afflicted areas of the body. For an athlete paralysed from the waist down, the leg muscles may atrophy, and there may be difficulty in developing the abdominal muscles. As a result, in tennis, the player may be reliant on upper body strength to produce tennis strokes with overuse injuries occurring; in wheelchair racing it may be impossible for the individual to bend into the most aerodynamic position.

14.2.2 Spinal cord injury

Spinal cord injury (SCI) affects movement, sensation and other nervous system functions that regulate body functions such as temperature, bladder and bowel control and heart rate. An SCI is often a result of a fall, vehicle accident or sports injury. The spinal cord may be severed, or, as a result of the trauma, bones in the spine may bruise or press on the spinal cord. This disrupts the transmission of nerve impulses between the brain and other parts of the body.

The Christopher and Dana Reeve Paralysis Resource Center (2002) has stated that there are 450000 Americans with SCI. The majority of injuries are the result of accidents and 82% are men. In the United States there are 10000 new cases of SCI per year.

Athletes with injuries in the cervical and thoracic areas may be more susceptible to respiratory infections (for example, due to the lack of ability to produce a strong cough). They may also experience spasticity (an increase in muscle tone), which can cause leg shaking or spasms.

Due to a lack of sensation, skin breakdowns are common. Sitting for long periods of time can produce pressure sores or ulcers which can become infected. Athletes may not even recognise when they have been injured due to lack of pain sensation.

14.2.3 Spina bifida

While SCI is normally caused by an accident or fall, spina bifida is a congenital (from birth, but not hereditary) condition incurred while the spinal cord developed during pregnancy. Individuals with spina bifida often experience many of the same limitations of an SCI (motor, sensatory, bladder and bowel function), although they do not generally experience body heat regulation and spasticity problems.

14.2.4 Polio

Polio or post-polio paralysis is caused by a virus that sometimes migrates to the body's neurological system. The virus attacks motor nerve cells resulting in permanent muscle weakness. Some individuals may experience more fatigue, pain or muscle weakness with age. This may be a normal ageing process (nerve cells change over time), but more attention is being given to the theory that the remaining functional nerve cells are overworked and begin to deteriorate. Of course, this would be a concern for athletes who train at high intensity.

14.2.5 Cerebral palsy

Cerebral palsy affects muscular coordination and voluntary movement. Either during development or at birth, areas of the brain which control coordination, motor function, speech or muscle tone may be affected.

14.2.6 Visual impairment

In some sports athletes are classified by perception of light, ability to recognise form and visual field. Dependent on the sport, athletes with visual impairments may have a seeing partner work alongside. Auditory or tactile cues are also used.

14.3 Considerations and limitations in design and materials based on Paralympic sport regulations

Regulation differences between the able-bodied sport and its Paralympic counterpart vary in commonality from sport to sport. For example, in wheelchair tennis, the only major difference is that wheelchair players are allowed to have two ball bounces before the player loses a point. Some sports allow assistants. Sports for visually impaired skiers, cyclists or runners allow sighted companions to aid manoeuvring over the course.

Some governing bodies specify equipment and adaptive devices that may be used. Other governing bodies evaluate assistive devices on a case by case basis. The most important factor is safety. In evaluating the device a technical advisor will determine if the equipment is safe for the athlete and the athletes and staff around them.

Athletes with motor disabilities often have stability, manoeuvrability and control issues. Governing bodies are concerned with allowing accessibility to athletes to participate balanced with inhibiting devices which would provide a skill advantage.

Once again, it is important to remember that the Paralympics deal with elite athletes and only a small portion of the sports and recreation activities available to individuals with physical challenges. While at the Paralympic level prosthesis or adaptive devices may be restricted, in other competitions or in recreation many more alternatives for assistive equipment are permitted.

Development and marketplace costs of assistive devices is an important consideration. Adaptive equipment in many sports is individualised, based on the specific capabilities of the athlete. Subsequently, the market is small or may even be just one individual. Subsequently, it is understandable that expensive design development costs may discourage high tech solutions.

14.4 Devices and materials used in Paralympic sports

The following sections provide a representative sample of the materials and types of wheelchairs, prosthetic and assistive devices allowed and used in Paralympic events. Again, it is important to note that many more solutions are available and used in other (non-Paralympic) competitions, in recreational use and for the many other sports, fitness and recreational activities in which individuals with physical and intellectual challenges participate. When applicable, it is noted that the equipment is the same as that which able-bodied athletes or participants would use.

14.4.1 Archery

There are currently three major classifications of archers at the Paralympic level. Classification ARW1 and ARW2 include wheelchair athletes, while classification ARST applies to standing athletes or athletes shooting from a sitting position in a chair. An ARW1 athlete may be a tetraplegic in a wheelchair or have a comparable disability such as triplegia or severe diplegia (two limbs affected), or a double amputee with additional upper body limitations. ARW2 archers may be paraplegic, have severe diplegia without control limitations in their upper limbs, significant trunk balance problems, spasticity in their lower limbs, or double amputation below the knee.

At the Paralympic level, assistive devices are permitted if the equipment is authorised by an International Paralympic Committee archery classifier. The equipment is intended to provide balance or to assist the archer in shooting and is not intended to provide a skill advantage.

Wheelchairs and chairs are allowed for athletes with lower limb and trunk disabilities. The wheelchair or chair may not support the bow arm and the bow cannot touch the wheelchair while the arrow is being released.

Archers with disabilities in the fingers of both hands are allowed to use a mechanical release device. Used to compensate for weak grasps or lack of muscle strength, these devices resemble gun triggers attached to small clamps called jaws. The mechanism may have a one-finger trigger with a nylon strap or padded fabric with Velcro for adjustment. The strap attaches the mechanism to the hand and wrist (Fig. 14.2). Alternatively, the mechanism may be a three-finger grip with a button release (Fig. 14.3). The release

14.2 Trigger releases assist archers with weak grasps or muscle strength. (Courtesy True-Fire Corp.)

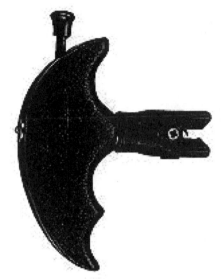

14.3 A three-finger grip with button release for archers. (Courtesy True-Fire Corp.)

mechanism itself is typically made from steel or aluminium (anodised in colours).

Note that these mechanical release devices are commonly found in archery and used by many able-bodied women, children and senior recreational archers as well as hunters, due to the need to hold the bowstring back for long periods of time.

Mouthpieces have been used by hemiplegics (individuals with paralysis on one side of the body) to release the bowstring. This can be a trigger mechanism released by a wired mouthpiece or as simple as a strip of nylon strap fabric or athletic tape looped around the bowstring. The archer is able to pull back the string with the material in their mouth.

Archers with upper limb disabilities may use a compound bow in certain divisions. A compound bow has a series of wheels, cams and cables that form a pulley system. As the archer pulls back on the string of a compound bow, the top and bottom limbs of the bow are pulled inward, which stores energy until the string is released. Upon release, the energy is transferred to the arrow (Fig. 14.4). Traditionally, the limbs of compound bows have been made from aluminium and fibreglass. As with the release mechanisms, these are standard archery equipment, not specific to archers with physical disabilities.

Prosthetic devices, which grip the bow, have been used to compensate for a missing upper limb. The device developed by TRS, Inc. is constructed from

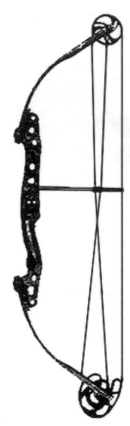

14.4 In certain divisions archers with upper limb disabilities may use a compound bow. (Courtesy Hunter's Friend)

titanium, surgical grade stainless steel, aircraft aluminum and polymers. To prevent damage to the bow, several layers of rubber are wrapped at the point where the prosthetic device grips the bow (Fig. 14.5).

At other levels of competition (national) and in recreation other innovative equipment is used to provide opportunities to archers with a wide variety of physical limitations. An avid archer and sound system repairman, Al Lefebvre, who lost his sight due to diabetes, developed the 'Sightless Sight System'. The archer uses earphones, which provides different tones for elevation and alignment. When the tones are synchronised, the archer shoots (Paciorek and Jones, 2001).

14.4.2 Athletics (track and field events)

Athletic events include track, throwing, jumping, the pentathlon and marathon. Athletes compete based on their functional classifications in each

14.5 Rubber covers an archery bow where a prosthetic device grips the surface. (Courtesy TRS, Inc.)

event. Materials science, technology and engineering design have positively affected Paralympic track and field events, making these activities inclusive to as many athletes as possible. Athletes compete using prostheses or wheelchairs. Visually impaired and blind athletes compete with a sighted companion.

Advanced materials and technology have made a tremendous contribution in the development of prosthetic devices used in track events. Once thought impossible, even above knee (AK) single and bilateral (both sides) amputees compete effectively. Improvements in design and utilisation of advanced materials have provided prostheses with a normal running heel to toe foot motion.

Today, the prosthetic devices use carbon fibre for strength and flexibility and titanium bolts for socket attachment. (The socket is the part of the prosthesis which fits around the remaining limb.) The heels absorb energy from the ground strike and release the energy into the foot of the prosthesis. Thereby, the athlete uses less of their own energy to move the leg forward (Fig. 14.6).

14.6 The prosthetic foot uses state-of-the-art materials and design to conserve the athlete's energy. (Courtesy *Ossur*)

14.7 Wheelchair racer. (Courtesy *Spokes n' Motion*)

Advances in the design and materials used in wheelchairs for track events have contributed to new speed records. Serious racers will invest in customised wheelchairs and stay informed of the latest technology (Fig. 14.7). The racers use a variety of lightweight metals (aluminium, titanium) and other materials (carbon fibres). *Sports 'n Spokes* magazine even includes an annual report of the wheelchair racers and latest technology.

Paciorek and Jones (2001) have noted that, in most field events, technique has provided competitive edge. However, in wheelchair field events, technique innovations have been the result of advancing technology in wheelchairs and wheelchair tie-down mechanisms.

In some instances wheelchairs have been replaced by special throwing chairs. The purpose of the chair is to provide a stable platform from which the athlete can throw. In the past, wheelchairs sometimes tipped over. Early versions of throwing chairs were composed of heavy metal. Today, lighter and less expensive PVC pipe (polyvinylchloride) is used in the construction. Chairs are often customised by the athletes.

Wheelchair tie-down systems used in the past were very time consuming to use during competition and did not provide adequate stability. One system utilised metal tracks with chains attached to the wheelchair. For the Barcelona Paralympic games the Olympic Stadium director and an industrial engineer developed a sophisticated, technologically advanced and stable tie-down system. A pressure gauge was attached to four large suction cups (typically used in the transportation of glass). The suction cups were attached to rods with rubber-covered hooks which attached to the wheelchair. The rubbed padding prevented damage to the wheelchair. The suction cups were applied to a metal base, although any non-porous, strong, clean dry surface could be used. This reduced set up time from 10–15 minutes to 10 seconds and provided a much more stable platform for the athletes (Paciorek and Jones, 2001).

Sighted guides can be used in certain events and for particular classifications of visually impaired runners. For short distances, auditory signals are yelled – spectators must remain silent. For longer distances, elbow leads or tethers connect the visually impaired to a sighted guide. The tether is a flexible material (but not allowed to be elastic) and may be no longer than 50 cm.

14.4.3 Boccia

Boccia, a ball game similar to lawn bowling, is one of the Paralympic sports played by individuals with a specific disability – cerebral palsy. The athletes throw, kick or use an assistive device to drive leather balls as close as possible to a small white target ball called a jack.

The boccia ball for Paralympic competition is different than a standard regulation ball. The Paralympic boccia ball is made of leather and filled with plastic granules. Since the balls are softer than standard balls, this allows the athletes to grasp the ball better.

Ramps and chutes may be used by specific classifications of Paralympic athletes. Typically, the ramps and chutes are individually constructed from hardware store materials and made from aluminum or plastic pipe. The pipe often is cut in half. The athletes must release the ball themselves, which can be accomplished through aids attached to the head, hand or even the mouth. Athletes in bowling also use ramps.

14.4.4 Cycling

The cycling events are divided into individual and three-cyclist teams and compete in road race and time trial events. Athletes with cerebral palsy use standard racing bikes or tricycles. Blind or visually impaired cyclists use tandem bicycles with a sighted teammate. Amputees and athletes with permanent motor disabilities compete in individual road race events using cycles specifically constructed for their needs.

Innovation in cycling technology for both competition and recreation has supported advances in equipment for individuals with physical challenges. The popularity of tandem cycling as a family recreational sport has increased availability of tandem bikes. Used in events for athletes with visual impairments, the cycles used are standard tandem cycles. The sighted pilot resides in the front seat with the visually impaired rider in the rear seat. Once again, observe that these are not cycles designed specifically for individuals with disabilities. The technology used in traditional racing cycles has migrated to tandem cycles including carbon fibre spoke wheels.

Cycling equipment, while adapted to meet the needs of the physically impaired, must also meet size clearance regulations implemented to prevent injuries in the event of a fall as well as to prevent an unfair advantage in crossing the finish line.

Standard dropped handlebars must be used although adaptations are allowed for athletes with upper limb disabilities, if needed to safely operate gear and brake levers. Handlebar adaptation may not provide any unfair aerodynamic advantage.

Artificial handgrips and prosthesis can be used by cyclists with an upper limb disability. For safety reasons (in case of a fall), these devices may not be mounted or fixed to any part of the cycle.

The regulations specify that the cyclist's position may only be supported by the handlebars, pedals and saddle. Although athletes with above knee amputations may use a thigh support, no thigh strapping devices are allowed and the thigh itself may not be fixed to the bicycle.

14.4.5 Equestrian

The only event including an athlete and animal, the equestrian competitions, may be one of the most beautiful events of the Paralympics. Adaptive equipment is developed to protect and provide safety and comfort for both the human and the horse. In riding events, due to the height of the horse, falls can be dangerous and even fatal. Helmet protection is absolutely required for anyone elevated.

Subsequently, in equestrian events adaptive devices often utilise as much 'breakaway' technology as possible. This may be manifested in simple rubber band latches or Velcro solutions.

However, the major challenge in design of adaptive equipment in equestrian events is providing a balance between stability for the rider while allowing the rider to separate from the horse (and not be dragged) in case

14.8 Hooded safety stirrups prevent the foot from sliding through the stirrup. (Courtesy Freedom Rider)

of a fall. Today, the design of a cost-effective paraplegic saddle remains a challenge.

Several types of stirrups are designed to accommodate riders with physical disabilities. Some are designed as safety stirrups which 'breakaway', to release the foot during a fall, preventing the rider from being dragged by the horse. One of the most popular designs has a rubber band latch. If a fall occurs, the weight of the rider is sufficient to release the rubber band. Leather hoods cover a stirrup to prevent the foot from sliding through (Fig. 14.8). Safety stirrups with thick sponge rubber foot cushion may also have a safety cage to prevent the foot from sliding through the stirrup. This stirrup is made from aluminum or lightweight nylon. Other stirrup configurations allow the foot to slide out by design and are made from stainless steel.

Reins may have knots, tabs or 'ladders' to prevent the hand from sliding and maintain the reins at the correct length to control and manoeuvre the horse as well as maintain balance for the rider (Fig. 14.9). Reins have been designed with suede or rubber tabs and in graduated colours as another feedback method to help the rider maintain proper rein length. Ladder reins make grasping reins easier for those who have less motor control. The

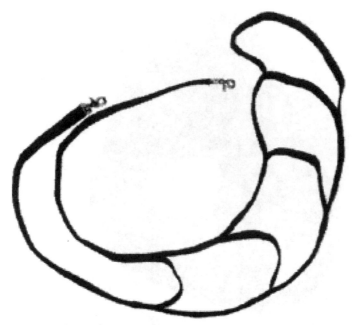

14.9 Ladder reins are easier to grasp for riders with less motor control. (Courtesy Freedom Rider)

rein lines have been made in traditional leather as well as in cotton webbing and synthetic fabric materials.

Equestrian saddles have been adapted for use for individuals with disabilities. One of the most technologically advanced has been developed by Supracor, Inc., which utilises its patented flexible, perforated honeycomb structure composed of thermoplastic elastomers. The cushioning material is highly effective in preventing and healing pressure sores. The ventilated honeycomb insert fits inside a cover of breathable wool (Fig. 14.10).

Other seat cushions use foam to alleviate the compacting and bunching of materials like fleece.

14.4.6 Sailing

Athletes from all disability classifications compete in sailing. Sailing classifications are based on the stability, hand function, manoeuvrability, vision and hearing.

Yachts with keels are used in Paralympics. Keel boats provide greater stability in the water. Specifically at this time Sonar class boats and 2.4 mR (Norlin Mark III) class boats are used in the Paralympics. The boats also have open cockpits allowing the sailors more room to manoeuvre and easy

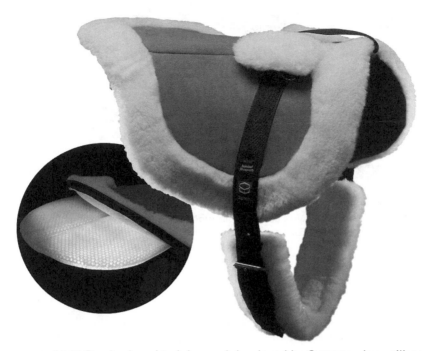

14.10 Bareback and training pad developed by Supracor, Inc. utilises its unique, patented Stimulite® honeycomb technology. Supracor and Stimulite are registered trademarks of Supracor, Inc. (Courtesy Supracor, Inc.)

access to controls. The 2.4 mR is designed for one individual sailing and the Sonar for a crew of three.

All sailors are required to wear personal floatation devices (PFD). Once again, this equipment (the boat and the floatation devices) is the same as that developed and used by able-bodied sailors.

Sailors without sensation in the buttocks generally use some type of cushion. Sitting for long periods of time and without sensation, pressure sores can form, without the sailor's recognition or feeling. Cushions also are used to assist sailors in maintaining positions while reaching for sheets or winches. There are several popular types of cushions, using waterproof materials. These range from gel-filled pads to inflatable rubber cushions which resemble the underside of an egg carton (Paciorek and Jones, 2001). Care must be taken to prevent punctures.

A vast array of very individualised equipment exists. Bars that cross from one side of the boat to the other provide a stable handhold (Fig. 14.11). Some sailors use track systems which allow their seats or chairs to swivel from side to side.

14.11 Bars provide sailors a handhold to manoeuvre. (Courtesy Gene Hinkel, IFDS)

14.4.7 Swimming

Swimming is one of the most inclusive Paralympic events. Paralympic swimmers may have hearing, visual or physical disabilities. An interesting aspect of swimming competition at this level is that while competing no limb prostheses can be used (an eye prosthesis can be worn). Devices that provide visual cues for the deaf or auditory signals for the visually impaired may be used. Safety is always of paramount importance both in the water and in terms of pool accessibility. In transferring individuals with lower body paralysis or weakness, sliding often occurs. Subsequently, attention is given to inhibiting sharp edges.

There are three major classifications for blind swimmers based on their visual acuity and light perception. S11 swimmers classified as having no light perception to some light perception but unable to recognise a hand shape from any distance or direction, must wear eye goggles made of opaque materials.

Tap-sticks, long thin poles with soft plastic bulbs at the end, underwater sound systems, and drip systems as athletes pass flags have been utilised. Swimmers who are deaf often use a strobe light or other visual signal to denote the start of a race.

14.4.8 Wheelchair tennis

A wheelchair tennis player pushes the chair with racquet in hand. Chair mobility is the first and foremost aspect of wheelchair tennis. With today's advances in chair technology (titanium chairs can be carried without effort in one hand), player mobility has vastly improved. This has led to the use

14.12 Paralympic gold medallist and 10-time US Open wheelchair tennis champion Randy Snow playing in a tennis wheelchair. (Courtesy Randy Snow)

of the chair as a power source as well as the advent of a more attacking style of play. Although two bounces are allowed in wheelchair tennis (the only major difference between wheelchair and able-body tennis), players are taking the ball on one bounce, with more and more players making contact inside the baseline as much as possible.

One of the fastest growing wheelchair sports, wheelchair tennis has only one major rule difference – the ball may bounce twice and the second bounce may be outside of the traditional court lines (Fig. 14.12).

There are two primary divisions separating the quadriplegics or equivalent disabilities from the paraplegics. The key equipment is the tennis wheelchair. Unlike a regular wheelchair, the wheels are cambered, which allows quick and pinpoint turns. Tennis wheelchairs traditionally have no brakes. The chairs have evolved using the latest manufacturing and materials. Although the lightest chair may not be the best alternative for each player, in general the industry and players have gone lightweight, using aluminum and titanium. Latex or rubber tires are used.

Based on the disability and the level of neurological injury of the player, players will strap themselves in their chairs, so that they are one with the chair. Straps may be plastic weight belts, rubber bands, bunge cords and pants belts.

Due to the neurological level of injury, some individual's bodies do not regulate heat well. Subsequently, any opportunity to keep cool through clothing is used. One player designed a hat with pockets he filled with ice. As the ice melted, it ran down his neck and back to keep him cool.

Quadriplegic (quad) players will have a disability or weakness in one of the upper limbs and will often use motorised powered wheelchairs with a joystick to manoeuvre. Quad players will either have a special prosthesis which holds the racquet and attaches the racquet to the arm or the racquet is taped into the hand.

Individuals with grip limitations typically attach the tennis racquet to their hand, wrist and arm. Although custom prosthetic devices are used, most players will simply use athletic tape or sometimes an athletic bandage. The prosthetic that holds the racquet slides over the wrist and is made of plastic.

14.4.9 Ice sledge hockey

Ice sledge hockey is played on a standard ice hockey rink; the athletes use an ice sledge and two sticks. Due to speed and player contact strict equipment regulations are implemented to ensure player safety (Fig. 14.13). In turn this affects the materials used.

Current rules regarding the disabilities of the Paralympic ice sledge hockey player allow those with permanent lower body locomotion disabilities including amputation, joint mobility, cerebral palsy and leg shortening. Athletes with restricted mobility such as hip disorders or individuals with chronic pain, hip disorders or knee or ankle disabilities do not meet the minimum disability level and are ineligible.

The ice sledge must meet specific design requirements. Height from the ice to the sledge frame, total blade (ice blade) length in proportion to the sledge (one-third of the total length), the diameter of the materials used in the construction of the sledge (3 cm maximum), puck clearance under the sledge, seat cushion height and any protrusion of equipment are regulated.

14.13 Ice sledge hockey (Courtesy *Spokes n' Motion*)

Straps are allowed but penalties can be assessed for repeated loss or delays. The ice sledge hockey stick may have a blade (similar to a standard hockey stick) or a pick end to grab the ice. The pick end (the lower end of the stick) cannot have a single, sharp end but must have at least six teeth. The teeth may not be sharp or conical to avoid the risk of injury to the players as well as to the ice-rink.

Ice sledge hockey is one of the few sports that specifies the use of materials. The handles, shafts and blades of the stick may be made of wood, plastic or aluminium/titanium.

Regulation helmets with a full cage or mask are mandatory. Athletes are also not allowed to remove the inside padding or ear protectors. Players may use a protective throat collar or bib. Illegal equipment can even result in assessment of penalties.

14.5   Resources

A six-part article series developed by Dr Kathleen Curtis appeared in *Sports 'n Spokes* over several years. It is an excellent introductory resource on physical disabilities and sports. Although not specifically focused on any one sport, the series provides an overview of the various medical conditions that affect wheelchair athletes and contains practical advice for accommodating wheelchair athletes.

Michael Paciorek and Jeffery Jones's book *Disability Sport and Recreation Resources* is the definitive text on sports for the physically and intellectually challenged. The book lists over 250 equipment suppliers and manufacturers that provide adaptive equipment for individuals with disabilities.

Palaestra is the journal and forum of sport, physical education and recreation for individuals with disabilities. The *Adapted Physical Activity Quarterly* presents the latest research on physical and activity and sports.

The International Paralympic Committee is the international governing body for the Paralympics. Each country has its own national governing body for different sports, which is the official body for information on acceptable adaptive equipment.

14.6   Future trends

The US Department of Health and Human Services has noted in its 'Healthy People 2010' report that the numbers of individuals with disabilities have increased. Among the 54 million Americans living with disability, more individuals report: no leisure time physical activity; less social participation; being overweight; more adverse effects from stress; lower employment rates; more environmental barriers; less access to preventive services; and more activity limitation than people living without disabilities. The Christopher Reeve Paralysis Foundation has noted that 2 million Americans live with some type of limb paralysis and that this is on the rise (Christopher Reeve Paralysis Foundation, 2002).

In contrast, it has been shown that paraplegic athletes are more successful than non-athletes in avoiding the major medical complications for which they are at risk (Stotts, 1986). Individuals with physical disabilities who participate in sports demonstrate increased mobility, improved self-image and a greater degree of independence in all aspects of personal health care (Schaefer and Proffer, 1989). Athletes with physical disabilities quickly moved on from primarily rehabilitation endeavours to organised competitive activities (Madorsky and Curtis, 1984). These studies confirm that sport and recreational activity for the individuals with physical disabilities is a key element to health and quality of life.

Adaptive equipment for individuals with physical disabilities will continue to utilise and take advantage of the materials, materials science and technology utilised in able-bodied sports.

As more physically challenged people participate in sports, recreation, fitness and conditioning activities, it is clear they will continue to seek out cost-effective and individualised solutions to improve their own accessibility. This is a tremendous opportunity for the materials science and engineering communities to contribute their expertise to provide accessibility to more individuals as well as to an ageing active senior population.

As opposed to 'following' the technology utilised in able-bodied equipment solutions, proactive design and evaluation of applicable materials that address the needs of individuals with physical disabilities are desirable. The materials and engineering community should embrace that solutions for individuals with physical disabilities often provide engineering solutions that are applicable to other able-bodied individuals. These adaptive devices can serve as training aids, or athletic equipment for women, children, seniors or anyone who may lack the physical strength of an adult male. Solutions that provide safer, more controllable and stable equipment are of great benefit to all athletes and recreational participants, not just to the individuals with physical disabilities.

In the future there must be an understanding, education and synergy between the athletic and engineering communities to develop cost-effective equipment that will assure inclusion, accessibility and life-long participation.

14.7   Acknowledgements

The author wishes to thank: Gene Hinkel of the International Federation of Disabled Sailors and Roger Cleworth, a member of the United States Disabled Sailing Team, for their assistance with the sailing section; Denise Avolio, Sports Manager of the National Disability Sports Alliance – Equestrian, for her information on equestrian Paralympic equipment; US Paralympic wheelchair tennis coach Dan James and three-time Paralympian and medal winner Randy Snow for their patience and openness over the past two years.

14.8   References

Christopher and Dana Reeve Paralysis Resource Center (2002). *Health: Basics By Condition – Spinal Cord Injury*,
http://www.paralysis.org/Health/HealthList.cfm?c=76, Christopher and Dana Reeve Paralysis Resource Center.
Christopher Reeve Paralysis Foundation (2002). *Quality of Life*,
http://www.christopherreeve.org/QLGrants/QLGrantsList.cfm?c=12,   Christopher Reeve Paralysis Foundation.
Curtis K A (1996). Health smarts, strategies and solutions for wheelchair athletes. *Sports 'n Spokes*, January–February, 1996, 25–31.
International Paralympic Committee (2001). *IPC Sport Classification Rules: Archery*. International Paralympic Committee.

International Paralympic Equestrian Committee (2002). *IPEC Rule Book*, International Paralympic Equestrian Committee.

ISAF/IFDS Sailing Committee (2003). *Sailing Manual*. ISAF/ISAF UK Ltd.

Madorsky J G and Curtis K A (1984). Wheelchair sports medicine. *American Journal of Sports Medicine*, **12** (2), 128–132.

Paciorek M J and Jones J A (2001). *Disability Sport and Recreation Resources*, 3rd edn, Traverse City, MI, Cooper Publishing Group.

Pallis J M and James D (2002), *Wheelchair Tennis, Coaching and Sports Science*, www.tennisserver.com/set/, The Tennis Server.

Schaefer R S and Proffer D S (1989). Sports medicine for wheelchair athletes. *American Family Physician*, **39** (5), 239–245.

Stotts K M (1986). Health maintenance: paraplegic athletes and nonathletes. *Archives of Physical Medicine and Rehabilitation*, **67** (2), 109–114.

US Department of Health and Human Services (2000). *Healthy People 2010*, Washington, DC, Office of Disease Prevention and Health Promotion.

Index